# 3rd International Workshop on Theoretical and Computational Physics: Complex Systems and Interdisciplinary Physics (IWTCP-3) and 40th National Conference on Theoretical Physics (NCTP-40)

Journal of Physics: Conference Series Volume 726



Da Lat, Vietnam
27 – 30 July 2015

**Editors:**

**Trinh Xuan Hoang**
**Hoang Anh Tuan**
**Nguyen Ai Viet**

ISBN: 978-1-5108-2612-0
ISSN: 1742-6588

**Printed from e-media with permission by:**

Curran Associates, Inc.
57 Morehouse Lane
Red Hook, NY 12571

CURRAN ASSOCIATES INC. proceedings.com

**Some format issues inherent in the e-media version may also appear in this print version.**



Printed by Curran Associates, Inc. (2016)

For permission requests, please contact the Institute of Physics
at the address below.

Institute of Physics
Dirac House, Temple Back
Bristol BS1 6BE UK

Phone: 44 1 17 929 7481
Fax: 44 1 17 920 0979

techtracking@iop.org

**Additional copies of this publication are available from:**

Curran Associates, Inc.
57 Morehouse Lane
Red Hook, NY 12571 USA
Phone: 845-758-0400
Fax: 845-758-2633
Email: curran@proceedings.com
Web: www.proceedings.com

# TABLE OF CONTENTS

## 3RD INTERNATIONAL WORKSHOP ON THEORETICAL AND COMPUTATIONAL PHYSICS (IWTCP-3): COMPLEX SYSTEMS AND INTERDISCIPLINARY PHYSICS



## 40TH NATIONAL CONFERENCE ON THEORETICAL PHYSICS (NCTP-40)

# Preface

This volume contains selected papers presented at the 3rd International Workshop on Theoretical and Computational Physics (IWTCP-3): Complex Systems and Interdisciplinary Physics and the 40th National Conference on Theoretical Physics (NCTP-40). Both the workshop and the conference were held from 27–30 July 2015 in Dalat Palace Hotel, Da Lat, Vietnam.

The NCTP-40 was held to mark the 40th anniversary of the National Conference on Theoretical Physics (NCTP), which was organized for the first time in Da Lat in August 1976, while the IWTCP-3 was also an External Activity of the Asia Pacific Center for Theoretical Physics (APCTP).

Both the IWTCP-3 and the NCTP-40 were organized under the support of the Vietnamese Theoretical Physics Society (VTPS), with a motivation to foster scientific exchanges between the theoretical and computational physicists in Vietnam and worldwide, and to promote high level research and educational activities in the country.

145 participants coming from 10 countries participated in the workshop and the conference. At the IWTCP-3 workshop, 14 invited talks were presented by international experts, together with 9 oral and 32 poster contributions. At the NCTP-40, 4 invited talks, 22 oral contributions and 63 posters were presented.

We would like to thank all invited speakers, participants and sponsors for making the workshop and the conference successful.

Trinh Xuan Hoang, Hoang Anh Tuan and Nguyen Ai Viet

Journal of Physics: Conference Series **726** (2016) 011001 doi:10.1088/1742-6596/726/1/011001

## Organizers

- Institute of Physics, Vietnam Academy of Science and Technology (IOP-VAST)
- Dalat University (DLU)
- Nuclear Research Institute, Da Lat (NRI)

## Sponsors

- Vietnam National Foundation for Science and Technology Development (NAFOSTED)
- Asia Pacific Center for Theoretical Physics (APCTP)
- Institute of Physics, Vietnam Academy of Science and Technology (IOP-VAST)
- Vietnam Academy of Science and Technology (VAST)

Journal of Physics: Conference Series **726** (2016) 011001 doi:10.1088/1742-6596/726/1/011001

# Committees

## IWTCP-3

### Honorary Chair

- Nguyen Van Hieu (Vietnam Academy of Science and Technology, Hanoi)

### Chair

- Nguyen Ai Viet (Institute of Physics, VAST, Hanoi)

### International Advisory Board

- Tae Soo Chon (Pusan National University, Pusan)
- Mukunda Das (Australian National University, Canberra)
- Hung T. Diep (University of Cergy-Pontoise, Paris)
- Peter Fulde (Asia Pacific Center for Theoretical Physics, Pohang)
- Mai Suan Li (Polish Academy of Sciences, Warsaw)
- Duc Nguyen-Manh (Culham Centre for Fusion Energy, Culham, UK)
- Andreas Savin (CNRS and Sorbonnes Universités, UPMC, Univ Paris 6, Paris)
- Dam Thanh Son (University of Chicago, USA)

### Organizing Committee

- Trinh Xuan Hoang (Institute of Physics, VAST, Hanoi), Chair
- Bach Thanh Cong (Hanoi University of Science, VNU, Hanoi)
- Hoang Dzung (Vietnam National University, Ho Chi Minh City)
- Phu Chi Hoa (Dalat University, Da Lat)
- Hoang Anh Tuan (Institute of Physics, Hanoi)

### Program Committee

- Nguyen Tri Lan (Institute of Physics, VAST, Hanoi), Chair
- Trinh Xuan Hoang (Institute of Physics, VAST, Hanoi)
- Nguyen Manh Duc (Culham Center for Fusion Energy, Culham, UK)
- Truong Minh Duc (Hue University, Hue)
- Nguyen Toan Thang (Institute of Physics, VAST, Hanoi)
- Vu Ngoc Tuoc (Hanoi University of Science and Technology, Hanoi)

Journal of Physics: Conference Series **726** (2016) 011001 doi:10.1088/1742-6596/726/1/011001

## NCTP-40

### Honorary Chair

- Nguyen Van Hieu (Vietnam Academy of Science and Technology, Hanoi)

### Chair

- Nguyen Ai Viet (Institute of Physics, VAST, Hanoi)

### Organizing Committee

- Trinh Xuan Hoang (Institute of Physics, VAST, Hanoi), Chair
- Bach Thanh Cong (Hanoi University of Science, VNU, Hanoi)
- Hoang Dzung (Vietnam National University, Ho Chi Minh City)
- Phu Chi Hoa (Dalat University, Da Lat)
- Le Van Hoang (Ho Chi Minh City Pedagogical University, Ho Chi Minh City)
- Nguyen Tri Lan (Institute of Physics, VAST, Hanoi)

### Program Committee

- Hoang Anh Tuan (Institute of Physics, VAST, Hanoi), Chair
- Phung Van Dong (Institute of Physics, VAST)
- Ho Trung Dung (Ho Chi Minh City Institute of Physics)
- Nguyen The Toan (Hanoi University of Science, VNU, Hanoi)
- Vu Ngoc Tuoc (Hanoi University of Technology)

### Secretariat

- Duong Thi Man (Institute of Physics, VAST, Hanoi)

# List of Participants

Đào Thị Thúy Nga
Bac Ninh Department of Education and Training
Email: daothuyngaht@gmail.com

Bach Thanh Cong
VNU University of Science
Email: congbt@vnu.edu.vn

Best B. Robert
National Institute of Diabetes and Digestive and Kidney Diseases, National Institutes of Health
Email: robert.best2@nih.gov

Bui Hong Nhung
Institute of Physics, VAST
Email: nhungqft@gmail.com

Bui Huu Nguyen
Hanoi University of Mining and Geology
Email: nguyencv90@gmail.com

Bui Phuong Thuy
Namdinh University of Technology Education
Email: buiphuongthuyphys@gmail.com

Bui Thi Ha Giang
Hanoi university of Education
Email: giang.giangsphn@gmail.com

Chon Tae-Soo
Pusan National University
Email: tschon.chon@gmail.com

Chu Thuy Anh
Institute of Physics, VAST
Email: ctanh@iop.vast.ac.vn

Dang Trung Si
NCS, Viện Vật lý, Viện Hàn lâm KHCNVN
Email: dangtrungsi@gmail.com

Dao Thu Hang
People's Police University of Technology and Logistics
Email: daohangkhtn@gmail.com

Dao Vong Duc
Institute of Physics, Vietnam Academy of Science and Technology
Email: dvduc@iop.vast.ac.vn

Journal of Physics: Conference Series **726** (2016) 011001 doi:10.1088/1742-6596/726/1/011001

Das P. Mukunda
The Australian National University
Email: mukunda.das@anu.edu.au

Dinh Nguyen Dinh
Institute of Physics, VAST
Email: dndinh@iop.vast.ac.vn

Dinh T Binh
Institute of Physics, VAST
Email: dtbinh@iop.vast.ac.vn

Do Chi Nghia
Hanoi Pedagogical University 2
Email: dochinghia.sp2@moet.edu.vn

Do Thi Huong
Viện Vật Lý, Số 10, Đào Tấn, Thủ Lệ, Ba Đình, Hà Nội
Email: dthuong@iop.vast.ac.vn

Do Thi Hong Hai
Department of Physics, Hanoi University of Mining and Geology
Email: dothihonghai@humg.edu.vn

Do Tuan Long
Faculty of Physics, Hanoi University of Science, Vietnam National University
Email: dotuanlong@hus.edu.vn

Doan Tri Dung
Ho Chi Minh City Institute of Physics
Email: dtdung@hcmip.vast.vn

Doan Quoc Khoa
Quang Tri Teacher Training College
Email: khoa_dqspqt@yahoo.com

Duong Thi Man
Institute of Physics, VAST
Email: duongthiman28@gmail.com

Duong Van Loi
Tay Bac University
Email: loidvtb@gmail.com

Dương Thị Diễm My
Khoa Vật lý, Trường Đại học Sư phạm
Email: myduong.phys@gmail.com

Giang Huong Bach
VNU University of Science
Email: gianghuongbach@gmail.com

Ha Thanh Hung
Ha Noi University of Education No.2
Email: hthung@grad.iop.vast.ac.vn

Ham Sihyun
Sookmyung Women's University
Email: sihyun@sookmyung.ac.kr

Ho Khac Hieu
Duy Tan University
Email: hieuhk@duytan.edu.vn

Ho Trung Dung
Theoretical Physics Research Group, Ton Duc Thang University
Email: htdung8@yahoo.com

Hoang Anh Tuan
Institute of Physics, VAST
Email: hatuan@iop.vast.ac.vn

Hoang Danh-Tai
Asia Pacific Center for Theoretical Physics (APCTP)
Email: danh-tai.hoang@apctp.org

Hoàng Dũng
Vietnam National University, Ho Chi Minh City
Email: hdung@vnuhcm.edu.vn

Hoang Ngoc Cam
Institute of Physics, VAST
Email: hncam@iop.vast.ac.vn

Hoang Van Ngoc
Truong Dai hoc Khoa hoc tu nhien Ha Noi
Email: hoangfvanwngocj@gmail.com

Hoang-Do Ngoc-Tram
Ho Chi Minh City University of Education
Email: ngoctramhd@gmail.com

Hoàng Ngọc Long
Viện Vật Lý, Viện Hàn lâm KHCNVN
Email: hnlong@iop.vast.ac.vn

Huỳnh Ngọc Toàn
Đại học Duy Tân
Email: hntoan1310@gmail.com

Hummer Gerhard
Max Planck Institute of Biophysics
Email: gerhard.hummer@biophys.mpg.de

Huynh Vinh Phuc
Dong Thap University
Email: hvphuc@dthu.edu.vn

Lê Hải Mỹ Ngân
HCM City University of Pedagogy
Email: myngan1988@gmail.com

Lê Đình
Hue University-College of Education
Email: dinhle52@gmail.com

Lê Thị Thanh Thủy
HCM City University of Education
Email: lethuy35102094@gmail.com

Lê Văn Hoàng
Ho Chi Minh City University of Education
Email: hoanglv2005@gmail.com

Le Cam Tu
HCMC University of Science
Email: letu106@gmail.com

Le Duy Manh
Institute of Physics, VAST, Hanoi
Email: ldmanh@iop.vast.ac.vn

Le Thi Quynh Huong
Department of Natural Sciences, Nha Trang
College of Education, Nha Trang City, Vietnam
Email: lethiquynhhuong@yahoo.com

Le Tho Hue
Institute of Physics, VAST
Email: lthue@iop.vast.ac.vn

Le Van Tan
Ho Chi Minh City Institute of Physics Vietnam
Academy of Science and Technology
Email: sobaobai@gmail.com

Le Van Vinh
Hanoi University of Science and Technology
Email: vinh.levan@hust.edu.vn

Le Van Xuan
Truong cao dang cong nghe va kinh te Ha Noi
Email: xuanlv.hn@gmail.com

Journal of Physics: Conference Series **726** (2016) 011001 doi:10.1088/1742-6596/726/1/011001

Liwo Adam Józef
Faculty of Chemistry, University of Gdansk
Email: adam@chem.univ.gda.pl

Luyện Thị San
Đại học Bách Khoa Hà Nội
Email: sancandy138@gmail.com

Lương Văn Tùng
Đại học Đồng Tháp
Email: lvt1961@yahoo.com.vn

Mai Suan Li
Institute of Physics, Polish Academy of Science
& Institute for Computational Science and Technology at Ho Chi Minh City
Email: masli@ifpan.edu.pl

Mẫn Văn Ngữ
Hung Yen Industrial Colleage
Email: nguqlkt@yahoo.com

Ngô Gia Vịnh
Bac Ninh Department of Education and Training
Email: ngogiavinh@bacninh.edu.vn

Ngô Thị Phương
HCM City University of Pedagogy
Email: phuongnt@hcmup.edu.vn

Ngo Son Tung
Institute of Physics, Polish Academy of Sciences
Email: nstung@ifpan.edu.pl

Ngo Thanh Cong
Ho Chi Minh City Institute of Physics
Email: thanhcong2491@gmail.com

Nguyễn Anh Kỳ
Institute of Physics
Email: anhky@iop.vast.ac.vn

Nguyễn Ánh Sáng
Vocational Training and comprehensive technical center of Phu Tho province
Email: anhsangk13@gmail.com

Nguyễn Đức Hiền
Sở GD & ĐT Gia Lai
Email: n.duchien@gmail.com

Nguyễn Minh Hoa
Đại học Y Duợc Huế
Email: minhhoa2806@gmail.com

Nguyễn Thị Hương
Đại học sư phạm Hà Nội
Email: huonghnue@gmail.com

Nguyễn Thị Kim Ngân
Đại học Cần Thơ
Email: ntkngan@ctu.edu.vn

Nguyễn Văn Hiệu
Vietnam Academy of Science and Technology
Email: nvhieu@iop.vast.ac.vn

Nguyen Chuong
Ton Duc Thang University
Email: nguyenhahungchuong@tdt.edu.vn

Nguyen Que-Huong
Marshall University
Email: nguyenh@marshall.edu

Nguyen Viet-Hung
Université Paris-Sud & Institute of Physics
Email: viet-hung.nguyen@u-psud.fr

Nguyen Ai Viet
Institute of Physics, VAST
Email: vieta@iop.vast.ac.vn

Nguyen Ba Hung
Vietnam Military Medical University & Institute of Physics, VAST
Email: hungnb80@yahoo.com.vn

Nguyen Ba Phi
Mien Trung University of Civil Engineering
Email: nguyenbaphi@cuc.edu.vn

Nguyen Dac Trung
Institute of Mechanics, Vietnam Academy of Science and Technology
Email: nguyentd@imech.ac.vn

Nguyen Dang Chien
Faculty of Physics, University of Da Lat
Email: chiennd@dlu.edu.vn

Nguyen Dinh Hien
Center for Theoretical and Computational Physics, Hue University
Email: dinhhienphysics@gmail.com

Nguyen Dinh Nam
Hanoi University of Sciences
Email: namdinhnguyen87@gmail.com

Journal of Physics: Conference Series **726** (2016) 011001 doi:10.1088/1742-6596/726/1/011001

Nguyen Hoang Linh
Institute for Computational Science and Technology (ICST)
Email: hoanglinh221191@gmail.com

Nguyen Hoang Vu
Institue of Physics
Email: nhvu@iop.vast.ac.vn

Nguyen Hong Son
Trade Union University
Email: atvsld@gmail.com

Nguyen Huy Thao
Hanoi University of Education No 2
Email: abcthao@gmail.com

Nguyen Huy Viet
Institute of Physics, Hanoi
Email: nhviet@iop.vast.ac.vn

Nguyen Mai Chung
Université Paris-Sud & Institute of Physics
Email: maichung.n@gmail.com

Nguyen Ngoc Hieu
Duy Tan University
Email: hieunguyenvly@yahoo.com

Nguyen Phuoc The
Duy Tan University
Email: thenpk14@gmail.com

Nguyen Quang Bau
Faculty of Physics, Hanoi University of Science, Vietnam National University
Email: nguyenquangbau54@gmail.com

Nguyen Quang Hung
Tan Tao University
Email: hung.nguyen@ttu.edu.vn

Nguyen Quoc Thai
Dong Thap University
Email: nqthai@dthu.edu.vn

Nguyen Quy Ngan
Department of Theoretical Physics, Faculty of Physics and Engineering Physics, University of Science
Email: 1113280@student.hcmus.edu.vn

Nguyen Quy Truong
University of Science, HCM City
Email: nqtruong@hcmus.edu.vn

Journal of Physics: Conference Series **726** (2016) 011001 doi:10.1088/1742-6596/726/1/011001

Nguyen The Toan
VNU University of Science
Email: toannt@vnu.edu.vn

Nguyen Thi Thao
Hanoi National University of Education
Email: ntthao.hnue@gmail.com

Nguyen Thi Trang
Ha Noi University of Sciences and Technology
Email: trangnt1980@gmail.com

Nguyen Thi Ha Loan
Hanoi Pedagogical University No.2
Email: hloansp2@gmail.com

Nguyen Thi Hai Yen
Institute of Physics
Email: yen.a3k44@gmail.com

Nguyen Thi Phuong Lan
Hanoi Pedagogical University 2
Email: lanntphuong@gmail.com

Nguyen Thi Thuy Hang
Ho Chi Minh City University of Technology
Email: hangthuynguyen2001@yahoo.com

Nguyen Thu Huong
Air Defence and Air Force Academy
Email: huong146314@yahoo.com

Nguyen Thu Nhan
Hanoi University of Science and Technology
Email: thunhaniep@gmail.com

Nguyen Tien Cuong
VNU University of Science
Email: ntcuong@vnu.edu.vn

Nguyen Tri Lan
Institute of Physics, VAST
Email: ntlan@iop.vast.ac.vn

Nguyen Trung Tin
Institute of Computational Science and Technology
Email: trung.tin.nguyen0309@gmail.com

Nguyen Truong Co
Institute for Computational Science and Technology
Email: truongcophysics@gmail.com

Nguyen Tu Niem
VNU University of Science
Email: nguyentuniem@gmail.com

Nguyen Van Hinh
Hanoi University of Industry
Email: nguyenvanhinh78@gmail.com

Nguyen Van Hop
Physics Department, Hanoi National University of Education, 136 Xuan Thuy, Hanoi
Email: hopnvdhsp@yahoo.com

O'Brien Edward
Pennsylvania State University
Email: epo2@psu.edu

On Van Vo
University of Thu Dau Mot
Email: vovanon1963@gmail.com

Pham Thoa
Department of Theoretical Physics, Faculty of Physics, University of Science, Vietnam National University, Ho Chi Minh City, Vietnam
Email: kimthoa.pro1989@gmail.com

Pham Dang Lan
Institute of Computational Science and Technology (ICST)
Email: l72mss@gmail.com

Pham Ngoc Son
Nuclear Research Institute
Email: pnson.nri@gmail.com

Pham Nguyen Thanh Vinh
Ho Chi Minh University of Pedagogy
Email: vinhpham@hcmup.edu.vn

Pham Thi Thanh Nga
Thuy loi University
Email: ngat493@gmail.com

Pham Tuan Vinh
Dong Thap University
Email: tuanvinhdhdt@gmail.com

Phan Ngọc Hưng
HCMC University of Education
Email: hung.catalunya@gmail.com

Phan Hong Lien
Military Technical Academy
Email: phlien.mta@gmail.com

Phan Minh Trường
Institute for Computational Science and Technology
Email: minhtruong1040300@gmail.com

Phan Van Cuong
Nha Trang University, Vietnam
Email: cuongpv@ntu.edu.vn

Phí Quang Văn
Institute of Physics, VAST
Email: pqvan@iop.vast.ac.vn

Phạm Khắc Hùng
Đại học Bách Khoa Hà Nội
Email: san.luyenthi@hust.edu.vn

Phung Van Dong
Institute of Physics, Vietnam Academy of Science and Technology
Email: pvdong@iop.vast.ac.vn

Podgornik Rudolf
J. Stefan Institute and University of Ljubljana
Email: rudolf.podgornik@fmf.uni-lj.si

Sasai Masaki
Nagoya University
Email: sasai@nuap.nagoya-u.ac.jp

To Duy Quang
Center for Computational Physics - Institute of Physics, VAST
Email: tdquang@iop.vast.ac.vn

Tran Cong Phong
Vietnam Institute of Educational Sciences
Email: tcphong@moet.edu.vn

Tran Lan Phuong
Ho Chi Minh University of Pedagogy
Email: tranlanphuongsp@gmail.com

Tran Minh Tien
Institute of Physics, VAST
Email: minhtien@iop.vast.ac.vn

Tran Thanh Dung
Thu Dau Mot University
Email: ttdung24@gmail.com

Tran Thi Thu Trang
Halong University
Email: trangtranthithu.c17@moet.edu.vn

Tran Viet Nhan Hao
Department of Physics, College of Education, Hue University
Email: tran.viet.nhan.hao@gmail.com

Trần Đình Thám
Trường Đại học Phạm Văn Đồng
Email: tdtham@pdu.edu.vn

Trần Minh Hiếu
Hanoi University of Science and Technology
Email: hieu.tranminh@hust.edu.vn

Trịnh Xuân Hoàng
Institute of Physics, Vietnam Academy of Science and Technology
Email: hoang@iop.vast.ac.vn

Trovato Antonio
University of Padua
Email: antonio.trovato@pd.infn.it

Truong Minh Duc
Hue University of Education
Email: tmduc2009@gmail.com

Trương Trọng Thúc
Trường THPT Phú Hưng, Xã Phú Hưng, huyện Cái Nước, tỉnh Cà Mau
Email: ttthuc@grad.iop.vast.ac.vn

Võ Văn Tài
College of Technology and Industrial Management (CTIM)
Email: vvtaiphys17@gmail.com

Viên Tuấn Anh
Can Tho University
Email: vien_tuananh@yahoo.com

Vo Quy
University of Stuttgart
Email: quy.vo@itb.uni-stuttgart.de

Vo Quoc Phong
The University of Sceince-Ho Chi Minh City
Email: vqphong@hcmus.edu.vn

Vo Van Vien
Tay Nguyen University
Email: wvienk16@gmail.com

Vu Ba Dung
Physics Faculty, Hanoi University of Mining and Geology, Ha Noi, Viet Nam
Email: vubazung305@gmail.com

Vu Ngoc Tuoc
Hanoi Univ. of Science and Technology
Email: tuoc.vungoc@hust.edu.vn

Vu Thi Mui
Institute for Computational Science and Technology (ICST)
Email: vuthimui132@gmail.com

Vuong Van Quan
The Institute for Computational Science and Technology (ICST)
Email: v2quan89@gmail.com

# Peer review statement

All papers published in this volume of *Journal of Physics: Conference Series* have been peer reviewed through processes administered by the proceedings Editors. Reviews were conducted by expert referees to the professional and scientific standards expected of a proceedings journal published by IOP Publishing.

# Revisiting the Fermi Surface in Density Functional Theory

**Mukunda P Das**[1] **and Frederick Green**[2]

[1] Department of Theoretical Physics, RSPE, The Australian National University, Canberra, ACT 2601, Australia

[2] School of Physics, University of New South Wales, Sydney, NSW 2052, Australia

E-mail: mukunda.das@anu.edu.au

**Abstract.** The Fermi surface is an abstract object in the reciprocal space of a crystal lattice, enclosing the set of all those electronic band states that are filled according to the Pauli principle. Its topology is dictated by the underlying lattice structure and its volume is the carrier density in the material. The Fermi surface is central to predictions of thermal, electrical, magnetic, optical and superconducting properties in metallic systems. Density functional theory is a first-principles method used to estimate the occupied-band energies and, in particular, the iso-energetic Fermi surface. In this review we survey several key facts about Fermi surfaces in complex systems, where a proper theoretical understanding is still lacking. We address some critical difficulties.

## 1. Introduction

Density Functional Theory (DFT) is the "model of choice" for understanding condensed matter at low energies. It has achieved a certain status as a standard first-principles method. Indeed for many, though not all, significant condensed-matter phenomena DFT is a powerful analytic tool for studying electronic, vibrational, magnetic, superconducting among others.

The basis of DFT (for example, see references [1]-[6]) rests upon two foundational theorems by Hohenberg and Kohn [7]. Over the last five decades great progress has been made on both the theory's fundamental aspects and its scope in application to various systems. It has a reputation as one of the most successful practical methods for treating many-body systems on a fully quantum-mechanical footing.

In physical terms, however, it is many-body microscopics that ultimately fixes the complex solid-state properties one seeks to explain and predict. This essential many-body aspect is not addressed by DFT in any frontal way. Its approach to many-body interactions is an implicit, and thereby indirect, one by its positing certain well-defined but nevertheless approximate "exchange-correlation" (XC) forces between electron pairs.

The classic Kohn–Sham version of DFT [8] generates effectively single-particle simulations of what are, in actuality, interacting correlated multi-particle systems. The theory's distinctive property is to recover, in principle, the real-space single-particle-number and energy densities of such a condensed system. Anything more than that cannot be inferred from the theorem, notably any phenomena involving excited states beyond the single-particle configuration. That is because such effects are irreducibly many-body in nature and, logically, demand an explicit many-body analysis.

To reinforce this fundamental point we recall the key physical reason why many-body mechanisms must, at some stage, appear explicitly in any description of interacting carriers. Landau, in his account of the weakly interacting Fermi liquid (see for instance Nozières and Pines [9]), showed that any energy change of an excited system away from the ground state takes the inevitable form

$$\delta E[f+\delta f] \equiv \sum_{\mathbf{k}} \varepsilon_{\mathbf{k}}[f]\delta f_{\mathbf{k}} + \frac{1}{2}\sum_{\mathbf{k},\mathbf{k}'} F_{\mathbf{k},\mathbf{k}'}[f]\delta f_{\mathbf{k}}\delta f_{\mathbf{k}'} \tag{1}$$

where $f_{\mathrm{k}}$ is the single-particle-state distribution function, and $\varepsilon_{\mathbf{k}}[f]$ is the corresponding single-particle spectrum in the interacting system. (The more general Landau formula describes the interacting free energy at finite temperature $T$; in this case $\delta f$ may as readily represent the effect of a thermal energy change, that is away from $T = 0$, as that of any other perturbation; a situation on which we have much more to say in the following.) Crucially, Landau's two-particle interaction potential $F_{\mathbf{k},\mathbf{k}'}[f]$ appears in equation (1). Such an interaction is not representable by *any* single-particle functional.

Although DFT, in principle, does guarantee replication of the exact total ground-state energy $E[f]$ over the occupied single-particle states, it cannot guarantee

(a) that its effective single-particle spectrum, say $\{\widetilde{\varepsilon}_{\mathrm{k}}[n]\}_{\mathbf{k}}$, as an auxiliary functional of the real-space density $n(\mathbf{r})$, is literally the replica of the actual one $\{\varepsilon_{\mathrm{k}}[f]\}_{\mathbf{k}}$, and

(b) that it will furnish a systematic estimate for the interaction potential.

Both (a) and (b) are essential to a description of the full behaviour of the particle assembly. For example, the physical response of the real system to any perturbation in the single-particle distribution is characterised directly by

$$\chi_{\mathbf{k},\mathbf{k}'}[f] \equiv \frac{\delta^2 E}{\delta f_{\mathbf{k}}\delta f_{\mathbf{k}'}} \equiv F_{\mathbf{k},\mathbf{k}'}[f].$$

Consequently the so-called Landau interaction parameters $F$ mediate, directly, all essential properties such as the dielectric response, spin susceptibility, compressibility, specific heat, and so on [9].

As we have remarked, the interaction parameters are inherently two-body properties that cannot be simulated merely in single-particle terms. This means that density functional theory for the ground state alone, viewed as the definitive and optimal *strictly single-particle* description of a many-body system, is insufficient to capture many significant response properties of a real, correlated system. Yet it is the response of a system over the entire arsenal of experimental probes that provides the structural information one needs to discover.

The inbuilt limitations of the Kohn–Sham formalism are inherited by any and all DFT-based machinery relying upon it exclusively. Owing to the essentially noninteracting, single-particle nature of the resultant models (and notwithstanding the popularity of DFT as a practical tool applied to actual many-body problems), a basic question of physics remains to be answered: How can one bring to experimental test any differences (at the very least, in the ground-state properties) between DFT as an effective one-body theory, and the established canon of many-body analysis?

In testing the limits of DFT the detailed measurable properties of the Fermi surface are particularly relevant. At the best of times, these are delicate qualities to predict. A widespread assumption still goes largely unquestioned: namely, that the Fermi surface, with all of its intricate topology, is uniquely a ground-state property and, uniquely, determinable via DFT. This cannot be the case in general, least of all for the complex interacting structures of current interest.

Journal of Physics: Conference Series **726** (2016) 012001 doi:10.1088/1742-6596/726/1/012001

For ideal free carriers in an ideally uniform three-dimensional sample, the Fermi sphere is bounded by an equi-energetic surface of constant curvature, of radius $k_F$, the Fermi momentum. Luttinger and Ward [10] showed that, in such a uniform metallic system, the Fermi surface will survive in the presence of inter-particle interactions. All of the low-lying excitations of the system live near the Fermi surface and account for much of the system's low-temperature behaviour.

Nevertheless, the real Fermi surfaces of real materials are not simply spheres or even quasi-spheres of at least positive curvature everywhere. Indeed they are topologically complicated, often multiply connected, and highly subject to the underlying crystalline geometry and interactions. They can even come as disconnected Fermi-surface pieces, as well as exhibiting Fermi-surface "nesting". The latter can induce unusual instabilities in certain metals: for example, at sufficiently low temperature a normal metal may move into a charge- or spin-density-wave state or other topological phase transition.

Present condensed-state research must deal extensively with novel and exotic materials. Invariably these are moderately to highly correlated in terms of their strong collective electron interactions. It follows that the relevant physics will necessarily display its many-body nature [11]-[13]. Equally it follows, in this context, that attempts to use any single-body description must be weighed with care.

In the next Section we outline the basic features of density functional theory. In view of very large growth in literature we present a brief picture of it, keeping only the salient points for our purpose. We discuss which ground-state properties within DFT can be trusted to give a reliable physical estimate, as well as what it is not designed to address. In Sec. 3 we survey various popular approximations to model the exchange-correlation functional, and present their relative merits or demerits. In this Section we provide a list of the most often-used computer codes that DFT practitioners rely upon.

In Sec. 4 we revisit a basic open question, still unresolved today: Is the Fermi surface uniquely a one-body ground-state property? From a many-body perspective this question has to be addressed carefully at zero temperature. Experimental observation of the Fermi surfaces from angle-resolved photo-emission spectroscopy (ARPES), Kohn anomaly, Shubnikov-de Haas and angle resolved magneto-resistance (ARMR) methods are discussed briefly and contrasted with calculated results of DFT, quasi-particle (GW) theory and DMFT. In Sec 5 finite-temperature aspects are discussed, since all actual experiments are done at $T > 0$. Our Summary with conclusions is presented in the final Section.

## 2. Outline of DFT

### *2.1. Hohenberg-Kohn theorem*

The Theorem for the Energy Functional of Hohenberg and Kohn [7] (an existence proof) states the following. Admitting certain general assumptions, any change in the ground-state density $n(\mathbf{r})$ of an interacting electron system subject to an adiabatically changing external potential $v_{\rm ext}(\mathbf{r})$, remains in one-to-one correspondence with that potential.

Since both $n(\mathbf{r})$ and the total number of particles, N, are uniquely tied to $v_{\rm ext}$ we can readily construct the full system Hamiltonian. From it, $n(\mathbf{r})$ can be calculated by solving a Schrödinger-like equation derived from a variational principle. We skip the many technical details (see references [1]-[6]) and proceed the popular Kohn-Sham method [8].

The total energy is given as a functional of the spatial density distribution $n(\mathbf{r})$ by

$$E[n] = T[n] + U[n] + V_{\rm ext}[n] = T[n] + U[n] + \int n(\mathbf{r})v_{\rm ext}(\mathbf{r})d\mathbf{r}, \qquad (2)$$

in which the first two terms on the right-hand side are the total kinetic energy $T$ and total inter-particle interaction energy $U$. The total potential energy $V_{\rm ext}$ includes the electron-ion electrostatic interaction, here considered to be external to the system of mobile electrons.

In the Kohn-Sham (KS) method we consider the kinetic energy to be exactly equivalent to that of a system of independent (noninteracting) electrons, with a basis of single-particle orbitals $\{\phi_i[n]\}_{i=1}^{N}$ that are themselves functionals of the exact density distribution for the $N$ electrons in the ground state of the system. The theorem guarantees these orbitals to be well-defined. They are determined systematically, in a self-consistent fashion.

The fundamental expression from the DFT analysis of the total ground-state energy now becomes more explicit in detail. In this setting we have

$$E[n] \equiv T[\phi_i] + U_{\rm H}[\phi_i] + E_{\rm xc}[\phi_i] + V_{\rm ext}[\phi_i]; \tag{3}$$

the interaction energy $U$ is first resolved into the mean-field Hartree component $U_{\rm H}$ and the exchange-correlation component $E_{\rm xc}$ with

$$U_{\rm H}[\phi_i] = \frac{1}{2}\int d\mathbf{r}\; v_{\rm H}(\mathbf{r})n(\mathbf{r}).$$

The set of orbitals is obtained through the Kohn-Sham equations (here indexed by the Brillouin-zone wavevector $\mathbf{k}$):

$$\begin{aligned}
\left[\frac{\hbar^2 k^2}{2m} + v_{\rm eff}(\mathbf{r})\right]\phi_{\mathbf{k}}(\mathbf{r}) &= \widetilde{\varepsilon}_{\mathbf{k}}\phi_{\mathbf{k}}(\mathbf{r}), \quad \text{where} \\
v_{\rm eff}(\mathbf{r}) &\equiv v_{\rm ext}(\mathbf{r}) + v_{\rm H}(\mathbf{r}) + v_{\rm xc}(\mathbf{r}), \quad \text{with} \\
v_{\rm H}(\mathbf{r}) &= \int \frac{e^2}{|\mathbf{r}-\mathbf{r}'|} d\mathbf{r}' n(\mathbf{r}'), \quad \text{and} \quad v_{\rm xc}(\mathbf{r}) \equiv \frac{\delta E_{\rm xc}}{\delta n(\mathbf{r})}
\end{aligned} \tag{4}$$

Two final steps are required to close this system self-consistently. First we define the constitutive relation for the density in terms of the single-particle KS orbitals:

$$n(\mathbf{r}) \equiv \sum_{\mathbf{k}} \theta(\mathbf{k})|\phi_{\mathbf{k}}[n(\mathbf{r})]|^2 \tag{5}$$

writing $\theta(k) = \theta(\widetilde{\varepsilon}_{\rm F} - \widetilde{\varepsilon}_{\mathbf{k}})$ for the carrier occupation number, equal to unity within the occupied Fermi sphere and zero for the empty momentum states $\mathbf{k}$ above it; here $\widetilde{\varepsilon}_{\rm F}$ is the corresponding DFT estimate of the Fermi energy. Second, an explicit form has to be decided for the exchange-correlation functional $E_{\rm xc}[\phi_i]$ so that $v_{\rm xc}$ can be constructed and fed into the KS equations. We defer this crucial point of principle to section 3.1 below.

By varying the total energy with respect to the undetermined orbitals $\phi_{\mathbf{k}}$ we obtain the $N$ Euler-Lagrange equations for the system. They are coupled nonlinear Schrödinger-orbital equations, conceptually very similar to Hartree-Fock but containing much more correlational input, over and above exchange.

The volume of $k$-space, enclosed by the Fermi iso-surface defined as $\{\mathbf{k}; \widetilde{\varepsilon}_{\mathbf{k}} = \widetilde{\varepsilon}_{\rm F}\}$, is exactly the number $N$ of mobile electrons physically occupying the lowest-lying states within the sample. Conservation of carriers requires this to be an invariant property of the closed system. In particular, any imposed external field may distort the surface and even change its topology substantially, but cannot alter the enclosed volume except possibly in a phase transition that totally re-orders the ground state.

### *2.2. Meaning of the Kohn-Sham eigenenergies*

Just as with Hartree-Fock theory, the early precursor of DFT, the total energy $E[n]$ is not the sum of all the single-carrier orbital energies $\widetilde{\varepsilon}_{\mathbf{k}}$. In fact, we know from the derivation that the $\widetilde{\varepsilon}_{\mathbf{k}}$ enter purely as Lagrange multipliers; variational parameters that are strictly artifacts from the physical standpoint. The set of $\widetilde{\varepsilon}_{\mathbf{k}}$ is simply the set of formal eigenvalues for the auxiliary one-body equations of DFT, whose eigenfunctions are guaranteed only to yield the correct local density $n(\mathbf{r})$; nothing else.

It is the net density profile, not its auxiliary contributions, that carries the genuine physical content of the KS equations. While the auxiliary KS eigenvalues may generally bear some qualitative resemblance to the true energy spectrum, there exists no guarantee that they form a trustworthy representation of the true single-particle spectrum.

There is one important exception to the caveat above. The value for the DFT Fermi energy $\widetilde{\varepsilon}_{\mathrm{F}}$ is, at least in principle, the actual Fermi level in the ground state. But this does not mean that the microscopic *topology* of the Fermi surface is at all reproduced by DFT.

We stress that, although there may be practical and even heuristic reasons to suggest that the set $\{\widetilde{\varepsilon}_{\mathbf{k}}\}_{\mathbf{k}}$ describes the "true" band structure, to date a basic microscopic justification of this hypothesis remains anything but settled. See for example references [14]-[16].

## 3. Implementation of the Kohn-Sham formulation

The only quantity that remains to be fixed is $E_{\mathrm{xc}}[n]$, the exchange–correlation energy functional. It is formally defined by the adiabatic connection formula [2], [5]. For computation, the expression for $E_{\mathrm{xc}}[n]$ has to be carefully constructed. This is where great efforts have gone into constructing several approximate expressions, namely the so-called LDA, GEA, GGA, hybrid functionals, ODF, and so on [4]-[6]. We now outline their main characteristics.

### *3.1. Approximations for the density functional*

(a) Local-Density Approximation (LDA): the general strategy of local-density approximations, as also those for local-spin-density (LSD) is to take known results for the XC potential $v_{\mathrm{xc}}[\nu]$ of a uniform system at density $\nu$ and apply them locally to an inhomogeneous system. This, in this model, $E_{\mathrm{xc}}[n(\mathbf{r})]$ becomes a sum of locally homogeneous (possibly spin-dependent) exchange-correlation energies of electrons over a small cell in real space, with a homogeneous density $\nu$ matching the local value $n(\mathbf{r})$.

Given this Ansatz, the total XC energy is approximated as

$$E_{\mathrm{xc}}^{\mathrm{LDA}}[n(\mathbf{r})] = \int d\mathbf{r}\; n(\mathbf{r}) v_{\mathrm{xc}}[n(\mathbf{r})]; \quad v_{\mathrm{xc}}[\nu] \equiv \frac{\delta E_{\mathrm{xc}}[\nu]}{\delta\nu}.$$

This formula works well when density gradients are small over the typical range of $k_{\mathrm{F}}(r)^{-1}$, the Fermi wavelength. Forms for $E_{\mathrm{xc}}[n]$ in LDA are often taken from parametrisation of highly precise Quantum Monte Carlo (QMC) calculations for the electron liquid.

(b) Gradient-Expansion Approximation (GEA): If the density variation is not small, one can try to include systematically the gradient corrections to the LDA expressions, going as $|n(\mathbf{r})|$, $|n(\mathbf{r})|^2$, etc.. In practice, low-order gradient corrections almost never improve the LDA results and higher-order corrections are exceedingly difficult to calculate. In any case, for real systems the results of GEA are worse than those of LDA [2].

(c) Generalised Gradient Approximation (GGA): Instead of finite-order, power-series-like gradient expansions one can use more general functionals of $n(\mathbf{r})$ and $\nabla n(\mathbf{r})$, which need not proceed order by order. Such functionals assume the general form

$$E_{\mathrm{xc}}^{\mathrm{GGA}}[n] = \int d\mathbf{r}\; f(n(\mathbf{r}); \nabla n(\mathbf{r})) n(\mathbf{r}),$$

where the $f(n(\mathbf{r}), \nabla n(\mathbf{r}))$, now non-local, is carefully constructed by satisfying at least the leading conservation sum-rules such as perfect-screening (of each carrier by its exchange-correlation hole). Various fitted forms are available in the current literature [6].

(d) Hybrid functionals: These form a set of approximate forms for the exchange-correlation energies, incorporating a portion of the exact exchange term via Kohn-Sham wave functions together with correlation estimates from empirical sources.

There are many parametrised hybrid forms, some of which are of use in atomic and molecular calculations. One of the forms is given here:

$$E_{\mathrm{xc}}^{\mathrm{hybrid}} \equiv aE_{\mathrm{x}}^{\mathrm{exact}} + (1-a)E_{\mathrm{x}}^{\mathrm{GGA}} + E_{\mathrm{c}}^{\mathrm{GGA}},$$

where $a$ is an adjustable mixing coefficient.

(e) Orbital-dependent functionals: This is known as the "third generation" of DFT. For details, see Engels' paper in chapter 2 of reference [2]. Here, instead of just density-dependent functionals, one uses orbital-dependent functionals. Since the orbitals will obviously embody more microscopic information, there are several advantages of this approach to highly correlated systems.

(f) Calculations by a sort of garden variety of techniques: these apply both to the methodology and to functional approximations too numerous to detail here: the so-called VASP, SIESTA, CRYSTAL, PAW, CASTEP, Quantum Espresso, FPLO, ABNIT. And so on. Some of these functionals have been devised for building into computer codes developed over many years by many people. These codes opened many new gates to the detailed computation of many physical quantities and were popularly adopted by large numbers of practitioners, even when the quantities calculated were being pushed somewhat beyond the advertised "fitness-for-use" of the codes.

### *3.2. Successes and failures*

It is difficult to describe in simple terms not only the noted successes of DFT but also, more to the point, the sometimes minimised cases of its failures. There are many reviews and texts to highlight fulsomely its manifold impressive successes.

Regarding the alleged outstanding success of the GGA, Perdew and Kurth [2] have written that "LSD has been so successful in SSP [solid state physics] and a small residue of GGA nonlocality in solids does *not* provide a universally better description than LSD."

Kokko and Das [17] have shown LDA and GGA do not always afford regular systematics, making it difficult to say objectively which one is better for ground-state properties of inhomogeneous systems, such as 3d and 4d transition metals. Many discussions have appeared in the past few years regarding the successes and failures of these DFT approximations [4], [6], [18]. The general belief seems to be that the approximations are systematically developed, despite their remaining always somewhat uncontrolled and therefore placing their reliability in question.

## 4. Fermi surfaces of metals

In keeping with the central issue of this paper, at this stage we are able to discuss questions on the nature of Fermi surfaces and the adequacy of their description.

### *4.1. Is the Fermi surface of a metal a ground-state property?*

It is widely held that the set of iso-energetic electron bands crossing the Fermi level $\varepsilon_{\mathrm{F}}$ define the Fermi surface (FS); a key quantity in understanding the electronic structure of any metallic material. This simple intersection may fix the true locus of the FS, or it may not (depending on

Journal of Physics: Conference Series **726** (2016) 012001 doi:10.1088/1742-6596/726/1/012001

how faithful is the model). But it certainly tells nothing of the physical properties, qualitative and quantitative over the entire surface itself.

The conventional way of mapping a Fermi surface is to measure the energy-distribution curves (EDC), for distinct k-points of the Brillouin zone, via angle-resolved photoemission spectroscopy (ARPES) and thus to ascertain the k-locations where the bands transect the Fermi energy. The connection between band structure and ARPES results is authoritatively discussed in references [19] (an especially clear review of ARPES techniques) and [20].

The main issue is that a metal hosting many-electron correlations will retain a well-defined Fermi surface. In relatively weakly correlated metals, the single-electron (i.e. independent-particle) band structure reproduces the measured FS reasonably faithfully. As mentioned before, certainly for strongly correlated systems and even when strongly correlated, the one-electron band structure is liable to become inadequate in providing a good picture of the FS. From equation (2) the introduction we recall the need to compute at least the two-body Landau interaction $F_{\mathbf{k},\mathbf{k}'}$ at the Fermi surface to determine the response properties.

It is then clearly necessary to check the results of the single-particle DFT for both band structure and the energy dispersion of the quasi-particle (that is, many-body-dominated) energies. In the Fermi-liquid picture, absent any inter-particle interactions, the band distribution of the carriers at $T = 0$ is the Fermi-Dirac step $f_{\mathbf{k}} = \theta(\varepsilon_{\mathrm{F}} - \varepsilon_{\mathbf{k}})$. This noninteracting distribution is precisely the formal situation required within KS theory.

By contrast, in the presence of interactions the would-be step function, even at zero temperature, becomes smeared around the Fermi energy. This is because minimisation of the total interacting ground-state energy favours a physical configuration in which a portion of the (otherwise noninteracting) one-body excitations is relocated in $k$-space from somewhat below to somewhat above the Fermi level. In other words: an increase of kinetic energy for the distribution, by promotion of lower-lying electrons to slightly higher states, is more than offset by a lowering of the collective correlation energy.

Owing to this interaction-induced rearrangement, the originally noninteracting single-particle modes about the FS map one-to-one to stable (but now correlated) quasi-particle (QP) state counterparts, whose distribution retains the step-function form. But within the noninteracting-basis picture, this induces the appearance of a renormalisation constant $Z_k$ at the Fermi surface; $Z_k$ is always less than unity and rescales the jump in occupancy of the underlying *non-interacting* states at the Fermi level. Its size is fixed by the self-energy correction at the FS, arising from the many-body correlation effects.

The self-energy is immediately related to the Landau two-particle interaction [9] and therefore cannot be extracted from a purely one-body analysis. For, while the knowledge of the particle-particle interaction is sufficient to determine the self-energy and thus its contribution to the (true) band spectrum $\varepsilon_{\mathbf{k}}$, any model spectrum $\widetilde{\varepsilon}_{\mathbf{k}}$ is not in itself sufficient to determine the interaction. That is the core difficulty in securing the adequacy of DFT to describe the Fermi-surface reconstitution, or "renormalisation".

We arrive at a crucial point. In its many-body setting, when described relative to the originally noninteracting single-particle basis, the Fermi surface is always "renormalised" by contributions that are explicitly many-body in nature. The FS is deformed away from its noninteracting profile depending on the strength of interaction. This highlights the central fact that the physical ground state need not at all conform to the intuitive picture provided by effective one-body formulations such as DFT or by its conceptual ancestor, Hartree-Fock.

For simple bulk metals the Fermi temperature $T_{\mathrm{F}} = \varepsilon_{\mathrm{F}}/k_{\mathrm{B}}$ can be large, on the order of $10^4$K and more. There, it is a good approximation to treat the distribution $f_{\mathbf{k}}$ in its zero-$T$ limit; namely, a step function with cut-off at $\varepsilon_{\mathrm{F}}$. However, for complex metals possessing various phase transitions at low temperatures, the $T = 0$ limiting approximation will be of no use in capturing the correct physics. That is because the interactions themselves, quite apart from any thermal

Journal of Physics: Conference Series **726** (2016) 012001 doi:10.1088/1742-6596/726/1/012001

effects, have already smeared it out.

Even when phase transitions are not in play, we have seen how many-body renormalisation can induce a radical change in the landscape of the critical states around a "normal" Fermi surface. In the potential presence of phase changes, one has even less option than to approach head-on the theory of finite-temperature effects on electronic structure. The quick fix does not exist here.

Energy bands as estimated by DFT differ substantially from the actual QP states described by many-body theory. This is evident when comparing the respective DFT and QP expressions

$$\begin{aligned} \left[\frac{\hbar^2}{2m}k^2 + v_{\rm ext}(\mathbf{r}) + v_{\rm H}(\mathbf{r}) + v_{\rm xc}(\mathbf{r})\right]\phi_{\mathbf{k}}(\mathbf{r}) &= \widetilde{\varepsilon}_{\mathbf{k}}\phi_{\mathbf{k}}(\mathbf{r}); \\ \left[\frac{\hbar^2}{2m}k^2 + v_{\rm ext}(\mathbf{r}) + v_{\rm H}(\mathbf{r})\right]\psi_{\mathbf{k}}(\mathbf{r}) + \int d\mathbf{r}'\Sigma(\mathbf{r},\mathbf{r}';\varepsilon_{\mathbf{k}})\psi_{\mathbf{k}}(\mathbf{r}') &= \varepsilon_{\mathbf{k}}\psi_{\mathbf{k}}(\mathbf{r}). \end{aligned} \tag{6}$$

In the latter expression arising from standard many-body analysis, $\Sigma(\mathbf{r},\mathbf{r}';\omega)$ is the nonlocal, irreducible electronic self-energy. This dynamical self-energy term is far more structured and internally complex than its static DFT analogue $v_{\rm xc}$.

There exists a wide range of systematic and consistent approximations for $\Sigma$. A relatively simple and successful such model is the "GW" approximation, in which the "G" stands for the one-body Green function and the "W" for the microscopically screened two-body interaction [21]. In the frequency domain $\Sigma(\mathbf{r},\mathbf{r}';\omega)$ is given by

$$\Sigma(\mathbf{r},\mathbf{r}';\omega) \equiv i\int d\omega' G(\mathbf{r},\mathbf{r}';\omega+\omega')W(\mathbf{r}',\mathbf{r};\omega'),$$

in which the screened electron-electron interaction is approximated by

$$W(\mathbf{r}',\mathbf{r};\omega') \equiv \int d\mathbf{r}''\epsilon^{-1}(\mathbf{r}',\mathbf{r}'';\omega)v(\mathbf{r}''-\mathbf{r}).$$

and here $\epsilon^{-1}(\mathbf{r}',\mathbf{r}'';\omega)$ is the inverse of the dynamic dielectric function.

If $v_{\rm xc}$ were a good approximation to $\Sigma$ the DFT band energies $\widetilde{\varepsilon}_{\mathbf{k}}$ would also be good. However, comparison of calculations performed within KS and GW show large differences near the Fermi wave vector. There are many examples of DFT Fermi surfaces which are manifestly at variance with the physical, many-body QP Fermi surface. This is particularly so in multi-band and correlated materials [22], [23].

### *4.2. Fermi surfaces from dynamical mean-field theory*

Dynamical mean-field theory (DMFT) is a new development that takes into account local correlations more faithfully [24]. It is an improved GW method, where local correlations are incorporated by treating them as effective impurities in a periodic system. In brief, the method produces an effective mass and the associated renormalisation for the QPs can be a substantial, consistent with observed ARPES [25] in correlated electron systems.

### *4.3. Is the DFT-Kohn-Sham Fermi surface a ground-state property?*

At the start of this Section we examined the wider issues of principle leading to this question. The answer is straightforward. The KS equations certainly reproduce the correct physical density distribution and the total ground-state energy. In the course of things, given that a band structure emerges out of KS, a FS must also exists for it.

Nevertheless, when it comes to real systems with their interactions and inhomogeneities, the FS of DFT-KS is already known to be inequivalent *in principle* to the physical FS obtained from the microscopic Dyson equation [26], [27], which codifies the many-body effects within the true QP distribution. A detailed theoretical argument demonstrates that the inaccuracy of the FS in DFT-KS theory comes from an inbuilt lack of sufficiently strict convergence for the gradient approximation implicit in DFT.

In an analysis using a time-dependent generalisation of DFT, Cohen and Wasserman [28] conclude that KS FS is identical to the QP FS in the sense of a "distributional" argument – that is, at best in some average sense.

### *4.4. Are explicit many-body effects seen in the FS properties of metals?*

We cite some apposite and fruitful observations on the issue.

- A.K. Rajagopal notes in his review [14]: "The eigenvalues [of the DFT equation] do not have any special significance. The equation is a mathematical artifact of the HKS [Hohenberg-Kohn-Sham] formalism. By taking a pragmatic point of view by treating $E_i$ [our $\widetilde{\varepsilon}_{\mathbf{k}}$] as a one electron eigenvalue in the one electron theory of band structure, one arrives at the HKS band structure. There is much controversy regarding the definition of "Fermi surface" whether it is a ground state property of the system or not?"
- Richard Martin, on p. 131 of his text [15], also visits this point. We quote: "Is the exact Fermi surface of a metal given by the exact ground state DFT? .... this is not a trivial question for two reasons (i) a many-body metal must have a well-defined FS - this is assumed for the purpose. (ii) It is not a-priori obvious that FS is a ground state property. One way to see if the FS is determined by a ground state property is to consider the susceptibility to static perturbations. The exact DFT must lead to the correct Kohn anomaly and Friedel oscillations of the density far from an impurity, which depend in detail on the shape of the FS of the unperturbed metal."

We return to Martin's last point in detail, in subsection 4.7 below; since a singularity in the dielectric susceptibility $\chi(q)$ is induced by electron-hole FS excitations when $q = 2k_{\rm F}$ spans the Fermi surface (the Kohn anomaly), one may reconstruct the physical FS from this information. The loci of the singularity trace out the FS. It is thus natural, and appropriate, to interrogate the effectiveness of DFT-based approaches as to how well their FS estimates match measured singularity data.

### *4.5. Instability of the Fermi surface*

If the FS is a robust ground-state property, exactly how stable is it against perturbations? Here are several examples.

- For a number of reasons including interactions, temperature and impurities, a system of metallic electrons can undergo a phase transition precipitated by superconducting or magnetic pairing; by a charge- or spin-density wave (CDW/SDW); by a nematic state; or a number of other many-body collective states.
- As mentioned in the Introduction, Landau's theory of the Fermi liquid of 1957 became the canonical description of the metallic state. It was realised, even at that time, that the FS is unstable against strong interactions; the Pomeranchuk instability. A simple example occurs when an initially isotropic FS loses its symmetry in the presence of a strong interaction. The system will minimise its free energy at the cost of the FS becoming distorted and anisotropic. Again, the gain in kinetic energy is more than offset by a lowering of correlation energy.

- During the past few years a number of models in 2D systems have been studied, both theoretically and experimentally. These have once more brought to the fore important questions in our understanding of the FS as an invariant ground-state property [29]-[34].

*4.6. How does one determine the FS?*
A good exposition of Fermi-surface analysis is given in the text by Ziman [34]. Currently one can try to map out the FS by the following means:

(i) Theoretically, by DFT through the naive KS band structure (with formally questionable status).

(ii) DFT augmented with *exact* response theory.

(iii) Empirically, by ARPES [19], [20] measurements: the ARPES spectrum is not just a set of one-electron bands, but measures directly quasi-particle spectral function $A(\mathbf{k},\omega)$, given theoretically by the imaginary part of the Green function for one-particle excitations, with interaction effects (self-energy) fully included:

$$A(\mathbf{k},\omega) = -\frac{1}{\pi}\mathrm{Im}\{G(\mathbf{k},\omega)\} = -\frac{1}{\pi}\mathrm{Im}\left\{\frac{1}{\hbar\omega - \varepsilon^0_{\mathbf{k}} - \Sigma(\mathbf{k},\omega)}\right\}$$

where $\varepsilon^0_{\mathbf{k}}$ is the noninteracting single-carrier energy and $\Sigma(\mathbf{k},\omega)$ can be calculated by GW or DMFT.

(iv) Magnetic resonance effects. Electronic Fermi surfaces are selectively measured by observing of the oscillation of transport properties in differently oriented magnetic fields $\mathbf{H}$. This approach results in, for example, the de Haas–van Alphen effect (dHvA) and the Shubnikov–de Haas effect (SdH). The former is an oscillation in magnetic susceptibility and the latter in resistivity. The determination of the periods of oscillation for various strengths and directions of $\mathbf{H}$ allows one to infer the size and shape of the Fermi surface [34].

(v) Mapping out the FS using Angle-Dependent Magneto-Resistance oscillations (ADMRO) [36], [37]. Reference [36] has a comparison of ARPES and the ADMRO measurements in $Sr_2RuO_4$. The authors point out certain obvious inconsistencies in the ARPES data.

*4.7. Fermi surface from canonical response theory*
The Lindhard formula for the charge-density response (see reference [34], p 129) is defined for noninteracting electron states, with occupancy $f^0_{\mathbf{k}}$ at band energy $\varepsilon^0_{\mathbf{k}}$. The dielectric function $\epsilon(\mathbf{q},\omega)$ of an electron liquid is then given by

$$\epsilon(\mathbf{q},\omega) = 1 - v(q)\chi(\mathbf{q},\omega) \quad \text{where} \quad \chi(\mathbf{q},\omega) = \sum_{\mathbf{k}} \frac{f^0_{\mathbf{k+q}} - f^0_{\mathbf{k}}}{\hbar\omega - \varepsilon^0_{\mathbf{k+q}} + \varepsilon^0_{\mathbf{k}}}. \tag{7}$$

Evaluating this quantity and taking its static limit $\omega \to 0$ at finite $q$ and $T = 0$, we have

$$\chi(\mathbf{q}) \equiv \chi(\mathbf{q},0) = \frac{3n}{4\varepsilon_{\mathrm{F}}}\left[1 + \frac{4k_{\mathrm{F}}^2 - q^2}{4k_{\mathrm{F}}q}\ln\left|\frac{2k_{\mathrm{F}}+q}{2k_{\mathrm{F}}-q}\right|\right]. \tag{8}$$

This formula is logarithmically singular at $q = 2k_{\mathrm{F}}$.

The dielectric function is continuous but its $q$-derivative has a logarithmic infinity at $q = 2k_{\mathrm{F}}$. A more realistic calculation shows that the singular logarithmic behaviour is not only evident in the static dielectric susceptibility for an isotropic FS, but indeed that the logarithmic anomaly occurs for any Fermi surface, where $\chi(\mathbf{q},\omega)$ assumes the quite general form displayed in the second of the expressions given in equation (7).

The Kohn singularity will manifest itself under very broad conditions. It demands only that there be a non-vanishing iso-energetic Fermi surface. If one selects a measurement with $\mathbf{q} = 2\mathbf{k}$

as a function of orientation for those Fermi momenta spanning the Fermi surface (that is, such that $\varepsilon_{\mathbf{k}} = \varepsilon_{\mathrm{F}}$) one maps out the physical FS in reciprocal space.

For a lattice system the inter-ionic potential is quasi-statically screened by the dielectric function. As a result the phonon frequency, which is much lower than the Fermi energy, is itself dependent on $\epsilon(\mathbf{q}) = 1 - v(q)\chi(\mathbf{q})$. Thus the singularity is reflected as a distinct "kink" in the phonon dispersion, known as the Kohn anomaly [38].

The dynamic dielectric function $\epsilon(\mathbf{q}, \omega)$ can be constructed from detailed many-body theory by systematic inclusion of the exchange-correlation forces. In this context two theories are at hand: (1) the Kohn-Sham theory of independent electrons (subsuming all many-body effects into an effective one-body potential) and (2) explicit many-body terms contributing directly to the Dyson equation for the quasi-particle excitations. One can then very well see that the Kohn singularity appears in both KS and QP Fermi surfaces, since both approaches lead to a well-defined estimate for it. They may then be compared as to their agreement with the Kohn-anomaly phonon data, among others.

## 5. Finite-temperature effects on the FS

DFT was formally extended by Mermin [37] to include finite temperatures, by considering a grand canonical ensemble of particles. Unlike the case of the theory for $T = 0$, no rigorous DFT functional is available for the grand potential. Nevertheless there are some approximate LDA-type functionals that have been invoked in plasma physics, nuclear physics and quantum-chemistry applications of DFT [39].

To our knowledge, finite-temperature band structures are rarely considered relevant to metal physics. This is understandable for simple metals, since the Fermi temperature far exceeds the scale of any experimental $T$. The electronic structure and FS at room temperature and below are reasonably well understood, taking into account the constraints previously discussed. Further, these systems do not exhibit any phase change at low temperatures. Here, therefore, one believes that $T = 0$ DFT is adequate and there are many calculations reported in the literature.

In reference [40] a calculation by the LDA+DMFT method is compared with ARPES and dHvA experimental results. It is shown there that DFT calculations, either with LDA or GGA, fail to reproduce the experimentally observed electronic structure of the multi-band material $KFe_2As_2$.

In this report our main attention falls on metals of multi-band type, with multiple FSs and many phase transitions at low temperatures. Fermi-surface instabilities at finite temperature form a topic of intense interest as of this writing [41]. Typical examples are $Sr_2Ru_2O_7$ and many crystalline pnictides [42]–[44]. In such cases, finite-temperature effects ought to be considered within the relevant theories to be applied. Since serious finite-temperature DFT has not been attempted – in contrast with rather naive extrapolations of DFT at $T = 0$ – many issues have now come to attention.

We first remark that $k_{\mathrm{F}}$ ceases to be a relevant parameter away from $T = 0$, for cases in which the condition $\varepsilon_{\mathrm{F}} \gg k_{\mathrm{B}}T$ ceases to hold. Rather, when Fermi and thermal energies become comparable, a different wave-vector $k_m$ comes into play, defined in terms of the actual chemical potential $\mu(T)$ by

$$\frac{\hbar^2 k_m^2}{2m^*[\mu(T)]} + U(k_m) \equiv \mu(T)$$

where $U$ is the momentum-dependent mean-field potential. For low temperature the sharpness of the FS still exists but only so long as $k_{\mathrm{B}}T \ll \mu$.

There is already a remarkable example of a large finite-temperature effect that is inaccessible to simplistic extrapolations of zero-temperature methodology. Backes et al. [46] performed a finite-temperature and -pressure DFT based on Born-Oppenheimer molecular dynamics for a Fe pnictide system. Their finite-$T$ results, in pronounced contrast with the electronic structure and FS computed at zero temperature, are new and entirely different [46]. The authors report that the band structure as well as the density of states, unlike those for $T = 0$, exhibit pronounced structural oscillations when measured at finite $T$. Oscillation of the band energies, particularly near the Fermi surface at ambient pressure, then leads to thermal broadening as expected.

The investigations by Backes et al. show that the FS at finite temperature has a conspicuously different behaviour in experimental observations, substantially at odds with DFT-based estimates relying on purely zero-temperature arguments. These need to be investigated further. Given these experimental findings, the finite-temperature aspects are undoubtedly missed by any effectively $T = 0$ density functional calculation.

In contrast with the Backes et al. work, a recent instance of a conventional DFT calculation taken outside its proper limit of validity is provided by Sen et al. [45], who have advanced a misleading picture of the FS structure for FeAs-based pnictide materials at finite temperature. Their estimate is improperly generated in that what should be, inherently, a fully finite-temperature calculation is treated by an unjustified extension away from a strictly $T = 0$ basis. We will expand this issue in a follow-up paper.

## 6. Conclusions

This brief overview deals with several nontrivial issues related to the ground-state predictions of density functional theory for metals. In particular we have reviewed what is needed for a clear and more faithful characterisations of the Fermi surface.

The FS is a central property of conducting electron systems, necessary for understanding their many physical properties. Transport and superconductivity are but two of them among a wide set of interesting and important many-body phenomena.

We started with a concise overview of the DFT, recognised as a first-principles theory for many-body systems. Since no real many-body problem can be solved exactly, the required systematic approximations to the many-body exchange-correlation functionals are outlined. Many computer codes, both general and specialised, rely on these approximations and are in widespread use in calculating almost every property, at least at zero temperature.

We have highlighted the Fermi surface as a special and especially delicate property of a Fermi system of carriers, and have looked at how one can use various experimental methods to map it: ARPES, SdH, dHvA, ARMRO and the interesting Kohn-anomaly technique. Since any measurements are carried out at finite temperature, we have emphasised the need for proper applications of finite-temperature theory rather then a too-casual stretching of strictly zero-temperature results. This becomes particularly evident when the Fermi energy of a metallic or semi-metallic material starts to enter the typical thermal scale. Uncritical extrapolations of $T = 0$ results cannot but fall short of physical accuracy. At best (in the circumstances relevant to contemporary investigations) the claim of first-principles reliability cannot be supported in their case.

The importance of finite-temperature theory is argued in the context of some novel correlated electron systems, notably for those with phase changes at low temperatures. We believe there are some important aspects of the physics of "fermiology" that need serious extension, in not indeed a thoroughgoing re-evaluation.

Finally, we emphasise that for a number of crucial material properties, density functional theory has proved itself one of the most fruitful and rigorous tools for furthering progress in condensed-matter physics over the last half-century. Our aim here is far from criticising the theory or belittling its manifold deep successes. Density functional theory, as with all theoretical innovations, has limits. It is rather with a view to upholding the true strengths of DFT that we have visited some delicate, nonetheless themselves crucial, aspects of the response of materials where the potential for uncritical misapplication of density functional arguments poses certain novel conceptual problems.

**Acknowledgment**
We thank Professor Reiner Dreizler for helpful correspondence.

**References**

[1] Engel E and Dreizler R M 2011 *Density Functional Theory* (Berlin: Springer)
[2] Fiolhais C, Nogueira F and Marques M A L eds 2003 *A Primer in Density Functional Theory, Lect. Notes Phys.* **620** (Berlin: Springer)
[3] Sholl D S and Steckel J JA 2009 *Density Functional Theory: A PracticalIntroduction* (New York: John Wiley and Sons)
[4] Capelle K 2006 *A Bird's-Eye View of Density-Functional Theory Brazilian J. Phys.* **36** 1318
[5] Dobson J F, Vignale G and Das M P eds 1997 *Electronic Density Functional Theory: Recent Progress and New Directions* (New York: Plenum)
[6] Burke K 2012 *Perspective on density functional theory J. Chem. Phys.* **136** 150901
[7] Hohenberg P and Kohn W 1964 *Phys. Rev* **136** B864
[8] Kohn W and Sham L J 1965 *Phys. Rev* **140** A1133
[9] Pines D and Nozières P 1966 *The Theory of Quantum Liquids Volume I: Normal Fermi Liquids* (New York: Benjamin)
[10] Luttinger J M and Ward J C 1960 *Phys. Rev.* **118** 1417; Luttinger J M 1960 *Phys. Rev.* **119** 1158
[11] Anisimov V and Izyumov Y 2010 *Electronic Structure of Strongly Correlated Materials* (Berlin: Springer) [Local density approximation and dynamical mean field theory compared]
[12] Cabra D C, Honecker H and Pierre Pujol P eds 2012 *Modern Theories of Many-Particle Systems in Condensed Matter Physics* (Berlin: Springer) ch 4
[13] Ylvisaker E R 2008 *DFT and DMFT: Implementations and Applications to the Study of Correlated Materials* (Davis: UC Thesis)
[14] Rajagopal A K 1980 *Adv. Chem. Phys.* **41** 164
[15] Martin R 2004 *Electronic Structure: Basic Theory and Practical Methods* (Cambridge: Cambridge University Press) p 137
[16] Giuliani G and Vignale G 2005 *Quantum Theory of the Electron Liquids* (Cambridge: Cambridge University Press) sec 7.24
[17] Kokko K and Das M P 1998 *J. Phys.: Condens. Matter* **10** 1285
[18] Kurth S, Perdew J and Blaha P 1999 *Inter J. Quantum Chem.* **75** 889
[19] Kordyuk A A 2014 *Low Temp. Phys.* **40** 286
[20] Damascelli A, Shen Z-X and Hussain Z 2003 *Rev. Mod. Phys.* **75** 473
[21] Hedin L 1965 *Phys. Rev.* **139** A790
[22] Svane A, Christensen N E, Cardona M, Chantis A N, van Schilfgaarde M and Kotani Y 2010 *Phys Rev B* **81**
[23] Aryasetiawan F 1992 *Phys RevB* **46** 13051
[24] Georges A, Kotliar G, Krauth W and Rozenberg M 1996 *Rev. Mod. Phys.* **63** 13
[25] Tomczac J M, Tomczac J M, van Schilfgaarde M and Kotliar G 2012 *Phys. Rev. Lett.* **109** 237010
[26] Mearns D 1988 *Phys. Rev. B* **38** 5906
[27] Schönhammer K and Gunnarsson O 1988 *Phys. Rev. B* **37** 3128 (1988)
[28] Cohen M H and Wasserman A 2004 in *Proceedings of the International School of Physics "Enrico Fermi": The Physics of Complex Systems (New Advances and Perspectives)* Mallamace F and Stanley H E eds (Amsterdam: IOS Press) pp 253-295
[29] Lamas C A Cabra D C and Grandi N 2008 *Phys. Rev. B* **78** 115104
[30] Jiang H-M, Jiang Yao Z-J and Zhang F-C 2012 *Eur. Phys. J.* **100** 47004
[31] Pourret A, Knebel G, Matsuda T D, Lapertot G and Flouquet J 2013 *J. Phys. Soc. Japan* **82** 053704
[32] Edalati M, Lo K W and Phillips P W 2012 *Phys. Rev. D* **86** 086003
[33] See reference [12] chapters 1 and 4.
[34] Ziman J M *Theory of Solids* 1965 (Cambridge: Cambridge Univesity Press)
[35] Ohmichi E, Adachi H, Mori Y, Meano Y, Ishiguro T and Oguchi T 1999 *Phys. Rev. B* **59** 7263
[36] Nam M S, Blundell S J, Ardavan A, Symington J A and Singleton J 2001 *J. Phys.: Condens. Matter* **13** 2271
[37] Mermin N D 1965 *Phys. Rev. A137* 1441
[38] Kohn W 1959 *Phys. Rev. Lett.* **2** 393
[39] Soitsov M V and Petkov I Z 1988 *Annals of Physics* **184** 121

[40] Backes S, Guterding D, Jeschke H O and Valenti R 2014 *New J. Phys.* **16** 083025
[41] Rodriguez Ponte P, Cabra D C and Grandi N 2013 *Eur. Phys. J. B* **86** 85
[42] Johnston D C 2010 *Adv. Phys.* **59** 83
[43] Canfield P C and Bud'ko S L 2010 *Ann. Rev. Condens. Matter Phys.* **1** 27
[44] Charnukha A *Charge Dynamics in 122 Iron Based Superconductors* 2014 (Berlin: Springer)
[45] Sen S, Ghosh H, Sinha A K and Bharathi A 2014 *Supercond. Sci. Technol.* **27** 122003
[46] Backes S and Jeschke H O 2013 *Phys Rev B* **88** 075111

# Theoretical evaluation of maximum electric field approximation of direct band-to-band tunneling Kane model for low bandgap semiconductors

**Nguyen Dang Chien[1,4], Chun-Hsing Shih[2], Phu Chi Hoa[3], Nguyen Hong Minh[1], Duong Thi Thanh Hien[1] and Le Hong Nhung[3]**

[1] Faculty of Physics, University of Dalat, Lam Dong 671460, Vietnam
[2] Department of Electrical Engineering, National Chi Nan University, Nantou 54561, Taiwan
[3] Faculty of Postgraduate Studies, University of Dalat, Lam Dong 671460, Vietnam

E-mail: chiennd@dlu.edu.vn

**Abstract.** The two-band Kane model has been popularly used to calculate the band-to-band tunneling (BTBT) current in tunnel field-effect transistor (TFET) which is currently considered as a promising candidate for low power applications. This study theoretically clarifies the maximum electric field approximation (MEFA) of direct BTBT Kane model and evaluates its appropriateness for low bandgap semiconductors. By analysing the physical origin of each electric field term in the Kane model, it has been elucidated in the MEFA that the local electric field term must be remained while the nonlocal electric field terms are assigned by the maximum value of electric field at the tunnel junction. Mathematical investigations have showed that the MEFA is more appropriate for low bandgap semiconductors compared to high bandgap materials because of enhanced tunneling probability in low field regions. The appropriateness of the MEFA is very useful for practical uses in quickly estimating the direct BTBT current in low bandgap TFET devices.

## 1. Introduction

The band-to-band tunneling (BTBT) has been proposed as the main mechanism to generate the conduction current in three-terminal tunnel devices which were initially named "surface tunnel transistors" [1] or tunnel field-effect transistors (TFETs) as widely known today. The most important innovation of employing the BTBT mechanism is to produce a sub-60 mV/decade subthreshold swing at room-temperature [2]-[4] which is a key requirement for scaling the threshold voltage and associated dynamic power consumption of transistors [5]. Furthermore, the off-state power dissipation of BTBT-based transistors is also relatively low since they are always operated in the highly reverse-biased condition of p-n junctions [6], [7]. With the breakthrough of the 60 mV/decade swing limit of conventional metal-oxide-semiconductor field-effect transistors (MOSFETs), TFETs have drawn much attention to serve as a promising candidate for energy-efficient applications [8], [9].

Properly calculating the BTBT generation rate in TFET devices is an important issue to study their physical operation, electrical characteristics and device design. Several full-quantum models based on

[4] To whom any correspondence should be addressed.

the k·p or tight-binding methods have been developed to accurately simulate the tunneling current in TFETs with different structures and materials [10], [11]. However, there are significant disadvantages of using full-quantum models due to their sophisticated mathematical formalisms such as huge memory space, long simulation time, big difficulties in extracting physics and design. Therefore, the two-band Kane model [12] has been popularly employed for TFET devices because of its simple formalism, applicability for both direct and indirect semiconductors, favorable integration in numerical simulation tools [13] and acceptable validity [4], [14], [15]. However, the Kane model was only derived for tunnel junctions with a uniform electric field. In cases of nonuniform electric field, an appropriate approach must be applied in the Kane model to calculate the tunneling generation rate. An intuitive approach by using nonlocal electric field in the Kane formula has demonstrated a good agreement with experiment data [14], [16], [17]. The electric field is nonlocally determined along the tunneling path of electrons. However, the nonlocal BTBT models may be relatively complicated for engineers who wish to quickly estimate the tunneling current in TFETs for design purposes. In this case, one can use another approximation in which the maximum electric field at the tunnel junction is used in the Kane model. The appropriateness of the maximum electric field approximation (MEFA) has been experimentally verified for reverse-biased p-n junctions with high bandgap semiconductor of silicon [18]. Regrettably, applying the MEFA of Kane model for TFET devices has not been clarified because previous researches have not realized local and nonlocal field terms in the Kane formula. In addition, it has been shown that low and direct bandgap semiconductors must be used in TFETs to achieve sufficient on-current for practical applications. The modeling and properties of the tunneling in low bandgap materials are quite different from those in high bandgap semiconductors [19], [20]. Therefore, the appropriateness of the MEFA of direct BTBT Kane model for low bandgap semiconductors remains questionable.

This study theoretically evaluates the maximum electric field approximation of direct BTBT Kane model for low bandgap semiconductors. Since important low bandgap semiconductors have a direct energy bandgap, the Kane model for direct semiconductors was considered. To compare the validity of the MEFA in low bandgap semiconductors with that in high bandgap materials, two fitting factors for the electric field terms in the pre-exponential and exponential terms were separately introduced. InSb with a direct bandgap of 0.17 eV was typified as a low bandgap semiconductor while Ge with a direct bandgap of 0.8 eV was considered as a high bandgap material. The paper consists of four sections, including the Introduction and the Conclusion. Section 2 first summaries the direct BTBT Kane model with decoupled the local and nonlocal electric field terms and then describes the device structure used for investigations. The analyses of the MEFA of direct BTBT Kane model for different bandgap semiconductors are subsequently presented in section 3.

## 2. Tunneling model and device structure

The nonlocal BTBT Kane model of which all electric field terms are identically specified by the nonlocal value of electric field along the tunneling path have been widely used to calculate the tunneling generation in the nonuniform field junctions of TFET devices. However, it is noted that this approach is an approximation because the Kane model was initially derived for a uniform electric field. For more accurate calculations, the physical origin and associated roles of each electric field term in the Kane model must be clarified to answer the question which value of electric field, the local or nonlocal field, should be used when considering nonuniform field tunnel junctions. This section first summaries the direct tunneling model to explicitly show the reason why the local and nonlocal electric field terms have to be decoupled appropriately. The device structure used in this study is then described and explained.

### *2.1. Direct tunneling model*

To illustrate and calculate the direct BTBT in semiconductors, Figure 1 shows the schematic energy-band diagram of a p-n junction under reverse-biased condition. The electric field is assumed to be uniform in the region of tunnel junction. In the quantum picture, it is physically allowable for electrons

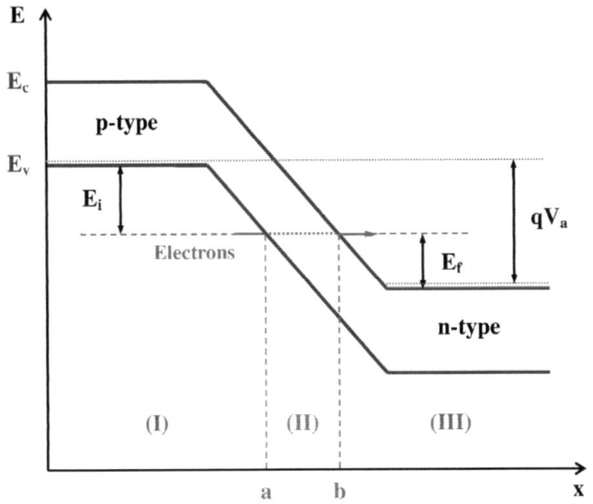


**Figure 1.** Band-to-band tunneling of valence electrons into conduction band in a reverse-biased p-n junction with the uniform electric field model.

to tunnel through the forbidden bandgap from the left-side valence band to the right-side conduction band. Generally, the tunneling generation rate per unit volume ($G_{BTBT}$) is obviously equal to the product of flux of electrons reaching to the high field junction for possible tunneling transitions, tunneling probability and transmission factor:

$$G_{BTBT} = \int dN_F \times P \times F_T \tag{1}$$

where $dN_F$ is the incident electron flux in a ring of perpendicular momentum between $k_\perp$ and $k_\perp + dk_\perp$ ($k_\perp$ is perpendicular component of wave vector), $P$ is the tunneling probability and $F_T$ is the transmission factor defined as the number of available conduction band states into which transmission of a valence electron can be performed. It is clearly that the transmission factor of direct BTBT is equal to one because both electron energy and momentum are strictly conserved.

The direct BTBT probability can be calculated by using the two-band Kane model [12]. In the Kane model, the tunnel electric field is treated as a constant perturbation field which causes the transition from the initial valence band state to the final conduction band state of tunneling electrons. The time-dependent perturbation theory limited to the first-order is employed to estimate the transition probability. Using the Fermi golden rule, the direct tunneling probability can be approximated as:

$$P = \frac{\pi^2}{9}\exp\left(-\frac{\pi m_r^{1/2} E_g^{3/2}}{2\hbar q \xi}\right)\exp\left(-\frac{2E_\perp}{\overline{E}_\perp}\right) \tag{2}$$

where

$$E_\perp = \frac{\hbar^2 (k_y^2 + k_z^2)}{4m_r},\ \overline{E}_\perp = \frac{\hbar q \xi}{\pi m_r^{1/2} E_g^{1/2}}, \tag{3}$$

$m_r$ is the reduced mass, $\xi$ is the electric field, $\hbar$ is reduced Plank's constant, $E_g$ is the semiconductor bandgap, $q$ is the elementary charge, $k_y$ and $k_z$ are perpendicular components of wave vector. Logically, the tunneling probability depends on the electric field at anywhere along the tunneling path. Therefore, the electric field terms in equations (2) and (3) must be nonlocally determined along this tunneling path. In other words, the electric field terms related to the tunneling probability should be specified to be the nonlocal electric field.

Intuitionally, the incident electron flux in a ring of perpendicular momentum is the product of the velocity in *k*-space, the density of states in *k*-space, the area of ring, and the occupancy of states. Using the expressions of the electron velocity and the density of state in *k*-space:

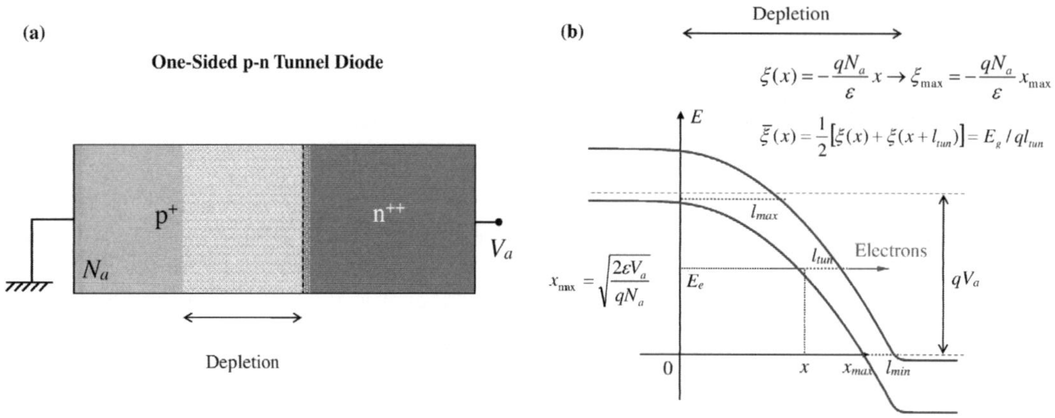


**Figure 2.** (a) Schematic structure of one-sided p-n tunnel diode and (b) energy-band diagram in reverse-biased condition of a one-sided p-n tunnel diode.

$$v_k \equiv \frac{dk}{dt} = \frac{q\xi}{\hbar}, \tag{4}$$

$$\rho(k) = \frac{1}{4\pi^3}, \tag{5}$$

the incident electron flux can be formulated as:

$$dN_F = v_k \times \rho(k) \times 2\pi k_\perp dk_\perp \times (F_v - F_c) = \frac{q\xi}{2\pi^2\hbar}(F_v - F_c)k_\perp dk_\perp, \tag{6}$$

where $F_v$ and $F_c$ are Fermi-Dirac distribution functions in valence and conduction bands, respectively. The presence of the conduction band Fermi-Dirac function in equation (6) is to exclude occupied conduction band states. From equation (6), we see that there is one term of electric field related to the electron velocity in *k*-space at the tunnel junction. This electric field term determines the number of electrons reaching to the tunnel junction for possible tunneling transitions. Therefore, the electric field term in equation (6) must be locally specified at the initial point of the tunneling. In other words, the electric field term related to the tunneling electron number has to be the local electric field.

Inserting equations (2) and (6) into equation (1) with denoting $\xi$ for local electric field and $\bar{\xi}$ for nonlocal electric field, we can express a full form of BTBT generation rate as:

$$G_{BTBT} = \frac{q^2 m_r^{1/2} \xi\bar{\xi}}{18\pi\hbar^2 E_g^{1/2}} \exp\left(-\frac{\pi m_r^{1/2} E_g^{3/2}}{2\hbar q\bar{\xi}}\right)\left[1 - \exp\left(-\frac{2E_e}{\bar{E}_\perp}\right)\right]. \tag{7}$$

By ignoring the negligible term of the exponential in square bracket in equation (7), the direct BTBT Kane model is simply formulated in a compact form [21]:

$$G_{BTBT} = A\frac{\xi\bar{\xi}}{E_g^{1/2}}\exp(-B\frac{E_g^{3/2}}{\bar{\xi}}), \tag{8}$$

where material dependent parameters $A$ and $B$ are given by:

$$A = \frac{q^2 m_r^{1/2}}{18\pi\hbar^2}, \quad B = \frac{\pi m_r^{1/2}}{2\hbar q}. \tag{9}$$

It is emphasized that only one electric field term in the prefactor of the Kane formula associated with the tunneling electron number is the local field whereas the remaining terms related to the tunneling probability are the nonlocal field.

### 2.2. Device structure

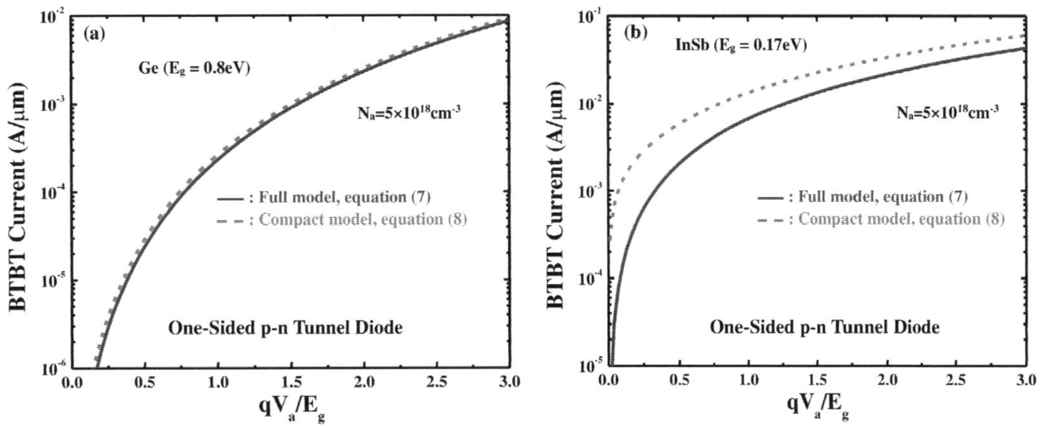


**Figure 3.** Current-voltage characteristics of reverse-biased one-sided p-n tunnel diodes using (a) high bandgap Ge and (b) low-bandgap InSb calculated by full and compact forms of the Kane model.

The direct BTBT Kane model presented in previous subsection is for tunnel junctions with a uniform electric field. In practical terms, the electric field profile of tunnel junctions in TFET devices are usually nonuniform. In this study, one-sided p-n junctions were chosen to represent nonuniform field tunnel junctions in TFETs, as schematically shown in figure 2(a). An important reason for this choice is that tunnel junctions in TFETs are all one-sided p-n junctions regardless of point- or line-tunneling TFET structure. Furthermore, the structure of one-sided p-n junction provides a simple calculation of nonuniform electric field profile [22]. Figure 2(b) shows the energy-band diagram in reverse-biased condition of one-sided p-n junction with lightly-doped p-type left side and heavily-doped n-type right side. The depletion region and associated nonzero electric field region almost penetrate into the lightly-doped p-side. The electric field is maximum at the most right of the depletion region and it nonlinearly decreases when going to the left.

## 3. Maximum electric field approximation (MEFA)

Clearly, the question arises of which value of electric field should be used in the Kane formula when the model is applied for a nonuniform electric field. There is no definite standard for choosing the value of electric field in the Kane model since we cannot know the exact and explicit expression of potential barrier in the classically forbidden region II in figure 1 [21]. Some authors have suggested the average field [12], [21], [23] but also another suggestion of using the maximum field has been made [24], [25]. Based on the investigation of the reverse-biased p-n junctions of silicon with a relatively high bandgap, Hurkx has experimentally shown that the maximum field is appropriate for use in the Kane model. In this section, the maximum electric field approximation of direct BTBT Kane model applied for nonuniform field junctions is discussed to theoretically evaluate its validity for low bandgap semiconductors.

### *3.1. MEFA of direct BTBT Kane model*

Starting from the Kane model of equation (8) with noting the local electric field $\xi = -(qN_A/\varepsilon)x$, the tunneling current in the one-sided p-n junction in reverse-biased voltages can be calculated by:

$$I = A\frac{qV_a\bar{\xi}}{E_g^{1/2}}\exp\left(-B\frac{E_g^{3/2}}{\bar{\xi}}\right), \quad (10)$$

where $V_a$ is the reverse-biased voltage.

Since the tunneling probability is strongly dependent on the length of tunneling path, the tunneling probability is only significant at the highest field region where the tunneling path is minimized. For

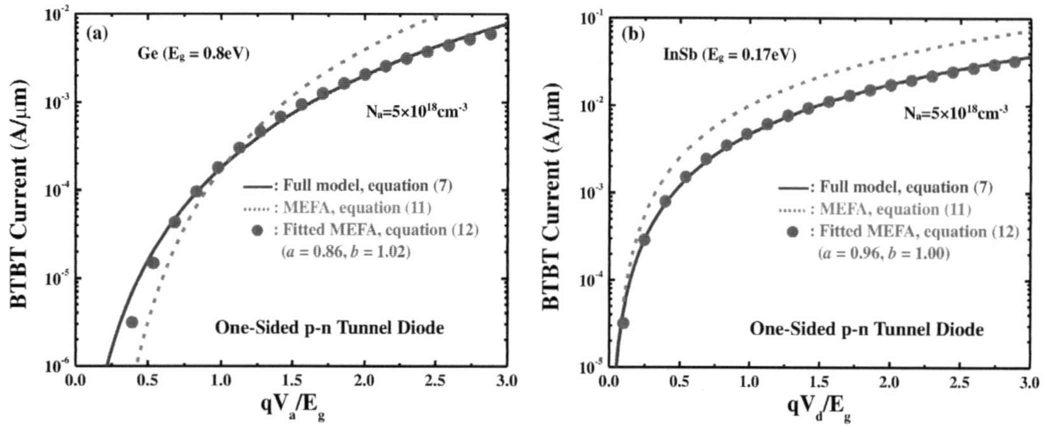


**Figure 4.** Calculated current-voltage curves of reverse-biased one-sided p-n tunnel diodes with (a) high bandgap Ge and (b) low-bandgap InSb using different forms of the Kane model.

simplicity, therefore, the nonlocal field terms in the Kane model are set with a maximum value of electric field at the tunnel junction. An important point in the MEFA has not been previously clarified that the local field term must be remained while the nonlocal field terms are replaced by the maximum electric field. This note is important because we usually use the Kane model in the form of equation (8), and thus it allows us to apply the MEFA for TFET devices. Using the MEFA, the tunneling current is written as:

$$I = A\frac{qV_a\xi_{\max}}{E_g^{1/2}}\exp\left(-B\frac{E_g^{3/2}}{\xi_{\max}}\right), \tag{11}$$

where $\xi_{\max}$ is the maximum electric field at the tunnel junction. In this study, we use the Hurkx's results to assume that the MEFA of direct BTBT Kane model is appropriate for high bandgap semiconductors.

### *3.2. Theoretical evaluation of MEFA for low bandgap semiconductors*

To evaluate the MEFA for low bandgap semiconductors, we introduce two separate fitting parameters *a* and *b* for the nonlocal field terms in the prefactor and the exponential of equation (11), respectively:

$$I_{fit} = A\frac{qV_a\xi_{\max}^a}{E_g^{1/2}}\exp\left(-B\frac{E_g^{3/2}}{\xi_{\max}^b}\right), \tag{12}$$

The fitting parameters are determined so that the tunneling current $I_{fit}$ is best fitted to the accurate solution of the Kane model. By comparing the values of fitting parameters for low bandgap semiconductors with those for high bandgap materials, the MEFA of direct BTBT Kane model for low bandgap semiconductors can be evaluated properly.

Although the full form of Kane model in equation (7) provides the more accurate solution of the tunneling generation rate, its compact form in equation (8) is much simpler for calculations. To decide which form of the Kane model should be used for the fitting procedure, figure 3 shows the calculated current-voltage characteristics of one-sided p-n tunnel diodes with different bandgap semiconductors. For high bandgap Ge in figure 3(a), the compact form of equation (8) is excellently fitted to the full model of equation (7). For low bandgap InSb in figure 3(b), the compact form presents a remarkable deviation from the full model because of very small applied voltages [19]. For exactly evaluating the MEFA, therefore, the tunneling current $I_{fit}$ is fitted to the full Kane model of equation (7) using the least-square method.

Figure 4 shows the current-voltage curves of one-sided p-n tunnel diodes using high bandgap Ge

Journal of Physics: Conference Series **726** (2016) 012002 doi:10.1088/1742-6596/726/1/012002

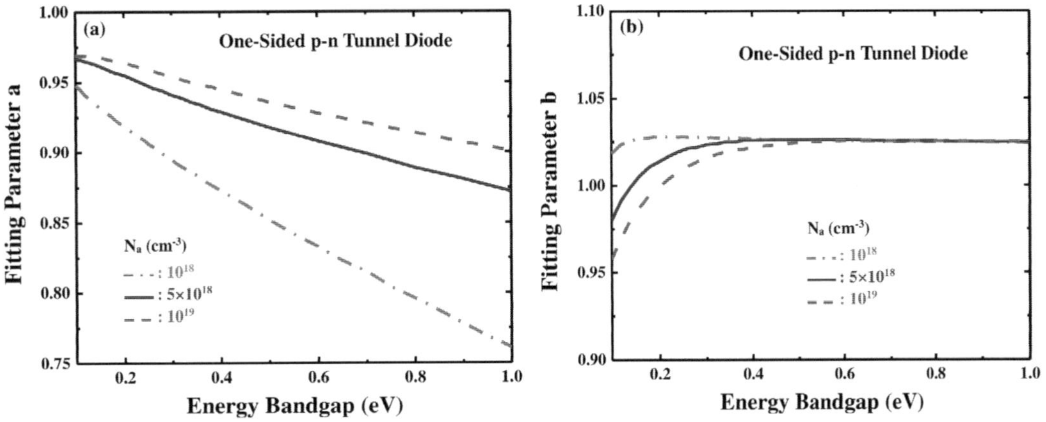


**Figure 5.** Fitting parameters *a* and *b* of the Kane model using the maximum electric field approximation with various doping concentrations.

and low bandgap InSb calculated by using the full Kane model, the MEFA without and with fitting parameters. Firstly, the deviation between the MEFA curve and the full model curve of low bandgap InSb tunnel diode is almost comparable to that of high bandgap Ge counterpart. It means that if the MEFA is considered as a good approximation for high bandgap materials, it is also a good approximation for low bandgap semiconductors. Basically in both high and low bandgap devices, the results of fitted MEFA are fitted well to those of the full model that confirms the appropriateness of the fitting procedure. The fitting parameter *a* in the case of low bandgap InSb is significantly larger than that in the case of high bandgap Ge, whereas the values of parameter *b* are slightly different between the two cases.

To further investigate the dependences of fitting parameters on semiconductor bandgap and doping concentration of tunnel junction, figure 5 shows the values of fitting parameters as a function of the energy bandgap with different doping concentrations. It is seen from figure 5(b) that parameter *b* does not depend on the bandgap except the region of very low bandgaps. The exponential change of parameter *b* when decreasing the bandgap in extremely low bandgap region is solely attributed to the contribution of the exponential in the square bracket in equation (7). The change of parameter *b* is more rapid in the tunnel junction with higher doping concentration because of the larger curvature of energy-band diagram. As shown in figure 5(a), in general, parameter *a* always increases with decreasing the bandgap, irrespective of doping concentration. The increase of parameter *a* approaching to one when decreasing energy bandgap implies that the role of maximum field in determining the tunneling probability approaches to that of the nonlocal field. This is because the tunneling is extended far from the maximum field region due to high tunneling probability in low bandgap semiconductors [19]. Moreover, for a high doping concentration, a large curvature of energy-band diagram results in a narrow tunneling path, a high tunneling probability and thus a large value of parameter *a*. In the mathematical point of view, therefore, the MEFA of direct BTBT Kane model is more appropriate for low bandgap semiconductors and high doping concentrations.

## 4. Conclusions

The maximum electric field approximation of direct BTBT Kane model has been theoretically elucidated to show that the local field term must be remained while the nonlocal field terms are replaced by the maximum electric field. It has been mathematically demonstrated that the MEFA can be applied more appropriately for low bandgap semiconductors and associated physical explanations have also been made adequately. The meaning of these conclusions is that the BTBT current in low bandgap TFET devices can be calculated easily and quickly for practical uses in device modeling and design.

**Acknowledgment:** This work was funded by the Vietnam National Foundation for Science and Technology Development (NAFOSTED) under grant number 103.02-2015.58.

**References**

[1] Baba T 1992 *Jpn. J. Appl. Phys.* **31**, L455
[2] Appenzeller J, Lin Y-M, Knoch J and Avouris Ph 2004 *Phys. Rev. Lett.* **93** 196905
[3] Zhang Q, Zhao W and Seabaugh S A 2006 *IEEE Electron Device Lett.* **27** 297
[4] Krishnamohan T, Donghyun K, Raghunathan S and Saraswat K 2008 *Tech. Dig. IEEE Int. Electron Devices Meeting* (15-17 December 2008, San Francisco) p 1
[5] 2013 *International Technology Roadmap for Semiconductor*
[6] Reddick W M and Amaratunga G A J 1995 *Appl. Phys. Lett.* **67** 494
[7] Wang P-F, Hilsenbeck K, Nirschl Th, Oswald M, Stepper Ch, Weis M, Schmitt-Landsiedel D and Hansch W 2004 *Solid-State Electron.* **48** 2281
[8] Ionescu A M and Riel H 2011 *Nature* **479** 329
[9] Zhao Q T et al. 2015 *IEEE Journal of Electron Devices Society* **3** 103
[10] Shin M 2009 *J. Appl. Phys.* **106** 054505
[11] Luisier M and Klimeck G 2009 *IEEE Electron Device Lett.* **30** 602
[12] Kane E O 1961 *J. Appl. Phys.* **31** 83
[13] 2010 *Synopsys MEDICI User's Manual* (California: Synopsys Inc.)
[14] Nayfeh O M, Hoyt J L and Antoniadis D A 2009 *IEEE Trans. Electron Devices* **56** 2264
[15] Shih C-H and Chien N D 2013 *J. Appl. Phys.* **113** 134507
[16] Peng J Z, Haddad S, Hsu J, Chen J, Longcor S and Chang C 1995 *Proc. Int. Conf. Solid-State and Integrated Circuit Technology* (24-28 Oct 1995, Beijing) p 141
[17] Kao K-H, Verhulst A S, Vandenberghe W G, Sorée B, Groeseneken G and Meyer K D 2012 *IEEE Trans. Electron Devices* **59** 292
[18] Hurkx G A M 1989 *Solid-State Electron.* **32** 665
[19] Shih C-H and Chien N D 2014 *J. Appl. Phys.* **115** 014507
[20] Shih C-H and Chien N D 2014 *IEEE Trans. Electron Devices* **61** 1907
[21] Moll J L 1970 *Physics of Semiconductors* (New York: McGraw-Hill) pp 252-255
[22] Pierret R F 1970 *Semiconductor Device Fuldamentals* (USA: Addison-Wesley Publishing Company) pp 209-225
[23] Logan R A and Chynoweth A G 1963 *Phys. Rev.* **131** 89
[24] Singh Tyagi M 1968 *Solid-State Electron.* **11** 99
[25] Fair R B and Wivell H W 1976 *IEEE Trans. Electron Devices* **23** 512

Journal of Physics: Conference Series **726** (2016) 012003 doi:10.1088/1742-6596/726/1/012003

# On the time-dependent extra spatial dimensions in six dimensional space-time

**Phan Hong Lien**[1] **and Do Thi Hong Hai**[2]

[1]Le Quy Don Academy of Technology, 236 Hoang Quoc Viet, Hanoi, Vietnam
[2]Hanoi University of Mining and Geology, Duc Thang, Bac Tu Liem, Hanoi, Vietnam

E-mail: phhlien.mta@gmail.com, dothihonghai@humg.edu.vn

**Abstract.** In this paper, we analyze the time-dependent extra spatial dimensions in six dimensional (6D) space-time. The 4-brane is assumed to be a de Sitter space. Based on the form of the brane-world energy-momentum tensor proposed by Shiromizu et al. and the five dimensions by Peter K. F. Kuhfittic, we extended the theory to the 2-codimension embedded in higher dimensions. The inflation scenario in 6D is investigated in two cases of cosmological constant: $\Lambda > 0$ and $\Lambda < 0$. The energy of two extra dimensions is calculated too.

## 1. Introduction

Nowadays, the models of extra dimensions introduce completely new perspectives of theoretical physics. Extra dimensions were proposed to assure a true solution for the hierarchy problem [1], and to explain the embedding of the Standard Model in a Grand Unified Theory [2] and the affecting of particle physics and cosmology. Particularly, in the brane-world theories, the D-brane inflation in warped compactification has attracted much attention recently [3]. In that case the inflation is identified as the separation between a brane and an anti-brane [4], which describes by 2-codimension branes couple to bulk metric in six and ten dimensions [5].

In this paper, we analyze the time-dependent spatial dimensions in six dimensional space time. The crossover distance $R_c$ between the two regimes is [6]

$$R_c = \frac{M_{PL}}{2M_6^2}.$$

Here the 4-brane is assumed to be a de Sitter space. Based on the form of the brane-world energy momentum tensor proposed by Shiromizu *et al.* and the five dimensions by Peter K. F. Kuhfittic [7], the theory to the 2-codimension embedded in higher dimensions is extended. The inflation scenario in 6D is investigated in two cases of cosmological constant: $\Lambda > 0$ and $\Lambda < 0$.

This paper is organized as follows. In Sec. 2 we present the formulation of Einstein field equation and the metric in 6D, where all spatial dimensions are time-dependent. The general solutions of these equations when $\Lambda > 0$ are investigated in Sec. 3. It shows that the extra dimensions expand by the factor $e^{Ht}$ in the branes. Sec. 4 is devoted to the case $\Lambda < 0$. The conclusion and discussion are given in Sec. 5.

Journal of Physics: Conference Series **726** (2016) 012003 doi:10.1088/1742-6596/726/1/012003

## 2. Formulation

We start from the bulk metric in six-dimensional space-time

$$ds^2 = -dt^2 + [R(t)]^2 \left[dr^2 + r^2 \left(d\theta^2 + sin^2\theta d\varphi^2\right)\right] + [\chi(t)]^2 \left[dy_1^2 + dy_2^2\right], \tag{1}$$

where $y_1, y_2$ are the coordinates in two extra dimensions. Note that the spatial dimensions are assumed to be time dependent by the scalar factors $R(t)$ and $\chi(t)$.

It is clear that the manifold (1) in six dimensions contains two five-dimensional space-time as its submanifolds at $y_1 = 0$ and $y_2 = 0$. The hypersurface $y_1 = 0$ and $y_2 = 0$ corresponds to 3-brane.

If we impose $y_1 = \rho sin\phi$, $y_2 = \rho cos\phi$, then the metric (1) takes the form

$$ds^2 = -dt^2 + [R(t)]^2 \left[dr^2 + r^2 \left(d\theta^2 + sin^2\theta d\varphi^2\right)\right] + [\chi(t)]^2 \left[d\rho^2 + \rho^2 d\phi^2\right], \tag{2}$$

where $\rho$, $\phi$ are the central symmetric coordinates in two extra dimensions, which could be compactified on a 2-dimensional sphere $(S^2/Z_2)$.

We assume that the energy-momentum tensor, which satisfies the six-dimensional Einstein field equations in the Early Universe

$$G_{MN} = k_6^2 T_{MN}, \tag{3}$$

could be writen in the form

$$T_{MN} = -\Lambda g_{MN} + \delta(\rho)\delta(\phi)\left(-\lambda q_{MN} + \tau_{MN}\right). \tag{4}$$

Here $G_{MN}$ is Einstein tensor, $g_{MN}$ is the metric tensor in six dimensions (6D), $\Lambda$ is the cosmological constant, the Latin letters denote the 6D indices $M, N = 0, 1, 2, 3, 5, 6$.

The $\delta$-function appears due to the boundary surface $\rho = 0$ and $\phi = 0$. $\lambda$ and $\tau_{MN}$ are the vacuum energy and the energy-momentum tensor of the 4-brane.

In this case the inflation is identified as the separation between a brane and an anti-brane like codimension-2 brane [3].

$$T_{\rho\rho} = -\Lambda g_{\rho\rho} + \delta(\rho)\left(-\lambda q_{\rho\rho} + \tau_{\rho\rho}\right), \tag{5}$$
$$T_{\phi\phi} = -\Lambda g_{\phi\phi} + \delta(\phi)\left(-\lambda q_{\phi\phi} + \tau_{\phi\phi}\right). \tag{6}$$

The 4-brane is assumed to be a de Sitter space, meaning $R(t) = e^{Ht}$ and $H = \sqrt{\Lambda_5/3}$, where $\Lambda_5$ is the cosmological constant of the 4-brane. The nonzero components of Einstein tensor in the orthonormal frame are determined from the Ricci tensor (see Appendix)

$$G_{tt} = 3\left(\frac{R'(t)}{R(t)}\right)^2 + 6\frac{R'(t)\chi'(t)}{R(t)\chi(t)} + \left(\frac{\chi'(t)}{\chi(t)}\right)^2, \tag{7}$$

$$G_{rr} = -2\left(\frac{R''(t)}{R(t)}\right) - \left(\frac{R'(t)}{R(t)}\right)^2 - 4\frac{R'(t)\chi'(t)}{R(t)\chi(t)} - 2\left(\frac{\chi''(t)}{\chi(t)}\right)^2 - \left(\frac{\chi'(t)}{\chi(t)}\right)^2$$
$$= G_{\theta\theta} = G_{\varphi\varphi}, \tag{8}$$

$$G_{\rho\rho} = -3\left(\frac{R''(t)}{R(t)}\right) - 3\left(\frac{R'(t)}{R(t)}\right)^2 - 3\frac{R'(t)\chi'(t)}{R(t)\chi(t)} - \frac{\chi''(t)}{\chi(t)}, \tag{9}$$

$$G_{\phi\phi} = -3\left(\frac{R''(t)}{R(t)}\right) - 3\left(\frac{R'(t)}{R(t)}\right)^2 - 3\frac{R'(t)\chi'(t)}{R(t)\chi(t)} - \frac{\chi''(t)}{\chi(t)}. \tag{10}$$

The time-dependent extra spatial dimensions are derived from the system of Eqs. (7), (8), and (9)

$$G_{tt}+G_{rr}-2G_{\rho\rho}=k_6^2\left(T_{tt}+T_{rr}-2T_{\rho\rho}\right),$$
$$8\left(\frac{R'(t)}{R(t)}\right)^2+8\frac{R'(t)\chi'(t)}{R(t)\chi(t)}+4\frac{R''(t)}{R(t)}=-2k_6^2T_{\rho\rho}. \tag{11}$$

Since $\delta(\rho),\delta(\phi)=0$ in the bulk, i.e $T_{MN}=-\Lambda g_{MN}$ and $R(t)=e^{Ht}$ with $H=\sqrt{\Lambda_5/3}$, we have the following equation for $\chi(t)$

$$12H^2+8H\frac{\chi'(t)}{\chi(t)}-2k_6^2\Lambda=0,$$
$$\frac{\chi'(t)}{\chi(t)}=-\frac{3}{2}H+\frac{1}{4H}k_6^2\Lambda. \tag{12}$$

Let $A_{in}$ be the initial value of $\chi(t)$ (at the onset of inflation), i.e. $\chi(0)=A_{in}$, then the solutions of Eq. (12) are

$$\chi(t)=A_{in}e^{-\frac{3}{2}Ht}e^{\frac{k_6^2\Lambda t}{4H}}. \tag{13}$$

In the next two sections, we will analyze two cases of the cosmological constant $\Lambda$.

## 3. The case $\Lambda>0$

From Eq. (9) we have

$$G_{\rho\rho}=-\frac{\chi''(t)}{\chi(t)}-3\left(\frac{R'(t)}{R(t)}\right)^2-3\frac{R'(t)\chi'(t)}{R(t)\chi(t)}=k_6^2T_{\rho\rho},$$

or, equivalently

$$\chi''(t)+3H\chi'(t)+\left(6H^2-k_6^2\Lambda\right)\chi(t)=0. \tag{14}$$

Its general solutions are given by

$$\chi(t)=C_1e^{-\frac{3}{2}Ht}e^{\frac{1}{2}\sqrt{4k_6^2\Lambda-15H^2}t}+C_2e^{-\frac{3}{2}Ht}e^{-\frac{1}{2}\sqrt{4k_6^2\Lambda-15H^2}t}, \tag{15}$$

where $C_1$ and $C_2$ are arbitrary constants.

The solutions (13) and (15) must agree, particularly at $t=\tau$, the end of inflation. It implies that

$$C_1=A_{in} \qquad \textit{and} \qquad C_2=0. \tag{16}$$

In fact, the second term $e^{-\frac{3}{2}H\tau}e^{-\frac{1}{2}\sqrt{4k_6^2\Lambda-15H^2}\tau}$ is many orders of magnitude below Planck scale, due to $H\tau=100$ [9], therefore it does not make physical sense and could be eliminated. So

$$\chi(t)=A_{in}e^{-\frac{3}{2}Ht}e^{\frac{1}{2}\sqrt{4k_6^2\Lambda-15H^2}t}. \tag{17}$$

Compare (13) with (15), we get

$$\frac{k_6^2\Lambda}{4H}=\frac{1}{2}\sqrt{4k_6^2\Lambda-15H^2}. \tag{18}$$

Two solutions of this equation are $\Lambda_1=2\Lambda_5/k_6^2$ and $\Lambda_2=10\Lambda_5/3k_6^2$.

It is easy to show that these solutions are consistent with $G_{\rho\rho}$ and $G_{\phi\phi}$. From Eq. (12) we can derive

$$\frac{\chi'(t)}{\chi(t)}=\frac{9}{4}H^2+\frac{3}{16}k_6^2\Lambda+\frac{k_6^4\Lambda^2}{16H^2}, \tag{19}$$

and Eq. (14) becomes

$$\frac{5\Lambda_5}{4} + \frac{3(k_6^2\Lambda)^2}{16\Lambda_6} = -k_6^2\Lambda. \tag{20}$$

This equation is satisfied, i.e. $\delta(\rho)\tau_{\rho\rho} = 0$ and $\delta(\phi)\tau_{\phi\phi} = 0$ in the brane and anti-brane if and only if $\Lambda = \Lambda_1$ or $\Lambda = \Lambda_2$.

We emphasize that due to the property of $\delta$-function for any continuos $\tau_{MN}$

$$\delta(\rho)\delta(\phi)\tau_{MN} = \delta(\rho)\delta(\phi)\left(\tau_{MN}\mid_{\rho=0,\phi=0}\right) \tag{21}$$

which describes the confinement to the brane and anti-brane, respectively.

For $\Lambda = 2\Lambda_5/k_6^2$, Eq (13) yields

$$\chi(t) = A_{in}e^{-\frac{3}{2}Ht}e^{\frac{3}{2}Ht} = A_{in}. \tag{22}$$

For $\Lambda = 10\Lambda_5/3k_6^2$, from both Eq. (13) and (15) we obtain

$$\chi(t) = A_{in}e^{-\frac{3}{2}Ht}e^{\frac{5}{2}Ht} = A_{in}e^{Ht}. \tag{23}$$

So, the extra dimensions expand by the factor $e^{Ht}$ in the branes, which is similar to the case of the four dimensional space.

## 4. The case $\Lambda < 0$

If $\Lambda < 0$, the solution (13) of Eq. (12)

$$\chi(t) = A_{in}e^{-\frac{3}{2}Ht}e^{\frac{k_6^2\Lambda}{4H}t}$$

is not acceptable, since at the end of inflation ($t = \tau, H\tau = 100$), $\chi(t)$ becomes many orders of magnitude below Planck scale.

In this case, the solution of (14) takes the form

$$\chi(t) = e^{-\frac{3}{2}Ht}\left(C_1 cos\frac{1}{2}\sqrt{15H^2 - 4k_6^2\Lambda}t + C_2 sin\frac{1}{2}\sqrt{15H^2 - 4k_6^2\Lambda}t\right). \tag{24}$$

It is plausible, but as in 5D [7], it also shows that $\chi(t)$ has shrunk significantly at the end of inflation.

## 5. Conclusion and discussion

In the above sections, the brane-world scenarios matter field are analyzed in the frame of the theory with two extra dimensions, where the spatial dimensions are time-dependent. It shows that when the cosmological constant $\Lambda > 0$, all spatial distance expands by the factor $e^{Ht}$ in the brane and anti-brane. When $\Lambda < 0$, the scalar factor $\chi(t)$ of external dimensions shrinks significantly at the end of inflation.

However, if the extra spatial are tightly curled up, as in conventional assumptions, a huge amounts of potential energy should be stored. It could be estimated, for example

$$R^{\rho}_{tt\rho} = R^{\phi}_{tt\phi} = |\frac{\chi''(t)}{\chi(t)}| = H^2 = \frac{\Lambda_5}{3}, \tag{25}$$

where the Hubble parameter is determined by [10]

$$H^2 = \frac{k_B^4 T^4}{(10^{19}GeV)^2}. \tag{26}$$

The source of this energy might just be the dark energy that accelerate the Universe's expansion rate [7].

Journal of Physics: Conference Series **726** (2016) 012003 doi:10.1088/1742-6596/726/1/012003

## Appendix

The components of the Ricci tensor are

$$
\begin{aligned}
R_{tt} &= -3\frac{R''(t)}{R(t)} - 2\frac{\chi''(t)}{\chi(t)}, \\
R_{rr} &= \frac{R''(t)}{R(t)} + 2\left(\frac{R'(t)}{R(t)}\right)^2 + 2\frac{R'(t)\chi'(t)}{R(t)\chi(t)}, \\
&= R_{\theta\theta} = R_{\varphi\varphi} \\
R_{\rho\rho} &= 3\frac{R'(t)\chi'(t)}{R(t)\chi(t)} + \frac{\chi''(t)}{\chi(t)} + \left(\frac{\chi'(t)}{\chi(t)}\right)^2 \\
&= R_{\phi\phi}.
\end{aligned}
$$

## References


[1] Randall L and Sundrum 1999 *Phys. Rev. Lett.* **83** 4690 [arXiv: help-th/9906064]
[2] Hall L J and Nomura Y 2001 *Phys. Rev.* D **64** 055003
[3] Lee H M and Papazoglou A 2009 *Phys. Rev.* D **80** 043506 [arXiv: 09014692]
[4] Cline J M, Descheneau J, Jovannini M and Vinat J 2003 JHEP **0306** 048 [arXiv: help-th/0304147]
[5] Bayntun A, Burgess C P, and van Nierop L 2010 *New Journal of Physics* **12** 075015
[6] Arkani - Hamed N, Dimopoulos S and Dvali G 1998 *Phys. Lett.* B **436** 257; Antoniadis I, Arkani - Hamed N, Dimopoulos S and Dvali G 1998 *Phys. Lett.* B **249** 263
[7] Shiromizu J, Maeda K and Sasaki M 2002 *Phys. Rev.* D **62** 024012; Kuhfittic Peter K F 2012 [arXiv: gr-qc/0608120]
[8] Kolb E W and Turner M S 1990 *The Early Universe* (Addison-Wesley, New York) pp 270-275
[9] Roman T A 1993 *Phys. Rev.* D **47** 1370 [arXiv: gr-qc/9211012]
[10] Liddle A 2003 *An Introduction to Modern Cosmology* (John Wiley and Sons Ltd) Advanced Topic 3 pp 141

Journal of Physics: Conference Series **726** (2016) 012004 doi:10.1088/1742-6596/726/1/012004

# Thermal neutron radiative capture cross-section of $^{186}$W(n, $\gamma$)$^{187}$W reaction

**V H Tan[1] and P N Son[2]**
[1] Agency for Radiation and Nuclear Safety, 113-Tran Duy Hung, Hanoi, Vietnam
[2] Nuclear Research Institute, 01-Nguyen Tu Luc, Dalat, Vietnam

E-mail: pnson.nri@gmail.com

**Abstract.** The thermal neutron radiative capture cross section for $^{186}$W(n, $\gamma$)$^{187}$W reaction was measured by the activation method using the filtered neutron beam at the Dalat research reactor. An optimal composition of Si and Bi, in single crystal form, has been used as neutron filters to create the high-purity filtered neutron beam with Cadmium ratio of $R_{cd}$ = 420 and peak energy $E_n$ = 0.025 eV. The induced activities in the irradiated samples were measured by a high resolution HPGe digital gamma-ray spectrometer. The present result of cross section has been determined relatively to the reference value of the standard reaction $^{197}$Au(n, $\gamma$)$^{198}$Au. The necessary correction factors for gamma-ray true coincidence summing, and thermal neutron self-shielding effects were taken into account in this experiment by Monte Carlo simulations.

## 1. Introduction

Thermal neutron radiactive capture cross sections are important data for safety analysis and design of reactors as well as in the studies of nuclear physics, and the nucleosynthesis process in astrophysics [1]. The previous measured values are quite discrepant from different experiments, and still not completely satisfied in quality and quantity for related applications. That is the reason why new experimental data and theoretical calculations to improve the neutron induced reaction cross-sections are necessary. In this study, the thermal neutron capture cross section for the reaction of $^{186}$W(n, $\gamma$)$^{187}$W has been measured by the activation method with the filtered neutron beam at the Dalat nuclear research reactor. A modern gamma-ray spectroscopy (ORTEC DSPEC) in compact with a high-efficiency HPGe detector was used to measure the specific radioactivity of the irradiated samples. The necessary correction factors for gamma true coincidence summing, and neutron self-shielding effects were taken into account in this experiment by Monte-Carlo simulations using the well know MCNP5 code. The new measured value of thermal neutron capture cross section for $^{186}$W nucleus was obtained relatively to the reference value of $\sigma_0 = 98.65 \pm 0.09$ (barn) for the $^{197}$Au(n, $\gamma$)$^{198}$Au standard reaction [2].

## 2. Experimental

The neutron filter technique [3] has been applied to derive a pure thermal neutron beam at the horizontal channel No.2 of the Dalat research reactor (DRR), Dalat Nuclear Research Institute. The single-crystals of 80 cm silicon and 6 cm bismuth were used as neutron filters, for which the epithermal component is absorbed or scattered out of the beam line. The measured thermal neutron flux at outer position of the beam line is $1.6\times10^6$ n/cm$^2$/s, and the value of Cadmium ratio $R_{cd}$(Au) is

420. Accordingly, the contribution of reaction rates by epithermal neutron component to the total reaction rate was less than 0.25%, which is much smaller than other sources of uncertainties, and may be negligible. The physical parameters of the thermal neutron beam line in comparison with others facilities are given in Table 1.

**Table 1.** Characteristics of the thermal neutron beam at the radial channel No.2 of DRR, in comparison with others facilities

| Parameters | Thermal neutron beams | | |
|---|---|---|---|
| | Channel No.2 of DRR *(this work)* | Channel No.4 of DRR | Ohio Reactor [3] |
| Thermal neutron flux $\phi_{th}$ | $1.6\times10^6$ n/cm$^2$.s | $1.7\times10^6$ n/cm$^2$.s | $8.5\times10^6$ n/cm$^2$.s |
| Cadmium ratio $R_{Cd}$(Au) | 420 | 112 | 266 |
| Filter materials | 6cm Bi + 80cm Si | 98cm Si + 1cm Ti + 35g/cm$^2$ S | 10.16cm Bi + 12.7cm Sapphire |
| Beam diameter | 3cm | 3cm | 3cm |

The samples were prepared from the high-purity metal foils of Au (99.99%, 1.27 cm in diameter, 0.05 mm in thickness) as a standard for neutron flux monitor and W (99.98%, 1.27 cm in diameter, 0.05 mm in thickness) as the target, by the SHIELDWERX company. Each W sample was sandwiched between two Au foils. The irradiation time was five hours for every sample group. The specific activities of the targets and the gold foils were measured by using the DSP-base gamma-ray spectrometer, in compacting with a HPGe detector (58% efficiency, 2.5keV energy resolution at 1.332 MeV of $^{60}$Co). The detector was calibrated by using a set of standard radioisotope sources: $^{152}$Eu, $^{60}$Co, $^{57}$Co, $^{137}$Cs, $^{241}$Am, $^{113}$Sn and $^{88}$Y. The absolute detection efficiency function for 5 cm distance from the end-cap of the detector was determined by Monte Carlo simulation method using the Geant4 toolkit [4]. The results of simulation and measurement of the efficiency curve are illustrated in Figure 1. The correction factors for neutron multiple scattering and self-shielding in the samples were also calculated by the Monte Carlo method, using the well-know MCNP5 code.

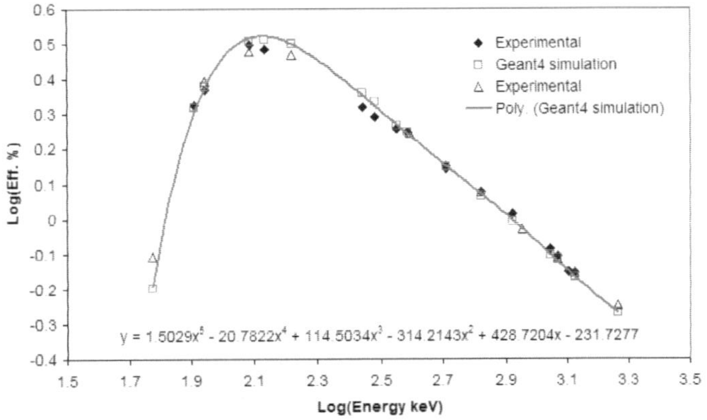


**Figure 1.** The absolute detection efficiency curve of the HPGe detector for foil sample at 5 cm from the detector end-cap

## 3. Data analysis

During irradiation in a neutron beam with energy spectrum $\phi(E)$, the capture reaction rate, R, of samples with neutron is defined as follows:

$$R = N\int \varphi(E)\sigma_a(E)dE \,, \quad (1)$$

where N is the number of nuclei in the sample, and $\sigma_a(E)$ is the neutron capture cross section as a function of incident neutron energy E. The average thermal neutron capture cross section, $\sigma_0$, and neutron flux, $\Phi$, are defined as following:

$$\sigma_0 = \frac{\int \sigma_a(E)\varphi(E)dE}{\int \varphi(E)dE}; \Phi = \int \varphi(E)dE. \quad (2)$$

Applying these average quantities, the integral equation (1) can be rewritten as follows:

$$R = N\sigma_0\Phi \quad (3)$$

The activity, A, of the irradiated sample at the end of neutron irradiation is calculated by expressions:

$$A = R(1 - e^{-\lambda t_1}), \quad \text{and} \quad A = \frac{Cf_c\lambda}{\varepsilon_\gamma I_\gamma e^{-\lambda t_2}(1 - e^{-\lambda t_3})}, \quad (4)$$

where C denotes the net counts of the corresponding gamma peak; $t_1$, $t_2$ and $t_3$ are irradiating, cooling and measuring times, respectively; $\lambda$ is the decay constant of the product nucleus; $\varepsilon_\gamma$ is the detection efficiency of the detector; $I_\gamma$ is the intensity of interesting gamma peak; $f_c$ is the correction factor for the effects of neutron multiple scattering and self-shielding in the irradiated samples. Finally, from equations (2-4), the thermal neutron capture cross sections $\sigma_0$ can be determined relative to that of $^{197}$Au standard by the following relations:

$$\sigma_0^x = \frac{C^x f(\lambda,t)^x f_c^x I_\gamma^{Au} \varepsilon_\gamma^{Au} N^{Au} \sigma_0^{Au}}{C^{Au} f(\lambda,t)^{Au} f_c^{Au} I_\gamma^x \varepsilon_\gamma^x N^x}; \quad (5)$$

$$f(\lambda,t) = \frac{\lambda}{(1 - e^{-\lambda t_1})e^{-\lambda t_2}(1 - e^{-\lambda t_3})}, \quad (6)$$

where the superscript 'x' denotes the nucleus of sample. The standard thermal capture cross section of $^{197}$Au is extracted from the reference [2]. The relevant decay data of product nuclei used in this work are given in Table 2.

**Table 2.** Decay properties of the activation product nuclei [5]

| Activation products | Half-life | γ-ray (keV) | Intensity (%) |
|---|---|---|---|
| $^{198}$Au | 2.6948 ± 0.0012 (d) | 411.8 | 95.62 |
| $^{187}$W | 24.000 ± 0.004 (h) | 479.53 | 26.6 ± 0.4 |

## 4. Results and Discussion

The thermal neutron radiative capture cross-section of $^{186}$W(n,γ)$^{187}$W reaction was obtained by the neutron filter technique and activation method. The Monte-Carlo simulation has been also performed for exact determinations of detector efficiency curve and correction parameters of neutron multiple scattering and self-shielding inside the irradiating samples. The new measured value of cross section is given in the Table 3, which is almost agreement with the previous experimental and evaluated values. In this experiment, the total uncertainty of cross-section value is about 3.5%, mainly due to the statistical errors (0.1-0.5%), the uncertainties of gamma-ray detection efficiency (1.5%), the reference cross-section of the monitor sample (1.5%) and the correction factors for self-shielding and multiple scattering effects (1.0%).

**Table 3.** The result of thermal neutron capture cross section for the $^{186}W(n, \gamma)^{187}W$ reaction, in comparison with previous measured and evaluated values

| Reaction | Neutron energy (eV) | Authors | $\sigma_0$ (barn) | Year |
|---|---|---|---|---|
| $^{186}W(n,\gamma)^{187}Mo$ | 0.0253 | **This work** | **37.5 ± 1.3** | 2015 |
| $^{186}W(n,\gamma)^{187}Mo$ | 0.0536 | M.S.Uddin [6] | 26.6 ± 1.6 | 2008 |
| $^{186}W(n,\gamma)^{187}Mo$ | 0.0253 | S.J.Friesenhahn [7] | 37.0 | 1966 |
| $^{186}W(n,\gamma)^{187}Mo$ | 0.0253 | JENDL 4.0 [8] | 38.099 | 2011 |
| $^{186}W(n,\gamma)^{187}Mo$ | 0.0253 | ENDF/B-VII.1 [9] | 38.0949 | 2011 |

### 5. Conclusion

The thermal neutron capture cross-section of $^{186}W(n, \gamma)^{187}W$ reaction at incident neutron energy of 0.0253 eV was determined to be 37.5 ± 1.3 (barn) by activation method using the new filtered thermal neutron beam at the radial channel No.2 of the Dalat research reactor. The Monte-Carlo simulation has been also performed for the experimental setups, detector efficiency determination, and correction factor calculations.

### Acknowledgments

This research is funded by Vietnam National Foundation for Science and Technology Development (NAFOSTED) under grant number "103.04-2012.59".

### References

[1] Agbemava S E, Nyarko B J B, Fletcher J J, Sogbadji R B M, Mensimah E and Asamoah M 2011 *Annals of Nuclear Energy* **38** 1616
[2] IAEA 2007 *"International Evaluation of Neutron Cross-Section Standards"* ISBN 92-0-100807-4
[3] Simpson O D and Miller L G 1986 *Nucl. Instr. and Meth.* **61** 245
[4] Nikolic J, Puzovic J, Todorovic D and Rajacic M 2015 *Nucl. Instrum. Methods Phys. Res. Sect.* **A799** 159
[5] Nudat 2.2 Database, http://www.nndc.bnl.gov/nudat2/
[6] Uddin M S, Chowdhury M H, Hossain S M, Latif S A, Hafiz M A, Islam M A, Zakaria A K M and Azharul I S M 2008 *Applied Radiation and Isotopes* **66** 1235
[7] Friesenhahn S J, Haddad E, Froehner F H and Lopez W M 1966 *Nuclear Science and Engineering* **26** 487
[8] Keiichi S et al 2011 *Journal of Nuclear Science and Technology* **48** 1
[9] Chadwick M B et al 2011 *Nuclear Data Sheets* **112** 2887

Journal of Physics: Conference Series **726** (2016) 012005 doi:10.1088/1742-6596/726/1/012005

# Fission Product Decay Heat Calculations for Neutron Fission of $^{232}$Th

**P N Son and N X Hai**
Nuclear Research Institute, 01-Nguyen Tu Luc, Dalat, Vietnam

E-mail: pnson.nri@gmail.com

**Abstract.** Precise information on the decay heat from fission products following times after a fission reaction is necessary for safety designs and operations of nuclear-power reactors, fuel storage, transport flasks, and for spent fuel management and processing. In this study, the timing distributions of fission products' concentrations and their integrated decay heat as function of time following a fast neutron fission reaction of $^{232}$Th were exactly calculated by the numerical method with using the DHP code.

## 1. Introduction

In a fission reaction, fission products (FP) are initially formed with known concentrations as independent fission yields, but this data still changes following times after the fission events. This is mainly because of the natural decay of radioactive fission products. In nuclear science and technology, the concentrations of fission products as functions of decay time are the key data required in aggregate decay heat calculations for designs and operations of nuclear-power reactors, fuel storage, transport flasks, and for spent fuel management and processing [1, 2]. Gamma and beta decay energies released from natural decay of the fission products contribute approximately 7% to 12% of the total energy generated through the fission process; this component of power is called "Decay Heat". After a reactor is shutdown, this source of radioactive decay energy still remains and cumulatively increases that necessary to apply a heat removal system for maintaining the safety level of temperature inside the reactor core.

In nuclear reactor technology, thorium is considering as a potential fuel material that could possibly supplement or even replace natural uranium [3]. $^{232}$Th is the naturally-occurring isotope with abundance of nearly 100%, and can be used as breeding-fuel material by capturing neutrons in a thermal and epithermal neutron flux to form $^{233}$Th isotope and subsequently decay to produce uranium-233 ($^{233}$U), which is an excellent fissile material for nuclear energy production [4]. For fast reactors, in which thorium-232 is used as the breeding-fuel component that driven by a fast neutron spectrum, $^{232}$Th is also fissionable with incident neutrons with energy above one MeV; and the decay energies of its fission products would significantly contribute to the decay heat power in the spend fuel. Accordingly, the updated knowledge of timing distributions or behaviour of concentrations and decay heat energies by fission products from fission reaction of $^{232}$Th is essential necessary in research and application of fast reactor technology.

In this work, the DHP program [5], a numerical calculation code, was used for calculation of fission products decay heat data for fast neutron fission of $^{232}$Th. The method used in this calculation is analysis procedure on the general solutions of the Bateman's Equations [6] for every full complex decay chain, in which the real-time build-up and decay of FP nuclides are determined. Based on the decay and fission yield data from JENDL 4.0 [7], the concentration of each FP nuclide is determined as functions of cooling time after a fission event. The difficulties of the complex systems of the decay-chains are solved by an additional computational algorithm implemented in the DHP code.

## 2. Analysis of Decay and Build-up Numbers

The number of $i^{th}$ nuclide at cooling time t after a fission burst can be calculated from the following equation:

$$N_i(t) = N_i(0)\exp(-\lambda_i t) + \sum_{j\neq i}^{M} N_{j\to i}(t) \tag{1}$$

in which, $N_i(0)$ is equal to the independent fission yield of the $i^{th}$ nuclide; the component of $N_{j\to i}(t)$ is the build-up number of nuclide $i^{th}$ at cooling time t, that was formed in the system of decay chains originated from the nuclide $j^{th}$. This term of the build-up number can be obtained by the general solutions of the Bateman's equation for every particular linear decay chain. In this work, we developed a numerical algorithm to calculate the term of $N_{j\to i}(t)$ in equation (1) directly by using the decay and fission yield data from JENDL4.0 [7] and/or the evaluated nuclear structure data file ENSDF [8]. The calculation procedure is generalized as the following steps [9]:

- For every nuclide $j^{th}$, the decay net that started from the nuclide $j^{th}$ is separated to form equivalent sub-branches of linear decay chains.
- For each linear decay chain, if the nuclide $i^{th}$ is a daughter nuclide in the decay chain, apply the general solution of Bateman's equation to calculate the build up number of nuclide ith due to the decay of nuclide $j^{th}$ through the current linear decay chain, as of the cooling time t.
- The total number $N_i(t)$ can be obtained by summing all the build-up numbers of nuclide ith from all the linear decay chains in which the build-up numbers are calculated by apply the general solution of Bateman's equation [6] for every liner decay branch in a differential period of time $\Delta t$. Let consider to a linear decay branch of n nuclides:

$$N_1 \to N_2 \to N_3 \to \ldots N_i \to \ldots N_n$$

The build-up number of nuclide $N_i$ due to the decay of nuclide $N_{i-1}$ is calculated by the differential equation:

$$\frac{dN_i(t)}{dt} = \lambda_{i-1}N_{i-1}(t) - \lambda_i N_i(t) \tag{2}$$

In case of $\{N_j(t) \neq 0$ and all of the remain nuclides $N_{1,2,\ldots j-1,j+1,\ldots i,\ldots M}(t) = 0$ at $t = 0\}$, the solution of build-up number for $N_i(t)$ contributed from nuclide $N_j$ at $t > 0$ is express as:

$$N_{j\to i}(t) = \sum_{m=j}^{i} C_m e^{-\lambda_m t} \tag{3}$$

$$C_m = \frac{\prod_{k=j}^{i-1}\lambda_k}{\prod_{k=j,\,k\neq m}^{i}(\lambda_k - \lambda_m)} . N_j(0) \tag{4}$$

In general case of $\{N_j(t) \neq 0 ; j = 1, 2, 3, \ldots n \text{ at } t = 0\}$, the total build-up number of time dependent $N_i(t)$, at $t > 0$, contributed from all of other nuclides in a liner decay branch can be estimated as the following general expression.

$$\sum_{j} N_{j\to i}(t) = \sum_{l=1}^{i} \prod_{k=l}^{i-1} N_l(0) \left( \sum_{m=l}^{i} \frac{e^{-\lambda_m t}}{\prod_{\substack{q=l \\ q\neq m}}^{i} (\lambda_q - \lambda_m)} \right) \tag{5}$$

A computational procedure for exact calculation of the time-dependent distribution of fission product concentration, $N_i(t)$, has been developed and integrated in the DHP code. In this calculation, all of decay chains and decay modes including $\beta^-$ decay to ground state, first and second isomer states, double $\beta^-$ decay, electron capture decay to ground and isomer states, alpha decay, delay $\beta^-$ decay, and internal transitions in the fission product system are taken into consider in this calculation. The block diagram of the DHP computational algorithm and window interface are shown in Figure 2 and Figure 3, respectively. The results of calculation for timing concentration functions of several fission product nuclides for times after a fission reaction of $^{232}$Th are shown in Figure 1, in which decay characteristics for every fission product and the independent neutron fission yields were extracted from JENDL4.0 [7].

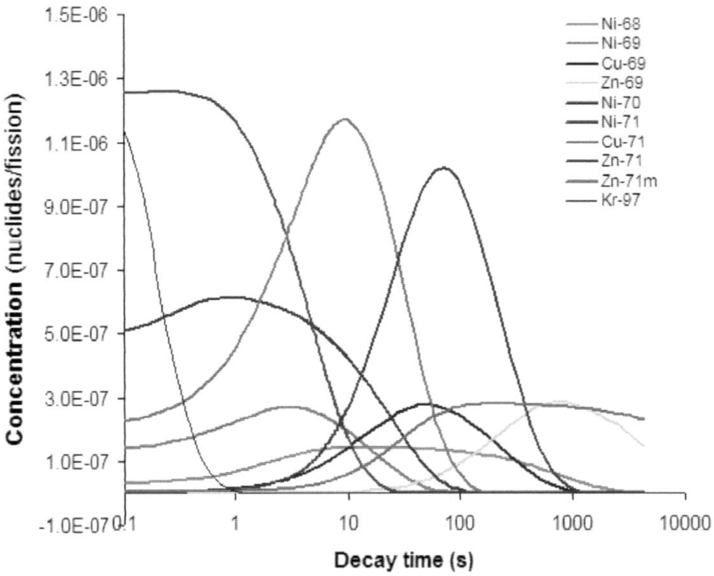


**Figure 1.** Calculated Results of concentration functions for a number selected fission products after a fast neutron fission reaction of $^{232}$Th

## 3. Beta Decay Average Energies Calculations

The average energy values of beta and gamma-rays released from beta decay of individual fission product are the most important quantities for nuclear decay heat summation prediction. The expressions for average beta and gamma-ray energies based on the Gross theory of beta decay were applied in these calculations [10, 11].

$$\overline{E}_\beta = T_{1/2} \int_{-Q}^{0} S_\beta(E) \int_{1}^{-E_g+1} mc^2(E-1)pE(-E_g+1-E)^2 F(E)dEdE_g \tag{6}$$

$$\overline{E}_\gamma = T_{1/2} \int_{-Q}^{0} S_\beta(E)mc^2(Q+E_g) \int_{1}^{-E_g+1} pE(-E_g+1-E)^2 F(E)dEdE_g \tag{7}$$

where: F(E) stands for the Fermi function. $E_g$ is related to the excitation energy $E_i$ as $E_g = -(E_i - 1)$; m, p and c denotes for electron rest mass, electron momentum and light velocity, respectively. The beta strength function $S_\beta(E)$ can be determined from the beta feeding function $I_\beta(E)$ as the following formula:

$$S_\beta(E) = \frac{I_\beta(E)}{f(Z,Q-E).T_{1/2}} \tag{8}$$

f(Z, $Q_\beta$-E) is the integrated Fermi function. Q, E and $T_{1/2}$ are the beta decay energy, excitation energy in the daughter nuclide and the half-life, respectively. The function $I_\beta(E)$ can be systematically normalized based on the beta branching intensities adopted from the ENSDF data file, or experimentally derived from measured distributions of beta decay intensity. The calculated results of average beta and gamma energies from beta decay for several selected fission products are shown in Table 1.

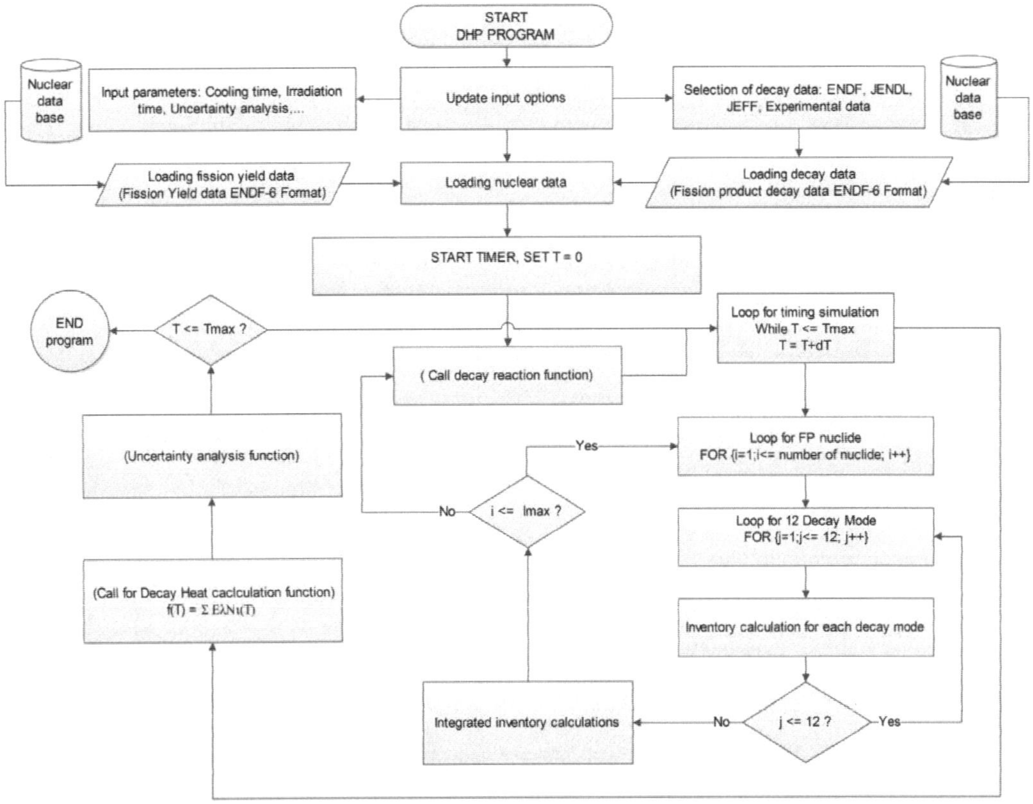


**Figure 2**. The block diagram of the computational procedure.

Journal of Physics: Conference Series **726** (2016) 012005 doi:10.1088/1742-6596/726/1/012005

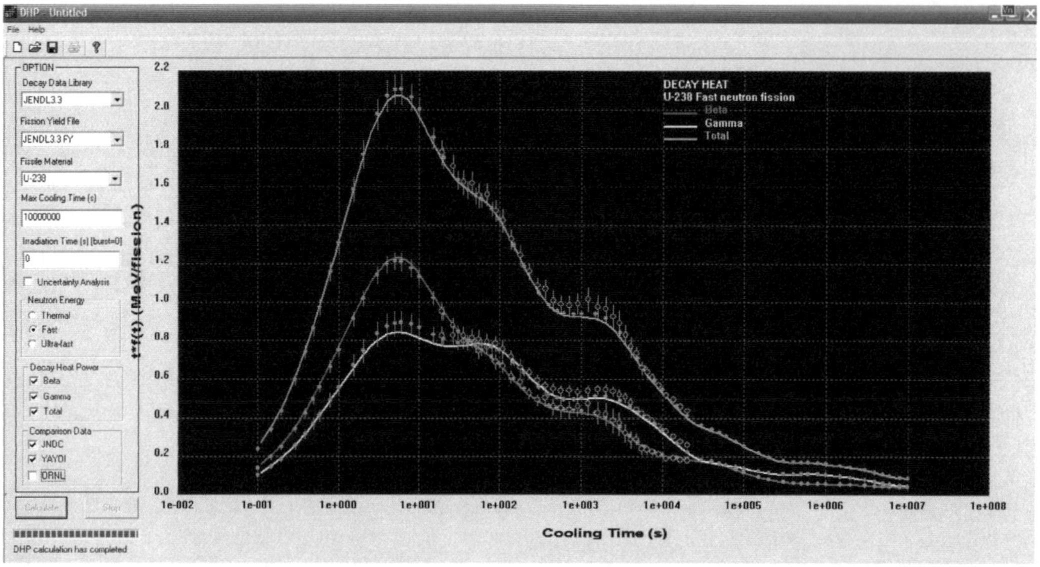


**Figure 3**. The window interface of the DHP program.

**Table 1.** The calculated average beta and gamma energies for selected nuclide.

| Nuclide | Q (MeV) | T1/2 (s) | (a) | | (b) | |
|---|---|---|---|---|---|---|
| | | | $\overline{E_\beta}$ (MeV) | $\overline{E_\gamma}$ (MeV) | $\overline{E_\beta}$ (MeV) | $\overline{E_\gamma}$ (MeV) |
| Rb-89 | 4.496 | 909.0 | 0.9970 | 2.2981 | 0.9303 | 2.2313 |
| Rb-90 | 6.587 | 158.0 | 2.0512 | 1.9996 | 1.9060 | 2.2712 |
| Rb-90m | 6.696 | 265.0 | 1.4047 | 3.2281 | 1.0811 | 3.9332 |
| Rb-91 | 5.891 | 58.4 | 1.6930 | 2.6675 | 1.3739 | 2.6876 |
| Rb-93 | 7.462 | 5.84 | 2.5226 | 2.225 | 2.1544 | 2.5402 |
| Sr-93 | 4.137 | 445.0 | 0.8197 | 2.2709 | 0.7860 | 2.1724 |
| Sr-94 | 3.508 | 75.2 | 0.8388 | 1.4503 | 0.8309 | 1.4380 |
| Sr-95 | 6.087 | 23.9 | 2.2151 | 1.1534 | 1.8928 | 1.7990 |
| Y-94 | 4.917 | 1120 | 1.8072 | 0.7987 | 1.8111 | 0.7875 |
| Y-95 | 4.453 | 618.0 | 1.4465 | 1.1285 | 1.3793 | 1.2471 |
| Cs-138 | 5.374 | 2010.0 | 1.2428 | 2.3646 | 1.2223 | 2.4047 |
| Cs-138m | 5.457 | 916.0 | 0.2700 | 0.4122 | 0.2565 | 0.4211 |
| Cs-139 | 4.213 | 556.0 | 1.6567 | 0.3406 | 1.6487 | 0.3451 |
| Cs-140 | 6.22 | 63.7 | 1.9708 | 1.7946 | 1.8399 | 1.9520 |
| Cs-141 | 5.251 | 24.9 | 2.0171 | 0.8087 | 1.5002 | 1.7041 |
| Ba-141 | 3.213 | 1100 | 0.9224 | 0.8679 | 0.9125 | 0.9108 |
| Ba-142 | 2.211 | 636 | 0.3740 | 1.0852 | 0.4032 | 1.0736 |
| Ba-143 | 4.246 | 14.3 | 1.2924 | 1.207 | 1.1835 | 1.3595 |
| Ba-144 | 3.119 | 11.5 | 0.9837 | 0.6239 | 0.9253 | 0.7921 |
| Ba-145 | 4.923 | 4.31 | 1.4714 | 0.3368 | 1.2999 | 1.7954 |
| La-142 | 4.503 | 5470.0 | 0.8682 | 2.2952 | 0.8184 | 2.4434 |
| La-143 | 3.426 | 848.0 | 2.9339 | 0.3468 | 1.2238 | 0.4618 |
| La-144 | 5.541 | 40.8 | 1.3639 | 2.3128 | 0.9867 | 3.0789 |
| La-145 | 4.108 | 24.8 | 1.5014 | 0.6509 | 0.7752 | 2.1097 |
| Ce-145 | 2.535 | 181.0 | 0.6366 | 0.8047 | 0.6268 | 0.8665 |
| Ce-146 | 1.026 | 811.0 | 0.2543 | 0.3647 | 0.2164 | 0.3520 |
| Ce-147 | 3.29 | 56.4 | 1.4929 | 0.2469 | 0.6905 | 1.4939 |
| Ce-148 | 2.06 | 56.0 | 0.5635 | 0.3416 | 0.5908 | 0.4598 |
| Pr-146 | 4.169 | 1450.0 | 1.3294 | 0.9969 | 1.2839 | 1.0685 |

| | | | | | | |
|---|---|---|---|---|---|---|
| Pr-147 | 2.686 | 804.0 | 0.7493 | 0.6836 | 0.6715 | 0.9207 |
| Pr-148 | 4.932 | 136.0 | 1.6738 | 0.9697 | 1.3111 | 1.7841 |
| Pr-148m | 5.05 | 121.0 | 1.6659 | 0.9128 | 1.1289 | 2.2984 |
| Pr-149 | 3.397 | 136.0 | 0.9161 | 1.0003 | 0.8588 | 1.3145 |
| Pr-151 | 4.102 | 18.9 | 1.8861 | 1.5435 | 1.1181 | 1.3521 |
| Nd-149 | 1.691 | 6220.0 | 0.5097 | 0.4342 | 0.4585 | 0.4109 |
| Nd-151 | 2.442 | 746.0 | 0.5926 | 0.8546 | 0.5763 | 0.8939 |
| Nd-153 | 3.336 | 32.0 | 1.1675 | 0.3636 | 1.1348 | 0.5452 |
| Nd-154 | 2.735 | 25.9 | 0.8878 | 0.4842 | 0.8573 | 0.5466 |
| Nd-155 | 4.222 | 8.9 | 1.5925 | 1.4987 | 1.0859 | 1.5365 |
| Pm-152 | 3.505 | 246.0 | 1.3034 | 0.3387 | 1.2696 | 0.4149 |
| Pm-153 | 1.881 | 324.0 | 0.6669 | 0.1311 | 0.6582 | 0.1329 |
| Pm-154 | 4.044 | 104.0 | 0.8482 | 1.7520 | 0.8527 | 1.8657 |
| Pm-155 | 3.224 | 48.0 | 2.7903 | 0.5935 | 1.0710 | 0.5639 |
| Pm-156 | 5.155 | 61.9 | 1.3029 | 1.7765 | 1.2024 | 2.2048 |
| Pm-157 | 4.514 | 27.69 | 1.4396 | 0.7605 | 1.5430 | 0.8389 |
| Sm-157 | 2.734 | 482.0 | 0.908 | 0.5180 | 0.8549 | 0.5558 |
| Sm-158 | 1.999 | 331.0 | 0.5167 | 0.5947 | 0.5155 | 0.5936 |
| Eu-158 | 3.485 | 2750.0 | 0.8503 | 1.2892 | 0.8241 | 1.3656 |

(a): Calculated with $I_\beta(E)$ from ENSDF [8]
(b): Calculation with $I_\beta(E)$ from experimental intensities [12]

The average energies shown in Table 1 are calculated in two cases: (a) calculated based on the input data of branching intensities extracted from the ENSDF data files; (b) calculated from the experimental beta decay intensities measured by Greenwood et al. [12] in which a total absorption gamma-rays spectrometer (TAGS) was used; TAGS is an efficient experimental method for average energies determinations, that overcome the problem of pandemonium effects [13].

## 4. Results of Calculations

The summation model for decay heat calculations is as the following function:

$$f(t) = \sum_{i=1}^{M} \overline{E}_i \lambda_i N_i(t) \tag{9}$$

where: M denotes the maximum number of FP nuclides; $E_i = E_{i\beta} + E_{i\gamma}$ stands for the mean energy per decay of the i$^{th}$ nuclide, $\lambda_i$ the decay constant, $N_i(t)$ the corresponding concentration function for cooling time f(t) is the burst function of decay heat (MeV/Fission/s). The physical quantity equal to t×f(t) is called decay heat power function (MeV/Fission).

In this work, the JENDL FP Decay Data File 2011 [14] and fission yield data file from JENDL 4.0 [7] have been applied for calculations of fission product concentrations as functions of cooling time after fast neutron fission reactions of $^{232}$Th, and these results were adopted in decay heat calculations. The results of calculations for t×f(t), as functions of times after a fission event, are shown in Table 2 and Figure 4 in comparison with measured values by Akizama 1982 [15].

**Table 2.** Calculated fission product decay heat for times after fast neutron fission of $^{232}$Th

| Time after fission burst (s) | Decay Heat t×f(t) (MeV/Fission) | | | Time after fission burst (s) | Decay Heat t×f(t) (MeV/Fission) | | |
|---|---|---|---|---|---|---|---|
| | Beta | Gamma | Total | | Beta | Gamma | Total |
| 1.10E-01 | 0.1497 | 0.1191 | 0.2688 | 3.00E+02 | 0.4812 | 0.5834 | 1.0646 |
| 1.60E-01 | 0.2065 | 0.1641 | 0.3706 | 4.00E+02 | 0.4419 | 0.5208 | 0.9627 |
| 2.10E-01 | 0.2579 | 0.2048 | 0.4627 | 5.00E+02 | 0.4225 | 0.4908 | 0.9132 |
| 2.60E-01 | 0.3047 | 0.2418 | 0.5464 | 6.00E+02 | 0.4121 | 0.4760 | 0.8880 |
| 3.10E-01 | 0.3475 | 0.2755 | 0.6231 | 7.00E+02 | 0.4061 | 0.4684 | 0.8745 |
| 4.10E-01 | 0.4234 | 0.3352 | 0.7586 | 8.00E+02 | 0.4027 | 0.4646 | 0.8673 |

| | | | | | | | |
|---|---|---|---|---|---|---|---|
| 5.10E-01 | 0.4886 | 0.3864 | 0.8750 | 9.00E+02 | 0.4008 | 0.4630 | 0.8638 |
| 6.10E-01 | 0.5455 | 0.4309 | 0.9764 | 1.00E+03 | 0.3998 | 0.4627 | 0.8625 |
| 7.10E-01 | 0.5956 | 0.4700 | 1.0656 | 1.50E+03 | 0.3983 | 0.4669 | 0.8652 |
| 8.10E-01 | 0.6401 | 0.5047 | 1.1448 | 2.00E+03 | 0.3922 | 0.4677 | 0.8599 |
| 9.10E-01 | 0.6799 | 0.5356 | 1.2156 | 2.50E+03 | 0.3811 | 0.4629 | 0.8440 |
| 1.01E+00 | 0.7158 | 0.5635 | 1.2792 | 3.00E+03 | 0.3679 | 0.4551 | 0.8230 |
| 1.51E+00 | 0.8509 | 0.6688 | 1.5196 | 4.00E+03 | 0.3430 | 0.4368 | 0.7797 |
| 2.01E+00 | 0.9377 | 0.7379 | 1.6757 | 5.00E+03 | 0.3234 | 0.4187 | 0.7421 |
| 2.51E+00 | 0.9954 | 0.7862 | 1.7816 | 6.00E+03 | 0.3091 | 0.4022 | 0.7112 |
| 3.01E+00 | 1.0342 | 0.8213 | 1.8554 | 7.00E+03 | 0.2987 | 0.3870 | 0.6857 |
| 4.01E+00 | 1.0772 | 0.8680 | 1.9452 | 8.00E+03 | 0.2912 | 0.3730 | 0.6642 |
| 5.01E+00 | 1.0938 | 0.8969 | 1.9907 | 9.00E+03 | 0.2857 | 0.3598 | 0.6455 |
| 6.01E+00 | 1.0967 | 0.9157 | 2.0124 | 1.00E+04 | 0.2814 | 0.3473 | 0.6287 |
| 7.01E+00 | 1.0922 | 0.9281 | 2.0203 | 1.50E+04 | 0.2678 | 0.2929 | 0.5607 |
| 8.01E+00 | 1.0838 | 0.9362 | 2.0200 | 2.00E+04 | 0.2565 | 0.2516 | 0.5081 |
| 9.01E+00 | 1.0734 | 0.9415 | 2.0148 | 2.50E+04 | 0.2446 | 0.2217 | 0.4662 |
| 1.00E+01 | 1.0620 | 0.9447 | 2.0067 | 3.00E+04 | 0.2321 | 0.1999 | 0.4320 |
| 1.50E+01 | 1.0061 | 0.9471 | 1.9532 | 4.00E+04 | 0.2064 | 0.1707 | 0.3771 |
| 2.00E+01 | 0.9604 | 0.9454 | 1.9058 | 5.00E+04 | 0.1815 | 0.1516 | 0.3331 |
| 2.50E+01 | 0.9265 | 0.9473 | 1.8737 | 6.00E+04 | 0.1593 | 0.1380 | 0.2972 |
| 3.00E+01 | 0.9025 | 0.9525 | 1.8550 | 7.00E+04 | 0.1406 | 0.1278 | 0.2684 |
| 4.00E+01 | 0.8726 | 0.9651 | 1.8377 | 8.00E+04 | 0.1255 | 0.1199 | 0.2454 |
| 5.00E+01 | 0.8515 | 0.9721 | 1.8236 | 9.00E+04 | 0.1135 | 0.1136 | 0.2270 |
| 6.00E+01 | 0.8308 | 0.9706 | 1.8013 | 1.00E+05 | 0.1039 | 0.1083 | 0.2122 |
| 7.00E+01 | 0.8085 | 0.9618 | 1.7702 | 1.50E+05 | 0.0767 | 0.0908 | 0.1675 |
| 8.00E+01 | 0.7850 | 0.9477 | 1.7327 | 2.00E+05 | 0.0641 | 0.0821 | 0.1461 |
| 9.00E+01 | 0.7612 | 0.9302 | 1.6914 | 2.50E+05 | 0.0573 | 0.0785 | 0.1359 |
| 1.00E+02 | 0.7377 | 0.9105 | 1.6482 | 3.00E+05 | 0.0539 | 0.0780 | 0.1320 |
| 1.50E+02 | 0.6363 | 0.8039 | 1.4402 | 4.00E+05 | 0.0520 | 0.0812 | 0.1332 |
| 2.00E+02 | 0.5643 | 0.7092 | 1.2735 | 5.00E+05 | 0.0528 | 0.0861 | 0.1388 |
| 2.50E+02 | 0.5149 | 0.6362 | 1.1511 | 6.00E+05 | 0.0544 | 0.0907 | 0.1451 |

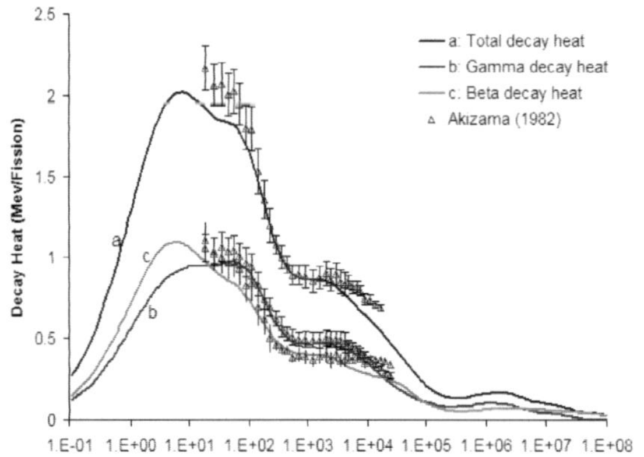


**Figure 4**. Calculated fission product decay heat after fast neutron fission of $^{232}$Th

## 5. Conclusions

In the present work, the computer code DHP has been improved for calculation procedure and updated for new available nuclear databases. Calculations have been carried out for the decay and growth distributions of concentrations for fission products from fast neutron fission of $^{232}$Th. The calculated data of fission product concentrations were then introduced into summation calculations of total and partial decay heat following times after fission burst. The decay and fission yield data used in this work is extracted from JENDL FP Decay Data File and fission yield data file 2011 [14]. As shown in Table 2 and Figures 4, within the decay time period from $1\times10^2$ s to $5\times10^3$ s, the present calculated results are agreements with the experimental values measured by Akizama [15]. It is also estimated that the updated DHP code can be used for calculations of fission product inventory concentrations and decay heat data exactly.

**Acknowledgments**
This research was partly funded by Vietnam National Foundation for Science and Technology Development (NAFOSTED) under grant number "103.04-2012.59".

**References**
[1] Nichols A L 2002 *"Nuclear Data Requirements for Decay Heat Calculations" Lectures given at the Workshop on Nuclear Reaction Data and Nuclear Reactors: Physics, Design and Safety*, Trieste, 2002
[2] Oyamatsu K 1999 *"Easy-To-Use Application Programs to Calculate Aggregate Fission Product Properties on Personal Computers"* JEARI-Conf 99-002
[3] Kazimi M S, Czerwinski K R, Driscoll M J, Hejzlar P and Meyer J E 1999 *"On the Use of Thorium in Light Water Reactors"* MIT-NFC-TR-016
[4] OECD 2015 *"Introduction of Thorium in the Nuclear Fuel Cycle"*, NEA No. 7224
[5] Son P N and Katakura J 2007 *"An Application Program for Fission Product Decay Heat Calculations"* JAEA-Data/Code 2007-018
[6] Tobias A 1980 *Prog. Nucl. Energy* **5** 1
[7] Shibata K O, et al 2011 *J. Nucl. Sci. Technol.* **48** 1
[8] Bhat M R 1992 *"Evaluated Nuclear Structure Data File (ENSDF)", Nuclear Data for Science and Technology*, Edited by Qaim S M, Springer-Verlag, Berlin, Germany, Page 817
[9] Son P N 2014 *Int. J. of Nucl. Ener. Sci. and Enginee.* **4** 75
[10] Katakura J, Yoshida T, Oyamatsu K, and Tachibana T 2001 "JENDL FP Decay Data File 2000" JAERI-1343 (2001)
[11] Yoshida T and Katakura J 1986, *Nucl. Sci Eng.* **93** 193
[12] Greenwood R C, Helmer R G, Putnam M H, and Watts K D 1997 *Nucl. Instr. Meth.* **A390** 95
[13] Hardy J C, Carraz L C, Jonson B and Hansen P G 1977 *Phys. Letts.* **71B(2)** 307
[14] Katakura J 2011 *"JENDL FP Decay Data File 2011 and Fission Yields Data File 2011"* JAEA-Data/Code 2011-025
[15] Akiyama M and An S 1982 *''Measurement of fission-product decay heat for fast reactors''* Proc. Int. Conf. on Nuclear Data for Science and Technology, Antwerp Belgium (1982): pp 237-244

# $\pi$-Plasmon model for carbon nano structures: Application to porphyrin

**Dao Thu Ha, Chu Thuy Anh, Do Thi Nga, Le Minh Thanh , Tran Thi Thanh Van, Nguyen Ai Viet**

Institute of Physics, Vietnam Academy of Science and Technology, 10 Dao Tan, Ba Dinh, Hanoi, Vietnam

E-mail: ctanh@iop.vast.ac.vn

**Abstract.** In traditional concept, the optical properties of semiconductors and semimetals near their fundamental optical band gaps are attributed to single excitations (such electron-hole pairs, excitons...). In our earlier article, we proposed the collective mechanism of $\pi$-plasmons for optical properties of low dimensional carbon nano structures. A simple way to calculate the peak positions of UV-vis absorption spectra was pointed out and gave a good agreement with experimental data. In this work we analyze different schemas to calculate the UV-vis absorption peaks. A new parameter $k$ which characterizes the dependence of schema on geometry and number of carbon sites is defined. As an example, the case of porphyrin was investigated.

## 1. Introduction

In recent years, optical properties of low dimensional systems, nano structures semiconductors and semimetals have attracted great intention due to their enormous, vast, and important application in modern nano technologies.

In traditional concept, the optical properties of semiconductors and semimetals near their fundamental optical band gaps are attributed to single excitations (such electron-hole pairs in one-particle picture, or excitons in two-particle picture ...). At high concentration, an electron-hole plasma in many particle-particle picture will formed, and the corresponding collective excitation is called plasmon.

Similarly, in carbon nano structures, such as graphene, graphene nano ribbons, carbon nanobots, Fullerton, $C_{60}$ molecules, etc, the optical properties determined by the interbank $\pi$e transitions (in the free $\pi$optoelectronic scheme), or from the ground sate to excited states (in the $\pi$exciton scheme), or by $\pi$-plasmas mechanism.

Porphyrins have been intensively investigated due to their wide range of applications such as for photon capture photons[1], photo dynamic therapy of tumors[2, 3], and selective catalysts[4]. Moreover, porphyrin-based materials also play substantial roles in biological systems as oxygen transporters[5], photoreceptor in green plants[6]. It is apparent that most applications of porphyritic are based on their optical properties. Understanding the nature of the visible light absorption of porphyrins, therefore, is of great importance to the advancement of technology.

The absorption peak of the Soret band of porphyritic is at 400nm with a strong intensity. The intensities of optical peaks decline when increasing wavelength. The maximum positions in the absorption spectra remain almost unchanged in different solvents with different kinds of porphyritic. In the previous work[7] we proposed simple model based on the Mie theory for an oblate spheroid allows us to understand the absorption spectra of porphyrin. The properties of the Soret band and Q bands can be explained by the plasmon mechanism. We have developed a theoretical model for the optical spectrum based on plasmonic excitation[8]. We explicitly point out the $\pi$-plasmonic feature of the Soret band. $\pi$-plasmon is the collective oscillations of $\pi$-electrons, which have recently been found to exist in low-dimensional systems such as graphene[9], graphite, and carbon nanotubes[10]. Because $\pi$-plasmon is caused by the excitation of a semilocalized electron, the properties of $\pi$-plasmon are quite different from the usual plasmons found in metallic materials. More insights into the nature of plasmon in the porphyrin molecule would open various new applications for such porphyrin-based materials.

In this work, we introduce a new concept: topological-dependent parameter $k$ of porphyrin, then investigate various models: between a porphyrin and a monolayer graphene plate which have the same density, or the same area, or the same atoms number.

## 2. $\pi$-Plasmon model for optical properties of flat carbon nano structures

To take in consideration, the chosen monolayer graphene, porphyrin, and flat bio-macro molecules are supposed to have the same thickness, $d = 3.4A^{\circ}$, equal to the diameter of a carbon atom. This assumption suggests that both graphene sheets, porphyrin, and flat bio-macro molecules can be treated as bulk systems. Plasmons in the bulk oscillate at plasma frequency $\omega_p$, determined by

$$\omega_P = \sqrt{\frac{ne^2}{m^*\varepsilon_0}}, \tag{1}$$

where $n$ is the density of electrons, $m*$ is the electron effective mass, $e$ is the electron charge, and $\varepsilon_0$ is the permittivity of the vacuum. The electron density $n$ is calculated by formula

$$n_i = \frac{N_C^i}{S^i d}, \tag{2}$$

where $N_C$ and $S$ are the number of carbon atoms and the considered area, respectively.

The distance between nearest-neighbor carbon atoms in graphene is $l = 1.24\,\mathring{A}$. The area of a hexagonal unit cell in graphene is $A = 3\sqrt{3}l^2/2$. Thus, the area of porphyrin corresponds to 11.5 hexagonal unit cells in graphene. Experimental data for $\pi$-plasmon are approximately $\omega_{\pi P} = 4.7$ eV.

## 3. $\pi$-Plasmon model for the Soret band of porphyrin

The structure of porphyrin can be seen in [7], and represented here in Fig.3 with 20 carbon atoms [(green) spheres], 14 hydrogen atoms [small (yellow) spheres], and 4 nitrogen atoms [(brown) spheres], and the reference compound [large (orange) spheres].

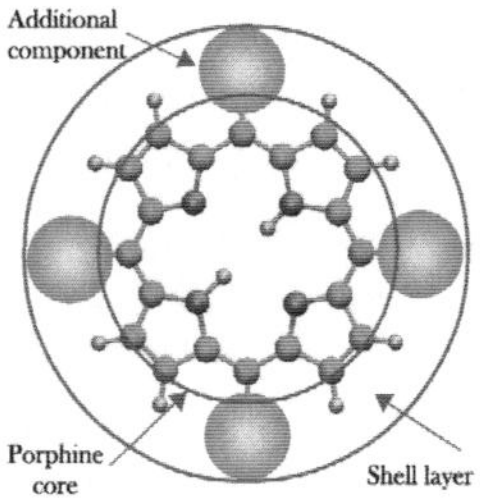


**Figure 1.** (Color online) Atomic structure of porphyrin, with 20 carbon atoms [(green) spheres], 14 hydrogen atoms [small (yellow) spheres], and 4 nitrogen atoms [(brown) spheres], and the reference compound [large (orange) spheres].

The typical absorption spectra of a porphyrin molecule is also described in our previous work which showed that the strongest absorption localized at around $414nm$ $(3eV)$, corresponds to the Soret band. The four weak absorption peaks are due to $Q$ bands.

The behaviors of the optical spectra are similar to those of other kinds of porphyrin family. However, the peak positions are shifted due to the additional components. This phenomenon can be explained by the conjugated systems creating a shell covering the porphyrin core. The core-shell interactions in porphyrins behave similarly to what we observe in the optical spectrum of core-shell nanoparticle systems. On the other side, plasmon is also found in low-dimensional materials. Based on these facts, they have made a suggestion of the possible exist of plasmon in porphyrins. They proposed an assumption for the nature of the largest peak, the Soret band or B band.

The ratio of the plasma frequency of porphyrin to that of graphene is given by[7]

$$\frac{\omega_P^{por}}{\omega_P^g} = k_{por}\sqrt{\frac{N_C^{por}}{N_C^g}\frac{S^g}{S^{por}}}, \tag{3}$$

where $k_{por}$ is a topological-dependent parameter of porphyrin, $\omega_P^{por}$ is the plasma frequency of porphyrin, $\omega_P^g$ is the plasma frequency of graphene, $N_C^{por}$ and $N_C^g$ are the number of carbon atoms in the area $S^{por}$ and $S^g$ of porphyrin and graphene, respectively. For porphyrin $N_C^{por} = 20$, the area of porphyrin $S^{por}$ corresponds to 11.5 hexagonal unit cells in graphene, so $N_C^g$=48 . Experimental data for $\pi$-plasmon are approximately $\omega_{\pi P} = 4.7$ eV. The our calculated $\omega_P^{por} \simeq 3.03$ eV, which is very closed to the position of the absorption peak of the Soret band $\omega_{Soret} \simeq 3$ eV. The value of topological-dependent parameter of porphyrin $k_{por} \simeq 0.64$. The position of the Soret band gives $\omega_{Soret} \simeq 3$ eV.

Generalize above result, we introduce a topological-dependent parameter $k_i$ of the nano-carbon or bio-macro-molecule structures $i$, which is defined by the equation

$$\frac{\omega_P^i}{\omega_P^g} = k_i\sqrt{\frac{N_C^i}{N_C^g}\frac{S^g}{S^i}}, \tag{4}$$

where $\omega_P^i$ is the plasma frequency, and $N_C^i$ is the number of carbon atoms in its area $S^i$of the nano-carbon or bio-macro-molecule structures $i$.

## 4. Topological-dependent parameter k of porphyrin:

### *4.1. The dependence of k on carbon atoms density:*

With $x, y$ the number of hexagonal unit cells of each edge, $N$ is the number of carbon atoms, $S$ is the considered area, $A$ is the area of a hexagonal unit cell, means $S$ is the number of hexagonal unit cells in the considered area.

Journal of Physics: Conference Series **726** (2016) 012006 doi:10.1088/1742-6596/726/1/012006

$$n = \frac{N}{S}. \tag{5}$$

As stated before, the thickness of porphyrins molecules and monolayer graphene are supposed to be the same, so we examine the variation of $k$ along with number of carbon atoms over the corresponding area ($n = N/S$), which can be considered as the carbon atom density. The investigation is made for three different monolayer graphene plate type: the parallelogram, the rhombus, and the equilateral triangle.

In the case of parallelogram:

$$N_{paral} = 2\,(xy + x + y)\,; S_{paral} = xy, n_{paral} = 2 + 2\frac{(x+y)}{xy}. \tag{6}$$

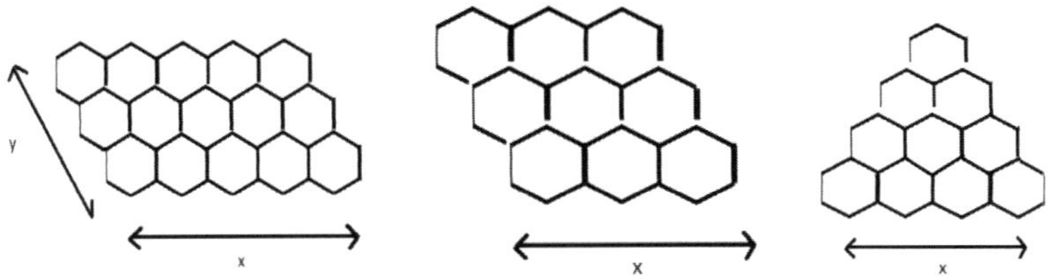


**Figure 2. Left**: Parallelogram molecule. **Center**: Rhombus molecule. **Right**: Equilateral triangle molecule.

Application of similar methods for the rhombus:

$$N = 2x(x+2), S = x^2, n = 2 + \frac{4}{x} \tag{7}$$

In case of equilateral triangle:

$$N = x^2 + 4x + 1, S = \frac{x+1}{2}x, n = 2\frac{x^2+4x+1}{x^2+x}. \tag{8}$$

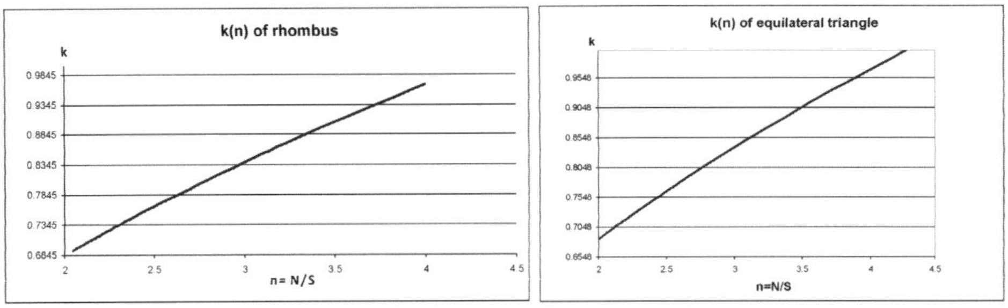


**Figure 3. Left**: The dependence of k on density of rhombus model. **Right**: The dependence of k on density of "equilateral" triangle model.

As shown in Fig.3, when the size of models increases, the value of $n = N/S$ decreases. As the size reaches to infinite (or the number of carbon atoms reach to infinite), $n \to 2$ in all three models. Then, $k$=0.6845 when $size \to \infty, N \to \infty, n \to 2$. $k$ increases linearly but differently in each case.

Journal of Physics: Conference Series **726** (2016) 012006 doi:10.1088/1742-6596/726/1/012006

*4.2. The dependence of k on the considered area*

We consider the case where porphyrin and graphene have the same number of carbon atoms $N_c = 20$, which is the number of carbon atoms of porphine core. Four different models of monolayer graphene are investigated in Fig.4 Fig.5.

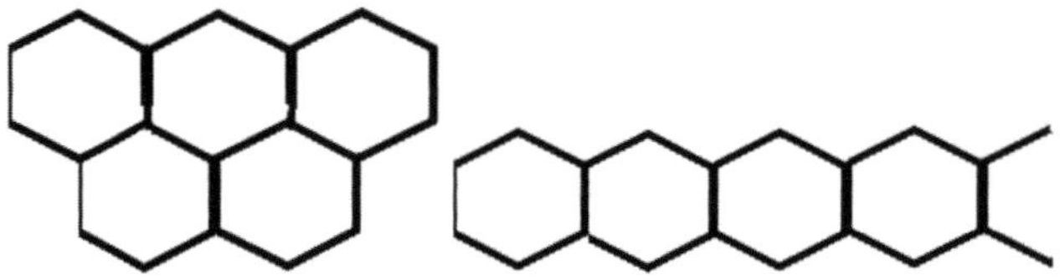

**Figure 4. Left**: With the arrangement of carbon atoms as shown, the respective area of graphene is S=5A, which lead to the result of $k = 0.968$.
**Right**: The area of graphene now is S=4A+0.5A=4.5A, so k is 1.02.

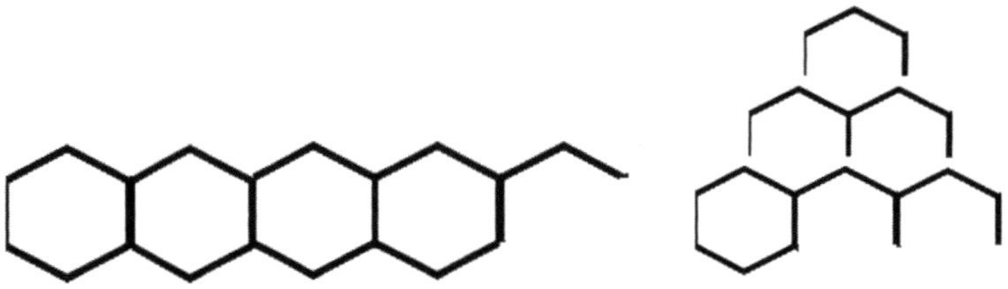

**Figure 5. Left**: S=4A+1/6A=25/6A. k= 1.06.**Right**: S=17/3A, $k = 0.91$.

*4.3. The dependence of k on the number of carbon atoms*

In case that considered area of porphyrin and monolayer graphene is the same (S=11.5A), four different models are investigated in Fig.6 and Fig.7.

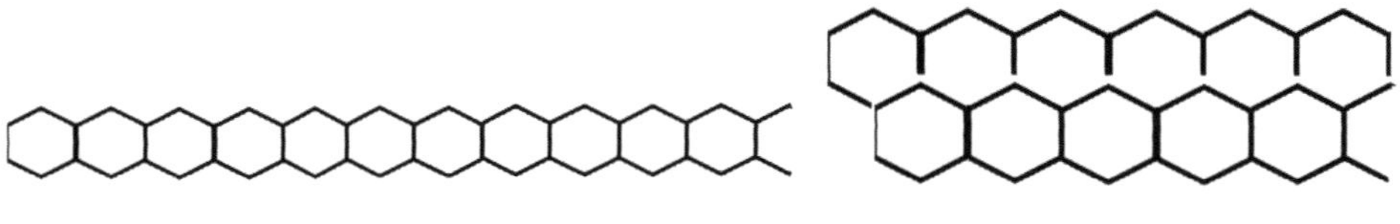

**Figure 6. Left**: The number of carbon atoms in this case is Ng=50, substitute the result in formula (3), we get k=1.009.
**Right**: Ng=42, so k=0.925.

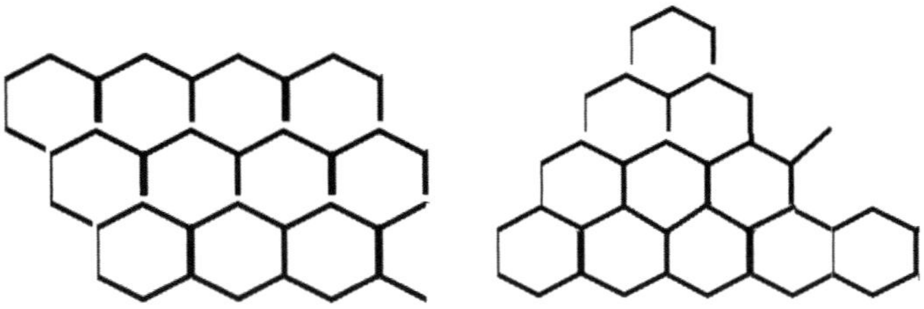

**Figure 7. Left**: The number of carbon atoms of graphene is 36, $k$ corresponding is 0.856. **Right**: Ng=38, k=0.88.

## 5. Conclusion

As the continuous results of our previous works on theoretical model for calculating the absorption spectrum of porphyrin and carbon nano structures, the topological dependence of $k$ has opened various possibilities on biological macromolecules research. It has been shown obviously that $k \simeq 1$, which helps to verify our proposed theoretical model on the $\pi$-plasmon nature of main optical properties of porphyrin and carbon nano structures. Above investigations showed that there exist each constant $k$ for each corresponding molecule type. It has important meaning that $k$ can be used to categorize biological macromolecules. The theoretical $k$ in this work will be compared with experimental one in our next investigations.

## Acknowledgement

We acknowledge financial support from the Protocol project No. VAST. HTQT. Belarus 03/14-15.

## References

[1] Fukuzumi S 2000 *The Porphyrin Handbook* (San Diego, CA: Academic Press)
[2] Batlle A M 1993 *J. Photochem. Photobiol. B* **20** 1744–1751
[3] Sugisaki K, Usui T, Nishiyama N, Jang W D, Y Yanagi S Y, Amano S and Kataoka K 2008 *Invest. Ophthalmol. Vis. Sci.* **49** 894–9
[4] Gharaati S, Moghadam M, Tangestaninejad S, Mirkhani V, Mohammadpoor-Baltork I, Barati B and Sadegh F 2013 *J. Organomet. Chem.* **78** 741–742
[5] Chen Y H, Yet S F and Perrella M A 2003 *Exp. Biol. Med.* **228** 447–53
[6] Storm C B, Corwin A H, Arellano R R, Martz M and Weintraub R 1966 *J. Am. Chem. Soc.* **88** 2525–32
[7] Phan A D, Nga D T, Phan T L, Thanh L T M, Chu T A, Bernad S and Viet N A 2014 *Phys. Rev. E* **90** 062707
[8] Chen Y C, Lee M W, Li L L and Lin K J 2008 *J. Macromol. Sci. Phys.* **47** 955–966
[9] Despoja V, Novko D, Dekanic K, Sunjic M and Marusic L 2013 *Phys. Rev. B* **87** 075447
[10] Kramberger C and et al 2008 *Phys. Rev. Lett.* **100** 196803

Journal of Physics: Conference Series **726** (2016) 012007 doi:10.1088/1742-6596/726/1/012007

# Generalized Bogoliubov Polariton Model: An Application to Stock Exchange Market

**Chu Thuy Anh, Truong Thi Ngoc Anh, Nguyen Tri Lan, and Nguyen Ai Viet**

Institute of Physics, Vietnam Academy of Science and Technology, 10 Dao Tan, Ba Dinh, Hanoi, Vietnam

E-mail: ctanh@iop.vast.ac.vn

**Abstract.** A generalized Bogoliubov method for investigation non-simple and complex systems was developed. We take two branch polariton Hamiltonian model in second quantization representation and replace the energies of quasi-particles by two distribution functions of research objects. Application to stock exchange market was taken as an example, where the changing the form of return distribution functions from Boltzmann-like to Gaussian-like was studied.

## 1. Introduction

Bogoliubov method[1] is frequently used in superconductivity[2, 3] systems and condensed matter physics. It is a well-known method, used as a filter for reducing special hamiltonians of interacting particles to almost noninteracting quasi-particles[4, 5, 6, 7]. In this investigation, Bogoliubov method is used to describe the transition of two branches polariton Hamiltonian model in second quantization. This method will be developed in the case of boson gases and plasmon-polariton. Then it will be applied in stock exchange market, as an example of complex system.

Quasi-particles are replaced by complex systems, described by statistical functions. It will appear then two branches as in boson or plasmon-polariton case. But in stock exchange market, we know only one system corresponding to one branch, the other branch is unknown. The transition of distribution function in stock exchange market corresponds to one branch. The unknown duality system remains as a challenge for our next investigation.

## 2. Bogoliubov transformation

Consider a system containing two quasi-particles $a$ and $b$, their interaction is described by second quantized Hamiltonian

$$H = \sum_k \left\{ \varepsilon_a(k) a_k^+ a_k + \varepsilon_b(k) b_k^+ b_k + g_k \left[ a_k^+ b_k + b_k^+ a_k \right] \right\}, \quad (1)$$

in which $\varepsilon_a(k)$ , $\varepsilon_b(k)$ are quasi-particle energy, represented by creation operators $a_k^+$, $b_k^+$ and annihilation operators $a_k^-$, $b_k^-$, interaction constant $g_k$ and pulse $k$. Interacting particle system could be transformed to non-interacting particle system via Bogoliubov transformation by annihilation operators $A$, $B$, with corresponding energy $E_A$ and $E_B$. For simplicity, we use only real parameters. After diagonalized, Hamiltonian of system becomes:

$$H = \sum_k \left\{ E_A(k)\, A_k^+ A_k + E_B(k)\, B_k^+ B_k \right\}. \tag{2}$$

Bogoliubov direct transformation

$$A_k = u_k a_k + v_k b_k, \; B_k = -v_k a_k + u_k b_k. \tag{3}$$

Bogoliubov inverse transformation

$$a_k = u_k A_k - v_k B_k, \; b_k = v_k A_k + u_k B_k. \tag{4}$$

Normalization condition

$$u_k^2 + v_k^2 = 1, \tag{5}$$

With given Hamiltonian with three parameters $\{\varepsilon_a, \varepsilon_b, g\}$, we need to transform $\{u, v\}$ in order to get Hamiltonian of non-interacting quasi-particle system

$$\{\varepsilon_a, \varepsilon_b, g\} \Longleftrightarrow \{E_A, E_B, u\}, \tag{6}$$

with $\{E_A, E_B\}$ parameters of non-interacting quasi-particle system.

*2.1. Boson case*

In case that $a$ and $b$ are bosons, commutation relation for bosonic creation and annihilation operators

$$\left[a_k, a_k^+\right] = \left[b_k, b_k^+\right] = 1, \tag{7}$$

other commutations of $a$ and $b$ are 0. It is easy to see that $A$ and $B$ are also bosons because

$$\left[A_k, A_k^+\right] = \left[B_k, B_k^+\right] = 1, \tag{8}$$

other commutations of $A$ and $B$ are 0.

Dispersion law of new quasi-particles is

$$E_{A,B}(k) = \frac{1}{2}\left\{ \left[\varepsilon_a(k) + \varepsilon_b(k)\right] \pm \sqrt{\left[\varepsilon_a(k) + \varepsilon_b(k)\right]^2 + 4g_k^2} \right\}. \tag{9}$$

Transformation parameters are

$$\begin{cases} u_k^2 &= \frac{1}{2}\left\{1 + \frac{|\varepsilon_a(k) - \varepsilon_b(k)|}{\sqrt{[\varepsilon_a(k)+\varepsilon_b(k)]^2 + 4g_k^2}}\right\}, \\ v_k^2 &= \frac{1}{2}\left\{1 - \frac{|\varepsilon_a(k) - \varepsilon_b(k)|}{\sqrt{[\varepsilon_a(k)+\varepsilon_b(k)]^2 + 4g_k^2}}\right\}. \end{cases} \tag{10}$$

*2.2. Plasmon polariton*

In case of plasmon-polariton, quasi-particle $a$ is photon $\gamma$, quasi-particle $b$ is plasmon, dispersion law

$$\varepsilon_a = \varepsilon_\gamma(k) = \frac{ck}{\sqrt{\epsilon_0}}, \; \varepsilon_b = \varepsilon_P(k) = \hbar\omega_P = \frac{ck_P}{\sqrt{\epsilon_0}}, \tag{11}$$

in which $c$ is speed of light, $\epsilon_0$is dielectric constant , $\omega_P$ is plasmon frequency. Consider $\hbar = c = 1$ and $\epsilon_0 = 1$.

Dispersion law of plasmon-polariton up and down are described in Fig.1.

Transformation parameters $u^2$, $v^2$ are described in Fig.2.

When $k$ increases, in the upper line plasmon becomes photon, while photon become plasmon in the lower line.

This property could be applied to complex systems.

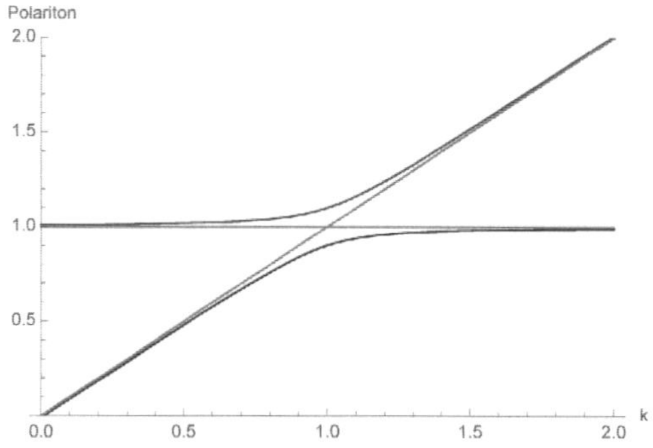


**Figure 1.** Plasmon polariton dispersion law.

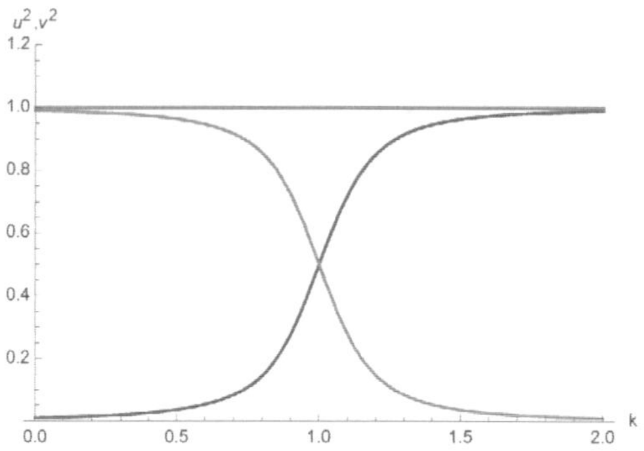


**Figure 2.** Transformation parameters.

## 3. Application for financial market

Consider two complex systems $\alpha$ and $\beta$ , their quasi-interaction is described by second quantized quasi-Hamiltonian

$$\hat{H} = \sum_{\kappa} \left\{ \left[ \varepsilon_{\alpha}(\kappa) + \varepsilon_{\alpha 0} \right] \alpha_{\kappa}^{+} \alpha_{\kappa} + \left[ \varepsilon_{\beta}(\kappa) + \varepsilon_{\beta 0} \right] \beta_{\kappa}^{+} \beta_{\kappa} + G_{\kappa} \left[ \alpha_{\kappa}^{+} \beta_{\kappa} + \beta_{\kappa} \alpha_{\kappa} \right] \right\}, \tag{12}$$

in which $\varepsilon_{\alpha}(\kappa)$, $\varepsilon_{\beta}(\kappa)$ are quasi-energy of complex systems, calculated from quasi-vacuum $\varepsilon_{\alpha 0}, \varepsilon_{\beta 0}$ ; quasi-annihilation operator $\alpha_{\kappa}$, $\beta_{\kappa}$ quasi-creation operator $\alpha_{\kappa}^{+}$, $\beta_{\kappa}^{+}$, action coordinates quasi-pulse $\kappa$ ; and $G(\kappa)$ quasi-interacting constant.

Statistical parameter $x$ is expended another space dimension. $P_{\alpha}(x)$ , $P_{\beta}(x)$ are $x$ distribution functions corresponding to complex systems, satisfy normalized condition on $X$

$$\begin{aligned} \textstyle\int_X P_{\alpha}(x)\, dx &= 1, \\ \textstyle\int_X P_{\beta}(x)\, dx &= 1, \end{aligned} \tag{13}$$

quasi- Hamiltonian is rewritten

$$\hat{H} = \sum_{\kappa} \int_X dx \left\{ \left[ \varepsilon_{\alpha}(\kappa) + \varepsilon_{\alpha 0} P_{\alpha}(x) \right] \alpha_{\kappa}^{+} \alpha_{\kappa} + \left[ \varepsilon_{\beta}(\kappa) + \varepsilon_{\beta 0} P_{\beta}(x) \right] \beta_{\kappa}^{+} \beta_{\kappa} + G_{\kappa}(x) \left[ \alpha_{\kappa}^{+} \beta_{\kappa} + \beta_{\kappa} \alpha_{\kappa} \right] \right\}. \tag{14}$$

Bogoliubov method could be applied to diagonalize quasi-Hamiltonian of interacting complex systems to non-interacting complex systems, described by quasi-annihilation operators $\hat{A}$ and

Journal of Physics: Conference Series **726** (2016) 012007 doi:10.1088/1742-6596/726/1/012007

$\hat{B}$ ($\hat{B}^+$, $\hat{B}^+$ are corresponding quasi-creation operators) at quasi-energy $\Omega_A$ and $\Omega_B$. In this investigation, all parameters are real. After diagonalization, Hamiltonian becomes

$$\hat{H} = \sum_\kappa \int_X dx \left\{ \Omega_A(\kappa, x) \hat{A}_\kappa^+ \hat{A}_\kappa + \Omega_B(\kappa, x) \hat{B}_\kappa^+ \hat{B}_\kappa \right\} = \sum_\kappa \int_X dx \hat{H}_\kappa(x) \tag{15}$$

Bogoliubov direct transformation

$$\hat{A}_\kappa = U_\kappa \alpha_\kappa + V_\kappa \beta_\kappa,\ \hat{B}_\kappa = -U_\kappa \alpha_\kappa + V_\kappa \beta_\kappa. \tag{16}$$

Bogoliubov inverse transformation

$$\alpha_\kappa = U_\kappa \hat{A}_\kappa - V_\kappa \hat{B}_\kappa,\ \beta_\kappa = V_\kappa \hat{A}_\kappa + U_\kappa \hat{B}_\kappa. \tag{17}$$

Normalization condition

$$U_\kappa^2 + V_\kappa^2 = 1, \tag{18}$$

For a given quasi Hamiltonian of interacting complex systems $\{\varepsilon_\alpha, \varepsilon_\beta, G\}$, $\{U, V\}$ need to be changed with normalization condition $U^2 + V^2 = 1$ to get the new quasi Hamiltonian of non-interacting complex system $\{\Omega_A, \Omega_B\}$

$$\begin{aligned} \hat{H}_\kappa(x) &= [\varepsilon_\alpha(\kappa) + \varepsilon_{\alpha 0} P_\alpha(x)] \alpha_\kappa^+ \alpha_\kappa + [\varepsilon_\beta(\kappa) + \varepsilon_{\beta 0} P_\beta(x)] \beta_\kappa^+ \beta_\kappa + G_\kappa(x) [\alpha_\kappa^+ \beta_\kappa + \beta_\kappa \alpha_\kappa] \\ &= \Omega_A(\kappa, x) \hat{A}_\kappa^+ \hat{A}_\kappa + \Omega_B(\kappa, x) \hat{B}_\kappa^+ \hat{B}_\kappa. \end{aligned} \tag{19}$$

In the mathematical point of view, it is a transformation

$$\{\varepsilon_\alpha, \varepsilon_\beta, G\} \Longleftrightarrow \{\Omega_A, \Omega_B U\}. \tag{20}$$

Dispersion law of the new system

$$\begin{aligned} \Omega_{\hat{A},\hat{B}}(\kappa, x) &= \tfrac{1}{2} [\varepsilon_\alpha(k) + \varepsilon_{\alpha 0} P_\alpha(x) + \varepsilon_\beta(k) + \varepsilon_{\alpha 0} P_\alpha(x)] \\ &\pm \tfrac{1}{2} \sqrt{[\varepsilon_\alpha(k) + \varepsilon_{\alpha 0} P_\alpha(x) - \varepsilon_\beta(k) - \varepsilon_{\alpha 0} P_\alpha(x)]^2 + 4G_\kappa^2(x)}, \end{aligned} \tag{21}$$

with transformation factor

$$\begin{aligned} U_\kappa^2(x) &= \tfrac{1}{2} \left\{ 1 + \frac{|\varepsilon_\alpha(k) + \varepsilon_{\alpha 0} P_\alpha(x) - \varepsilon_\beta(k) - \varepsilon_{\alpha 0} P_\alpha(x)|}{\sqrt{[\varepsilon_\alpha(k) + \varepsilon_{\alpha 0} P_\alpha(x) + \varepsilon_\beta(k) + \varepsilon_{\alpha 0} P_\alpha(x)]^2 + 4G_k^2(x)}} \right\} \\ &= 1 - V_\kappa^2(x). \end{aligned} \tag{22}$$

Increasing $\kappa$ in the lower line, $\alpha$ becomes $\beta$, while in the upper line, $\beta$ becomes $\alpha$.

In financial market, returns distribution function $x$ transform from Boltzman-like $P_B(x) = (1/2\pi\lambda)\, exp(-x/\lambda)$ to Gaussian-like $P_G(x) = \left(1/\sqrt{2\pi\sigma}\right) exp\left(-x^2/4\sigma^2\right)$ with corresponding statistical factors $\lambda$, $\sigma$, in this case, action coordinates $\kappa$ is time $t$. Apply Bogoliubov method, Hamiltonian can be written as

$$\begin{aligned} \hat{H}_t(x) &= [\varepsilon_\alpha(t) + \varepsilon_{\alpha 0} P_\alpha(t)] \alpha_t^+ \alpha_t + [\varepsilon_\beta(t) + \varepsilon_{\beta 0} P_\beta(t)] \beta_t^+ \beta_t + G_t(x) \left[\alpha_t^+ \beta_t + \beta_t \alpha_t\right] \\ &= \Omega_U(t, x) \hat{A}_\kappa^+ \hat{A}_\kappa + \Omega_D(t, x) \hat{B}_\kappa^+ \hat{B}_\kappa, \end{aligned} \tag{23}$$

in which $\alpha_t, \beta_t$ are quasi-annihilation operators, which describes impaction with quasi-energy $\varepsilon_\alpha(t) = bt$ to financial market at quasi-static energy $\varepsilon_\beta(t) = \varepsilon_M$. The evolution of returns distribution from Boltzmann like to Gaussian like is described in Fig. 3 and Fig. 4

In econophysics systems, quasi-energy of complex systems is identified by market returns. As observed in our previous works[8, 9], there exists the fluctuation of market returns distribution. Most of the published works investigated only on Gaussian-Boltzmann distribution transformation but not the reverse process. This works showed that there might exist Boltzmann-Gaussian distribution transformation also.

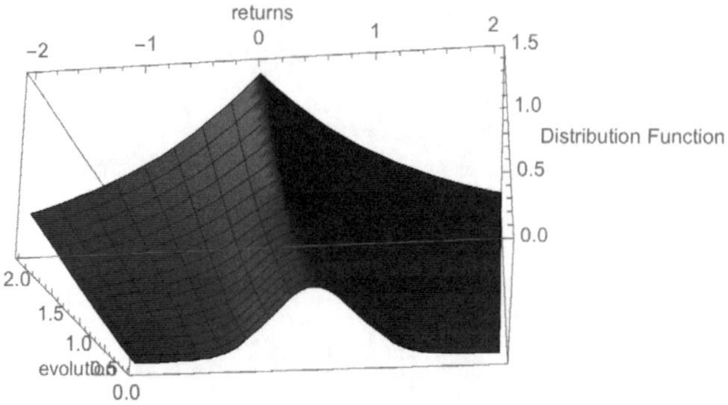


**Figure 3.** Returns distribution evolution.

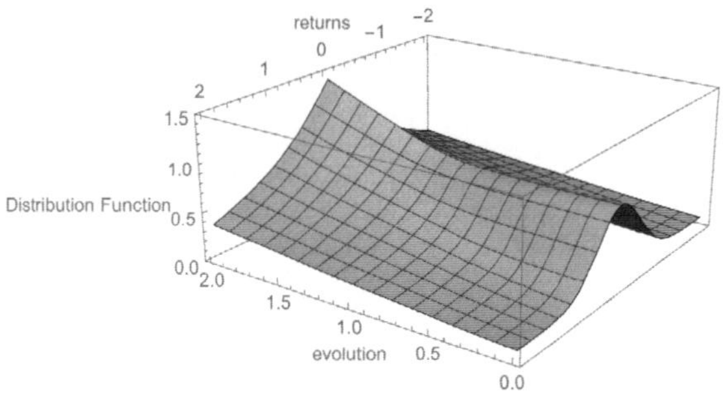


**Figure 4.** Returns distribution evolution.

## 4. Conclusion

After a brief presentation of general Bogoliubov method, investigations in Bosonic system and plasmon-polariton system are made. A transition between two lines of dispersion is observed.

In previous publications, a transition of market returns distribution function is realized in some investigations. Some explanations are made, with the hope to understand the dynamic of this transition. In this investigation, Bogoliubov method is used to describe the transition, as another view of this phenomenon. Beyond that, the result of this work opens a possibility to understand that there exist the vice-versa transformation, which is never ever been mentioned before.

## References

[1] Bogolubov N 1947 *J. Phys. USSR* **11** 23–32
[2] Bogoliubov N 1958 *Sov. Phys. - JETP* **7** 41–46
[3] Bogoliubov N 1958 *Nuovo Cim.* **7** 794–805
[4] Dodonov V, Dremin I, Man'ko O, Man'ko V and Polynkin P 1998 *J. Russ. Laser Res.* **19** 427–463
[5] Honegger R and Rieckers A 1996 *J. Math. Phys.* **37** 4292–4309
[6] Nieto M and Truax D 1997 *Fortsch.Phys.* **45** 145–156
[7] Shumovskii A 1991 *Theor. Math. Phys.* **89** 1323–1329
[8] Chu T A, Do H L and Nguyen A V 2013 *Communications in Physics* **23** 185–191
[9] Anh C T, Lan N T and Viet N A 2014 *Journal of Physics: Conference Series* **537** 012005

# A model of optical trapping cold atoms using a metallic nano wire with surface plasmon effect

**Nguyen Thi Phuong Lan**[1], **Do Thi Nga**[2] **and Nguyen Ai Viet**[2]
[1] Hanoi Pedagogical University 2, Nguyen Van Linh Street, Xuan Hoa Ward, Phuc Yen Town, Vinh Phuc Province, Vietnam
[2] Institute of Physics, 10 Dao Tan, Thu Le, Ba Dinh, Ha Noi, Vietnam

E-mail: lanntphuong@gmail.com

**Abstract.** In this work, we construct a new model of optical trapping cold atoms with a metallic nano wire by using surface plasmon effect generated by strong field of laser beams. Using the skin effect, we send a strong oscillated electromagnetic filed through the surface of a metallic nano wire. The local field generated by evanescent effect creates an effective attractive potential near the surface of metallic nano wires. The consideration of some possible boundary and frequency conditions might lead to non-trivial bound state solution for a cold atom. We discus also the case of the laser reflection optical trap with shell-core design, and compare our model with another recent schemes of cold atom optical traps using optical fibers and carbon nanotubes.

## 1. Introduction

In the last twenty years the problem of surface plasmon and surface plasmon polariton become one of the hottest topics, attract great intention, and their investigation grow up very quickly in both theoretical and experimental sides due the very large potential and vast application in modern nano and bio technologies [1-23].

Recently, study properties of cold neutral trapped atoms have attracted much interest because of their significance in both fundamental research and potential technological applications. So far, the investigation of hybrid systems that integrate isolated atoms with solid devices has become a hot research subject in both atomic physics and nano-optics [24-31]. The crucial property of these devices is the ability of the designed system to trap and manipulate cold atoms near the surface of the device in a nanoscale region. To trap, guide,and manipulate cold atoms in a nanoscale region, several different methods have been proposed [24,26] in which metallic nanometer structures have been paid more attention due to the property of plasmon resonances and exhibiting surface plasmons (SPs). The trapping methods of neutral atoms are of interest in applications in many domains of physics. Cold and trapped radioactive atoms can be used in the fundamental symmetry experiments, including the experiments on nuclear $\beta$ -decay, atomic parity non conservation and the search for parity and time-reversal violating electric dipole moment. Trapping of ultracold atoms also gives us an opportunity to study the collisional processes in cold atomic samples. Trapped cold atoms can be used in the formation of cold molecules and in studies of quantum statistical effects in atomic ensembles at low temperatures, such as the Bose-Einstein condensation (BEC). Since the cold molecules trap lifetime was approximately half a second, so the production of cold molecules opens up new ways of research in molecular spectroscopy. There are different schemes for trapping and storing cold atoms [24-31] such as optical trapping using the forces of electric dipole interaction between atoms and laser fields, magnetic trapping based on the use of the forces of magnetic dipole interaction, mixed magneto-optical trapping using simultaneous interaction between atoms and magnetic and laser fields, and also mixed gravito-optical and gravito-magnetic trapping.

In the previour works [28-31], we have shown that the neutral atoms and molecules can be captured at temperature down to nano-Kelvin by an optical trap, based on the use of an evanescent wave around a thin optical fiber, a single wall carbon nanotube (SWCNT) or other nano structures.

In this work we study the optical trap using the surface plasmon effect. This scheme gives strong effect on cold neutral atom, which need to be shown experimentally. Plasmon and plasmon polaritons theories always are difficult to understand for numerous experimental physicists and technicians, a simple model will be useful and needed.

## 2. Surfacre plasmon SP and surface plasmon polariton SPPs in cylindical metallic nanowires

Owing to the fact that the modified Drude model (MDM) is based on the Sommerfeld theory of metal, Maxwells equations and Newtons theory can be used to deal with the interaction between the free electrons in the metal and the penetrated electromagnetic fields (EM): under the excitations of the penetrated fields, the free electrons will execute a collective motion. For an single metal wire (SW), in particular a thin SW, subjected to the penetrated fields, the uniformly distributed free electrons within the wire are re-arranged to form a bunch-like state. On the other hand, due to the space charge effect, the electrons tend to return to their original uniform spatial distribution. These processes are the general physical basis of the following analyses.

When the axial symmetric mode ($TM_0$ mode) is excited (symmetric excitation) on the SW, the free electrons in the metal can oscillate longitudinally only in the vicinity of the surface of the wire, which is just the case of surface plasmon polariton (SPPs). So, the $TM_0$ mode is always the surface propagating mode . Yet, when the wire is excited by an asymmetric mode (asymmetric excitation) , the natural uniformly distributed free electrons in the metal are forced to change into electron bunches along the angular direction within the wire to match the azimuthal variation of the penetrated fields; the electron bunches will execute spiral motion along the wire by combining the longitudinal and transversal motions of the electrons. According to the electrodynamics theory, the transversal rotation of the electron bunches will emit EM radiation, which is just the physical mechanism of the radiation of the asymmetric mode.

The mechanism of the mode transformation for SPPs propagating along SW can be explored. When an asymmetric mode is excited on a very thin metal wire, the field penetration into the wire is very strong. The free electrons are forced to change their original uniform spatial distribution state, and the distribution re-arrangement of free electrons occurs in the wire to form electron bunches, which will also execute a special spiral motion along the wire. This motion causes the asymmetric mode to rapidly disappear through radiation. On the other hand, the variation of the spatial distribution of the electrons will result in a strong space charge force, which leads to a reverse trend for the re-arranged electron distribution returning to the original uniform state gradually.

To calculate the group velocity $d\omega = dk$ as a function of a metallic nano wire radius $R$, a model consisting of a metallic cylinder with a dielectric constant $\epsilon_m$ surrounded by a dielectric medium of dielectric constant $\epsilon_d$ was used (see the figure 1).

For the special case with TM mode ($H_z = 0$) and fundamental mode with no winding $m = 0$, continuity of the remaining tangential components $E_z$ and $H_\phi$ at the boundary lead to the equation for SPPs dispersion relations.

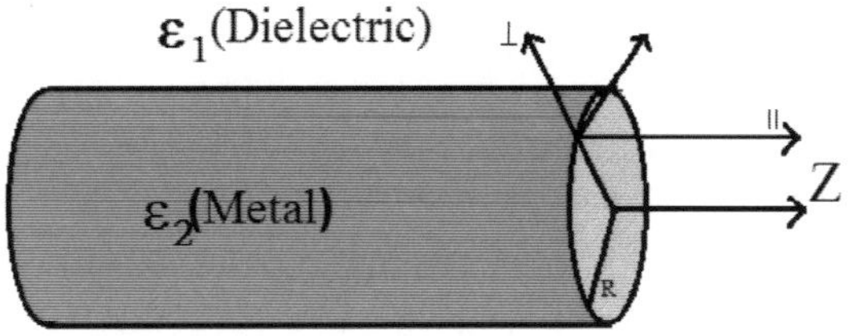


**Figure 1.** Model metallic nano wire

The surface plasmon propagation in this case is governed by the dispersion relation for the fundamental transverse magnetic mode, which is given by [16, 22]

$$\frac{k_2^2}{k_{2\perp}}\frac{J_0'(k_{2\perp}R)}{J_0(k_{2\perp}R)}-\frac{k_1^2}{k_{1\perp}}\frac{H_0'(k_{1\perp}R)}{H0(k_{1\perp}R)}=0, \tag{1}$$

where $k_i=\sqrt{\epsilon_i}\omega/c=\sqrt{k_{i\perp}^2+k_{i\|}^2}$ and $J_m$, $H_m$ are Bessel and Hankel functions of the first kind, respectively. By numerically solving the above equation (1), we can compute the group velocity as a function of $R$ for a given frequency $\omega$ and given $\epsilon_d$ and $\epsilon_m$.

In the limit of $k_{i\perp}=\sqrt{k_i^2-k_{i\|}^2}\simeq ik_\|$ we have

$$\frac{\epsilon_m}{\epsilon_d}=\frac{K_0'\left(k_\|R\right)I_0\left(k_\|R\right)}{K_0\left(k_\|R\right)I_0'\left(k_\|R\right)}, \tag{2}$$

with $K_m$, $I_m$ are modified Bessel and Hankel functions , respectively.

Taking the case of metal/vacuum interface $\epsilon_d=1$, and using the relation $\omega^2=\omega_P^2\left(\epsilon_\infty-\epsilon_m\right)$, the SSP dispersion relation for metallic nano wires is

$$\omega_{DW}\left(k_\|,R\right)=\omega_P\left\{\epsilon_\infty-\frac{K_1\left(k_\|R\right)I_0\left(k_\|R\right)}{K_0\left(k_\|R\right)I_1\left(k_\|R\right)}\right\}^{-1/2}, \tag{3}$$

here $k_\|$ is the in-plane (parallel to metal/dielectric surface) part of the wave vector.

In the case of small radius nanowires $k_\|R\leq 1$ we obtain the dispersion relation for SPPs

$$\omega_{DW}\left(k_\|,R\right)\simeq\omega_P\left\{\epsilon_\infty-\frac{2\epsilon_d}{\left[a+ln\left(k_\|R\right)\right]\left(k_\|R\right)^2}\right\}^{-1/2}, \tag{4}$$

where $a=\gamma_E-ln2$ , $\gamma_E=0.577$ is the Euler's constant.

Because in the case of plasmon-plariton the longitudinal-transversal splitting is zero $\omega_{LT}=0$, so we can consider the dispersion relation (3, 4) also for $k_\perp$ the off-plane (orthogonal to metal/dielectric surface) part of the wave vector. In the following parts of this work, we will ommit the sub-index $\|,\perp$ of the $k$ wave vectors.

## 3. Simple SPPs model for cylindical metallic nanowires

In analogy the cases of exciton plariton and phonon polariton, we consider a model Hamiltonian in 2nd quantization form for surface plasmon polariton [23] in metallic nano wires

$$H = \sum_k H_k = \sum_k \left\{ \omega_{\gamma k} a_k^+ a_k + \omega_{P1} b_k^+ b_k + g_k \left( a_k^+ b_k + b_k^+ a_k \right) \right\}, \tag{5}$$

with $a_k(a_k^+)$ and $b_k(b_k^+)$ are the annihilation (and creation) photon and plasmon operators correspondingly with momentum $k$, $\omega_{P1}$ is the surface plasmon energy with $m = 0$, the fundamental model will be denote by the symbol 1, and $\omega_{\gamma k} = ck/\sqrt{\epsilon_d}$ is the photon dispersion law.

We denote $g_k$ the plasmon-photon transition vertex (or coupling constant), this vertex is absence in traditional plasmon theory $g_{Bk} = 0$ because the bulk plasmon is longitudinal excitation, while photon is transversal excitation. For the case of surface plasmons due to the existence of the boundary conditions that electromagnetic waves must be satisfied at the interface, the plasmon-photon transition vertex might not be zero, and being the main parameter of our theory.

We use the Bogoliubov transformation technique taken from theory of superconductivity for diagonalization plasmon polariton Hamilatonian

$$H_k = \sum_k \left\{ \Omega_{Uk} \gamma_{Uk}^+ \gamma_{Uk} + \Omega_{Lk} \gamma_{Lk}^+ \gamma_{Lk} \right\}, \tag{6}$$

where $\gamma_{ik}$ (and$\gamma_{ik}^+$ ) are the annihilation (and creation) operators of the surface plasmon polarition SPPs with momentum $k$, and $i$ is the branch number $i = L = 1$ for lower and $i = U = 2$ for upper branch.

The transformations with unita condition $u_k^2 + v_k^2 = 1$ are

$$\gamma_{1k} = u_k a_k + v_k b_k, \ \gamma_{2k} = -v_k a_k + u_k b_k. \tag{7}$$

Using the commutative relations for annihilation and creation operators $\left[a_k, a_k^+\right] = 1$, $[b_k, b+] = 1$, $\left[\gamma_{ik}, \gamma_{ik}^+\right] = 1$, and equal zero for others, by standard calculation we obtain the surface plasmon polariton SPPs dispersion relations

$$\Omega_{ik}(R) = \frac{1}{2} \left\{ \left[\omega_{\gamma k} \pm \omega_{P1}(R)\right] - \sqrt{\left[\omega_{\gamma k} - \omega_{P1}(R)\right]^2 + 4g_k^2} \right\}. \tag{8}$$

Note that in the case of planar geometry, the upper branch is lying in the energy gap where the damping is too high so only the lower branch exist, while in the case of metallic nano wires both SPPs branches could be exist.

We take the Dispersion relations of the lowest energy SPPs modes on the silver nanowire of the metallic nanowires in the form

$$g_{Wk}^2(R) = \left[\omega_{k+} - \omega_{DW}(k, R)\right]^2 - \omega_{k-}^2, \tag{9}$$

where

$$\omega_{k\pm} = \frac{1}{2}\left(\omega_{P1} \pm \omega_{\gamma k}\right), \tag{10}$$

and $\omega_{DW}(k, R)$ is the Drude-like SPPs dispersion relation

$$\omega_{DW}(k, R) = \left\{ \omega_{\gamma k}^2 + \omega_{P1}^4(R) - \sqrt{\omega_{\gamma k}^4 + \omega_{P1}^4(R)} \right\}^{1/2}. \tag{11}$$

Note here our plasmon-photon effective coupling constant is well defined via $\omega_{\gamma k}$ and $\omega_{P1}(R)$.

The value of the plasmon-photon coupling constant $g_k$ depends on wave vector $k$ at the metal/vacuum interface of metallic nanowires ($\epsilon_d = 1$) is plotted in the figure 2

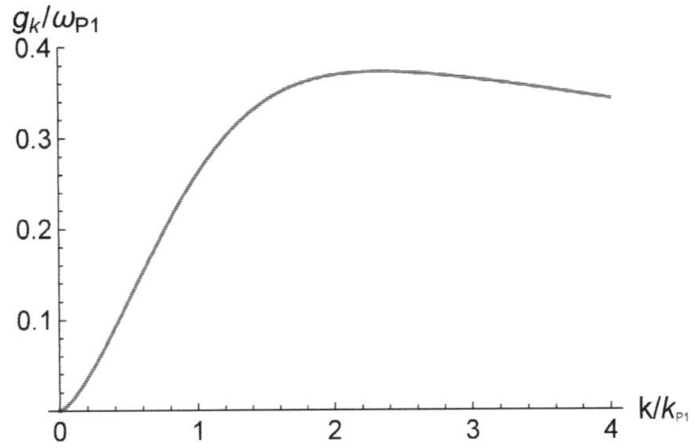


**Figure 2.** The value of the plasmon-photon coupling constant $g_k$ depends on wave vector $k$ at the metal/vacuum interface of metallic nanowires

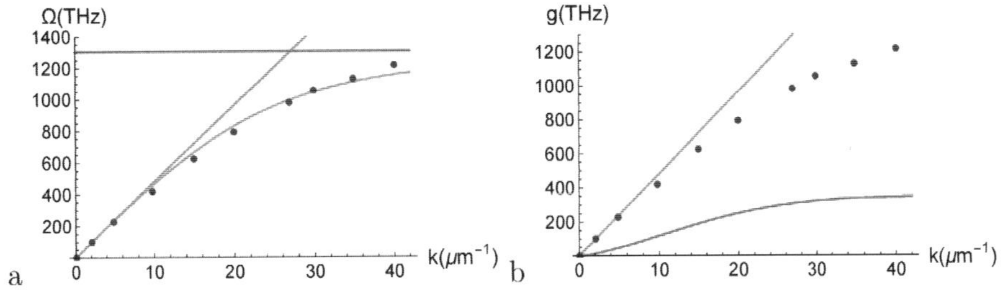


**Figure 3.** The calculated from Eqs. (8, 9) dispersion relation $\Omega_k$ and plasmon-photon coupling constant $g_k$ of the lowest energy SPPs modes with experimental data (blue dots)

The bottle neck region $k \simeq k_p$ is most important for all polariton problems, from this picture we can approximately take $g_{P1} = g_{k=k_P} \simeq 0.3\omega_{P1}$ for simple calculation and estimation.

For comparison, we take the case of Ag cylindrical nano wires with the data set $\omega_{P1} = 1370$ THz, $R = 100$ nm, $\epsilon_\infty = 2.9$. The calculated from Eqs. (8, 9) dispersion relation $\Omega = \Omega_L(k)$ and plasmon-photon coupling constant $g_k$ of the lowest energy SPP modes with experimental data (blue dots) are presented in the figure 3. The plamon energy 1370 THz was shown also for guidance. We can see the good agreement between theoretical and experimental results.

As mentioned above we can approximately take $g_{P1} \simeq 0.3\omega_{P1} = 450$ THz for simple estimations.

## 4. Cold atom surface plasmon polariton trapp with cylindical metallic nanowires

The schema of our model is an analogy of the optical trap model [ 24-31], but the optical fiber is replaced by a metallic nano wire with radius $R$.

For a clear comparison with the optical trap model using a silica fiber [29] and metallic nano wire with evanescent effect, we use evanescent wave filed by plasmon surface effects induced, the wave decays away from the metallic nano wire surface producing an attractive potential for the neutral atom. The atom needs to be kept away from the metallic nano wire for a stable trapping, the waveguide possibility of metallic nano wire is observed experimentally [21]. We assume that the input light fields are circularly polarized to produce the cylindrically symmetric optical potential. The optical potential of the atom is cylindrically symmetric, that depends on the radial distance $r$ from the atom to the carbon nanotube axis $Z$, but not on two other cylindrical coordinates $\varphi$ and $z$. The optical field generates an evanescent wave around the metallic nano wire to capture cold atoms. Assume that the time scale of atomic motion is much slower than the beating period of the light field, that is, the inverse of their frequency difference. The component $L_z$of the angular momentum of the atom is conserved, $L_z = m\hbar$ in the eigenstate, where $m$ is an integer, called the rotational quantum number. The centrifugal potential of the atom is repulsive and can be written as [24, 25]

$$U_{cp} = \hbar^2(m^2 - 1/4)/2\mu^2, \tag{12}$$

where $\mu$ is the atomic mass. A circularly polarized light was sent through the wire to produce the cylindrically symmetric potential $U$. The radial motion of the atom can be treated as the one-dimensional motion of a particle in the effective potential

$$U_{eff} = U_{cp} + U. \tag{13}$$

A circularly polarized light was sent through the fiber to produce the cylindrically symmetric potential $U$. The optical field generates an evanescent wave around the metallic nano wire. The metallic nano wire mode characteristic is determined by the radius $R$ and light wavelength $\lambda$. The parameter $q = 1/\Lambda$ is the inverse of the characteristic decay length $\Lambda$ of the evanescent-wave field. The modified Bessel function $K_0(qr)$ is used to describe the spatial dependence of the amplitude of the field outside the carbon nanotube in the linear-polarization approximation. $G$ is the coupling constant for the interaction between the evanescent wave and the atom, then $U = -GK_0^2(qr)$ is the optical potential outside the metallic nano wire. The effective potential for the radial motion of the atom can be written as

$$U_{eff} = \theta_{rec}\left[\frac{m^2 - 1/4}{k^2r^2} - gK_0^2(qr)\right], \tag{14}$$

where $\theta_{rec} = \frac{(\hbar k)^2}{2\mu}$ is the recoil energy, $g = G/\theta_{rec}$is the normalized coupling parameter and $k$ is the wave vector of the field. The potential $U$ is attractive, opposite to the centrifugal potential $U_{cp}$. The atom can have a stable bound state only if $U_{eff}$ has a local minimum at distance $r = r_m$ and allow at least one bound state exist.

The Schrodinger equation of a cold atom around the cylindrical metallic nanowire is written in the cylindrical system of coordinates $(r, \varphi, z)$ is

$$\left[-\frac{\hbar^2}{2\mu}\nabla^2 + U_{eff}\right]\psi(r,\varphi,z) = E\psi(r,\varphi,z), \tag{15}$$

where $E$ is the energy of atom in the optical potential, the wave functions are taken in the form $\psi(r,\varphi,z) = \frac{1}{\sqrt{r}}R(r)e^{im\varphi}e^{ikz}$ . The Schrodinger equation for radial motion of the atom in the presence of effective optical potential $U_{eff}(r)$ is

$$\frac{d^2R}{dr^2} + \frac{2\mu}{\hbar^2}\theta_{rec}\left[\varepsilon - \left(\frac{m^2 - 1/4}{k^2r^2} - gK_0^2(qr)\right)\right]R = 0, \tag{16}$$

with $\varepsilon = E/\theta_{rec}$.

The Schrodinger equation (16) can be solved numerically only. In the previous works [28-31], we replaced the "exact" potential $U_{eff}$ by a Morse-like simple effective potential. The set of parameters of this Morse-like effective potential was found by fitting with the "exact" potential or experimental data. With the Morse-like effective potential, the new Schrodinger equation now can be solved analytically. By that the lowest bound energy (ground state) is obtained and the efficiency of trap was estimated.

In the case of surface plasmon mechanism, the normalized coupling parameter is too large that instead of Morse-like effective potential we choose the exponential effective potential

$$V_{eff}\left(r\right) = -V_0 exp\left(-r/\delta\right), \tag{17}$$

where $V_0$ and $\delta$ are parameters of our model. We take $V_0 = g_P \simeq 0.3\omega_P$, and $\delta$ is the plasmon penetration length generated by evanescence field of surface plasmon in vacuum or air

$$\delta\left(\omega\right) = \frac{1}{\mid k_{d\perp} \mid} = \frac{c}{\omega}\sqrt{\frac{\mid \epsilon_m + \epsilon_d \mid}{\epsilon_d^2}}. \tag{18}$$

The binding energy (ground state) of the Schrodinger equation with the exponential potential (17) is

$$E = -\frac{\hbar^2}{8\mu\delta^2}\beta^2, \tag{19}$$

where $\beta$ is defined by the equation

$$J_\beta\left(\frac{2\delta}{\hbar}\sqrt{2\mu V_0}\right) = 0, \tag{20}$$

and $J_\beta$ is the Bessel function. Because the argument of the Bessel function is too large $\frac{2\delta}{\hbar}\sqrt{2\mu V_0}) \gg 1$, so in practical the equation (20) is not easy to solve.

For estimation we take the $V$-potential

$$V\left(r\right) = ar - b, \tag{21}$$

with the conditions that the two potential have the same deep and the same average efficiency

$$V\left(0\right) = V_{eff}\left(0\right), \int_0^{r_{max}} V\left(r\right)dr = \int_0^\infty V_{eff}\left(r\right)dr, \tag{22}$$

where $r_{max} = b/a$. From these conditions we get $b = V_0 = g_P$, and $a = V_0^2/\left(2\delta\right) = g_P/\left(2\delta\right)$. The new effective $V$-potential equals

$$V\left(r\right) = \frac{g_P}{2\delta}r - g_P. \tag{23}$$

The values of $V$-potential (yellow) and exponent potential (blue) are presented in the figure 4.

The energy spectra of Schrodinger equation with the $V$-potential can be obtained exactly by WKB method. The basic idea of the old quantum theory is that the motion in an atomic system is quantized, or discrete. The system obeys classical mechanics except that not every motion is allowed, only those motions which obey the old quantum condition

$$\int_{r_1}^{r_2} pdr = nh, \tag{24}$$

where $h$ is the quantum constant, $r_1, r_2$ are the turning points derived from equation $H\left(p,r\right) = E$. For $V$-potential $r_1 = 0$, $r_2 = b/a = \delta$, the quantum condition reads

$$\int_0^\delta dr\sqrt{2\mu\left(E - \frac{g_P}{2\delta}r + g_P\right)} = nh, \tag{25}$$

so

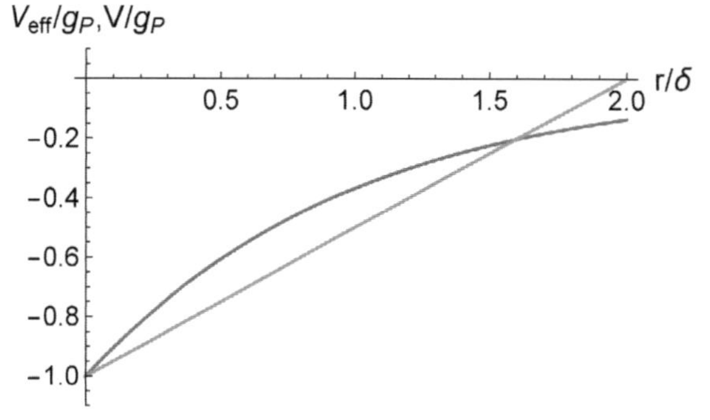


**Figure 4.** The values of $V$-potential (yellow) and exponent potential (blue)

$$\frac{4}{3}\sqrt{2\mu}\frac{(E+g_P)^{3/2}}{(g_P/(2\delta))} = nh = n2\pi^{-}h, \tag{26}$$

which determines the energy levels

$$E_n = -g_P + \left(\frac{3\pi^{-}h\wp}{4\delta\sqrt{2\mu}}\right)^{2/3} n^{2/3}, \tag{27}$$

and the lowest energy level (ground state)

$$E_1 = -g_P + \left(\frac{3\pi^{-}h\wp}{4\delta\sqrt{2\mu}}\right)^{2/3}. \tag{28}$$

The condition exist at least one bound sate reads $E_1 < 0$, or $g_P > (3\pi^{-}h\wp/4\delta\sqrt{2\mu})^{2/3}$, or

$$\delta\sqrt{g_P} > \frac{3\pi^{-}h}{4\sqrt{2\mu}}. \tag{29}$$

Now we estimate our obtained results. For estimation we take Ag nanowire with $\omega_{P1} = 1370$ THz, $R = 100$ nm, $\epsilon_\infty = 2.9$, $g_P = g_{P1} \simeq 0.3\omega_{P1} = 450$ THz, $\delta \simeq 50$ nm, the cold atoms are cerium Ce atoms with $\mu = 140M_0$, $M_0 \simeq 1$ GeV is the neutron mass. We use the thermal QED unit system with $\hbar = c = k_B = 1$, in which all physical values have to converted to eV energy unit, $\omega_{P1} = 5.6$ eV, $g_P \simeq 0.3\omega_{P1} = 1.7$ eV, $\delta = 7.4\ eV^{-1}$, $\mu = 1.4 \times 10^{11}$eV, $k \simeq 2$ eV (UV and optical range). We got that the condition (29) always satisfied, a bound state always exit, the SPPs mechanism trap factor $g_P/\theta_{rec} \simeq 1.5 \times 10^{10}$. Compare with non-SPPs trap mechanism factors $G/\theta_{rec} \sim 6 \times 10^3$[24 − 31], the SPPs enhancement factor in trapping efficiency $\eta = g_P/G \sim 10^6$ in agreeing with the very large enhancement factors in another phenomena with SPPs mechanism, such as SERS [1-3], GFRET [18], ... The bounding energy of the the SPPs trap mechanism is $E_B = -E_1 \simeq g_P = 1.7$ eV, which corresponds to the temperature $T_{SPPs} \sim 2 \times 10^4$ K, comparing with traditional optical trap temperature $T_t \sim 1$ mK [24-31], the SPPs enhancement factor in temperature is $\sim 10^7$. Trap with the SPPs mechanism, the "cold" condition of atoms seems not to needed anymore.

## 4. Conclusion

We have shown that the neutral cold atoms moving near a metallic nano wire can be captured by surface plasmon effect. The main idea is the considering a design model of trapping cold atoms with a metallic nano wire, by that the both two trapping effects can be used: optical evanescence and plasmon effects. Using surface plasmon effect, there are more choices of light sources and the estimated trapping efficiency could be in 6-7 orders stronger than efficiency of traditional trap models without plasmon effect. Since a metallic nano wire offer the possibility of engineering evanescent fields and surface plasmon effect., the system can be used to store, move, and manipulate cold atoms in a controlled manner. The our proposed model might be useful in practical design of cold atom SPPs trap with metallic nanowires.

## Acknowledgment

This research is funded by Vietnam National Foundation for Science and Technology Development (NAFOSTED) undergrant number 103.01-2015.42.

## References

[1] Garcia M A 2011 Surface plasmons in metallic nanoparticles: Fundamentals and applications *J. Phys.* D **44** 28
[2] Homola J 2006 *Surface plasmon resonance based sensors* (Springer Series on Chemical Sensors and Biosensors vol 4) ed J Homola (Berlin: Springer-Verlag) chapter 1 pp 3–44
[3] Xie W, Qiu P and Mao C 2011 Bio-imaging, detection and analysis by using nanostructures as SERS substrates *J. Mater. Chem.* **21** 5190
[4] An N B, Hieu N V, Thang N T, Thang T L and Viet N A 1981 Polariton theory of scattering of light beams in semiconductors *Ann. Phys. (NY)* **131** 149
[5] Kottmann J P and Martin O J F 2001 Plasmon resonant coupling in metallic nanowires *Opt. Exp.* **8** 655
[6] Schmeits M 1989 Surface-plasmon coupling in cylindrical pores *Phys. Rev.* B **39** 7567
[7] Wang K and Mittleman D M 2006 Dispersion of surface plasmon polaritons on metal wires in the terahertz frequency range *Phys. Rev. Lett.* **96** 157401
[8] Laroche T and Girard C 2006 Near-field optical properties of single plasmonic nanowires *Appl. Phys. Lett.* **89** 233119
[9] Ashley J C and Emerson L C 1974 Dispersion relations for non-radiative surface plasmons on cylinders *Surf. Sci.* **41** 615
[10] Liu L, Han Z and He S 2005 Novel surface plasmon waveguide for high integration *Opt. Express* **13** 6645
[11] Oulton R F, Sorger V J, Genov D A, Pile D F P and Zhang X 2008 A hybrid plasmonic waveguide for subwavelength confinement and long-range propagation *Nat. Photon.* **2** 496
[12] Tian J, Ma Z, Li Q, Song Y, Liu Z, Yang Q, Zha C, Akerman J, Tong L and Qiu M 2010 Nanowaveguides and couplers based on hybrid plasmonic modes *Appl. Phys. Lett.* **97** 231121
[13] Sun S, Hung-Ting C, Wei-Jin Z and Guang-Yu G 2013 Mode conversion of surface plasmon polaritons in silver double-nanowire systems *Opt. Exp.* **21** 14591
[14] Delga A, Feist J, Bravo-Abad J and Garcia-Vidal F J 2014 Quantum emitters near a metal nanoparticle: Strong coupling and quenching *Phys. Rev. Lett.* **112** 253601
[15] Teichroeb J H, Forresta J A, Ngai V and Jones L W 2006 Anomalous thermal denaturing of proteins adsorbed to nanoparticles *Eur. Phys. J. E.* **21** 1924
[16] Chang D E, Sorensen A S, Hemmer P R and Lukin M D 2007 Strong coupling of single emitters to surface plasmons *Phys. Rev.* B **76** 035420
[17] Schider G, Krenn J R, Hohenau A, Ditlbacher H, Leitner A, Aussenegg F R, Schaich W L, Puscasu I, Monacelli B and Boreman G 2003 Plasmon dispersion relation of Au and Ag nanowires *Phys. Rev.* B **68** 155427
[18] Ha C V, Nga D T, Viet N A and Nhung T H 2015 The local field dependent effect of the critical distance of energy transfer between nanoparticles *Opt. Comm.* **353** 49
[19] Alpeggiani F and Andreani L C 2014 Quantum theory of surface plasmon polaritons: Planar and spherical geometries *Plasmonics* **9** 965
[20] Ren-Bin Z, Wei-Hao L, Jun Z and Sheng-Gang L 2012 Surface plasmon wave propagation along single metal wire *Chin. Phys.* B **21** 117303
[21] Sun S, Hung-Ting C, Wei-Jin Z and Guang-Yu G 2013 Dispersion relation, propagation length and mode conversion of surface plasmon polaritons in silver double-nanowire systems *Opt. Exp.* **21** 4591
[22] Kolesov R, Grotz B, Balasubramanian G, Rainer J S, Nicolet A, Hemmer P R, Jelezko F and Wrachtrup J 2009 Waveparticle duality of single surface plasmon polaritons *Nat. Phys.* **5** 470
[23] Nga D T T, Lan N T P and Viet N A 2015 Second quantization model of surface plasmon polariton at metal planar surface *J. Phys.: Conf. Series* **627** 012018

[24] Kien F L, Balykin V I and Hakuta K 2006 Light-induced force and torque on an atom outside a nanofiber *Phys. Rev.* A **74** 033412.
[25] Balykin V I, Hakuta K, Kien F L, Liang J Q and Morinaga M 2004 Atom trapping and guiding with a subwavelength-diameter optical fiber *Phys. Rev.* A **70** 011401
[26] Giri P R 2008 Atom capture by nanotube and scaling anomaly *Int. J. Theor. Phys.* **47** 1776
[27] Garca-Segundo C, Yan H and Zhan M S 2007 Atom trap with surface plasmon and evanescent field *Phys. Rev.* A **75** 030902
[28] Anh C T, Nga D T, Thao T T, Thanh N V and Viet N A 2011 Traping cold atoms by a carbon nanotubes *Mod. Phys. Lett.* B **25** 979
[29] Anh C T, Nga D T, Lan N T and Viet N A 2012 Optically trapping cold atoms by using a silicon nanopillar *Photonics Global : Conf. Series* **C12a467** 6457989
[30] Nga D T, Viet N A, Nga D T T and Lan N T P 2014 Optical trapping of cold neutral atoms using a Two-Color evanescent light field around a carbon nanotube *J. Phys.: Conf. Series* **537** 012014
[31] Lan N T P, Nga D T T and Viet N A 2015 Trapping cold atoms using surface plasmons with phase singularities generated by evanescent Bessel beams *J. Phys.: Conf. Series* **627** 012017

# Simple Model for Gold Nano Particles Concentration Dependence of Resonance Energy Transfer Intensity

N M Hoa[1], C V Ha[2], D T Nga[3], N T Lan[3], T H Nhung[3] and N A Viet[3]

[1] Hue University of Medicine and Pharmacy, Hue, Viet Nam
[2] Thai Nguyen University of Education, Thai Nguyen, Vietnam
[3] Institute of Physics, Vietnam Academy of Science and Technology, Ha Noi, Vietnam

Email: minhhoa2806@gmail.com

**Abstract.** Gold nano particles (GNPs) concentration dependence of the energy transfer occurs between the fluorophores and GNPs is investigated. In the case of theses pairs, GNPs can enhance or quench the fluorescence of fluorophores depending upon the relative magnitudes of two energy transfer mechanisms: i) the plasmonic field enhancement at the fluorophores emission frequencies (plasmon coupled fluorescence enhancement) and ii) the localized plasmon coupled Forster energy transfer from fluorescent particles to gold particles, which quenches the fluorescence. The competition of these mechanisms is depending on the spectral overlap of fluorophores and GNPs, their relative concentration, excitation wavelength. Simple two branches surface plasmon polariton model for GNPs concentration dependence of the energy transfer is proposed. The experimental data and theoretical results confirm our findings.

## 1. Introduction

Fluorescence or Forster resonance energy transfer (FRET) has been widely used in biology and chemistry for measuring the distance or between two fluorophores to detect molecular interaction in a number of systems, thanks to their distance dependent dipole-dipole interaction mechanism. FRET can be used as spectroscopic ruler in various areas such as the interaction of biological molecules in vitro. In vivo assays in cellular research, nucleic acid analysic, signal transduction, light harvesting and metallic nanometarial etc. based on the mechanism of FRET a variety of novel chemical sensors and biosensors have been developed [1-14].

FRET is a distance dependent radiationless transfer energy from an excited donor fluorophore to a suitable acceptor through long-range Coulomb dipole-dipole interactions. There are few criteria must be replaced satisfied with met in order for FRET to occur: overlap between emission spectrum of donor (D) and absorption spectrum of acceptor (A), removed it parallel dipole orientations of D and A, suitable fluorescence lifetime of D, and suitable D-A separation (typically up to 10 *nm*).

Forster showed that the efficiency of traditional FRET process is on the inverse sixth power of the distance between D and A pair (or distance dependence of traditional FRET mechanism). Recent experiment argue that the Resonance Energy transfer (RET) between donor and metallic nanoparticle acceptor more look like the energy transfer to a metallic surface (dipole- Surface Energy Transfer (SET)) with the inverse four power of the distance between D and A pair (or distance dependence of SET mechanism) [3-4].

On the other hand, there are many works that demonstrated the non-Forster distance dependence of energy transfer. Bagchi et al. showed the $R^{-2}$ dependence rule for electronic excitation transfer (EET) from a segment of polyfluorene to tetraphenylporphyrin. The Forster expression seems to be inappropriate for the condensed phase systems where D and A can be closely packed [7].

In the our previous work [16] we studied the energy transfer between various type of D and A pairs; dye molecules, quantum dots (QDs), fluorescent nanoparticles are used as D and dye molecules, gold nanoparticles (GNPs) are used as A. For the first time, It is observed by experiments un-expected giant resonance energy transfer (Gi-RET) phenomenon is observed by experiments with very large critical transfer distance $d_0$, which increases from few ten nanometers to micrometers when donors are fluorescent and acceptors are GNPs. It was found that the critical transfer distances depends on the local field of both donor and acceptor when they cannot be considered as the point dipoles.

Additionally the conventional only quenching intensity depending on the concentration of GNPs in the solution, the unconventional with both fluorescence enhancement and quenching depending on the concentration of GNPs is observed. This effect plays an important role in the un-expected Gi-RET phenomenon. A model "nano-wave emitter station and antenna" is given to explain this phenomenon, but it has some limitations for explaining other experimental results.

In this work GNPs concentration dependence of the energy transfer occurs between the fluorophores and GNPs is investigated. When explaining the observed unconventional with both fluorescence intensity enhancement and quenching depending on the concentration of GNPs, a simple two branches surface plasmon polariton model is proposed. Our theoretical results has a good agreement with experimental studies.

## 2. The experimental result

D and A used for experiments are fluorophores and GNPs. The fluorophores are dye molecules, fluorescent nanoparticles and QDs (Fig.1). The Cyanine dyes (Cy5) weoe purchased from Invitrogen. The fluorescent nanoparticles are 100 *nm* Cy3 orange beads (OBs) (Invitrogen) and laboratory manufactured Rhodamine B (RB)-doped silica nanoparticles (RBDSNPs). Each OB contains about ~ 3000 Cy3 dye molecules. There are about ~ 4000 RB molecules doped in each 100 *nm* RBDSNP and ~ 150 RB molecules in 20 *nm* nanoparticles. The 4 *nm* CdTe QDs were provided by Institute of Materials Science, VAST. The GNPs were purchased from BB International with 20 *nm* in size. The energy transfer experiments were carried out for five types of D-A pairs: OBs – Cy5 dyes, Cy5 dyes – GNPs, CdTe QDs – GNPs , RBDSNPs – GNPs and OBs - GNPs.

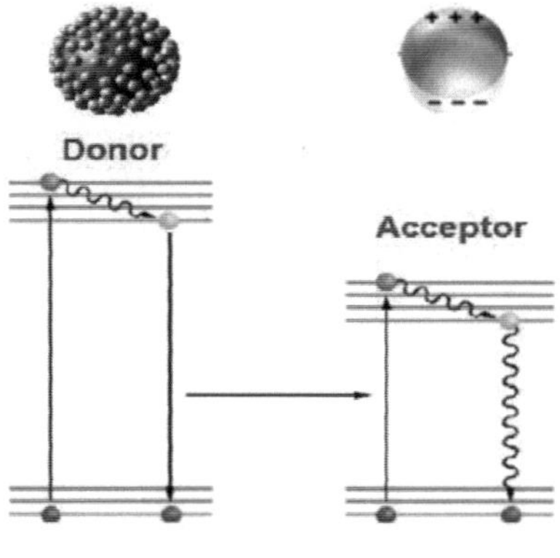


**Figure 1.** The Forster resonance energy transfer between the fluorophores and gold nano particles.

Fig.1 presents the energy transfer occurs between the fluorophores and GNPs, which are metallic nanoparticles. In theses pairs, the GNPs can enhance or quench the fluorescence of fluorophores depending upon the relative magnitudes of two energy transfer mechanisms: the plasmonic field enhancement at the fluorophores emission frequencies (plasmon coupled fluorescence enhancement); and the localized plasmon coupled Forster energy transfer from fluorescen particles to gold particles, which quenches the fluorescence [18].

Journal of Physics: Conference Series **726** (2016) 012009 doi:10.1088/1742-6596/726/1/012009

The experiments were conducted at room temperature and at 532 *nm* excitation for mixture of D-A pairs in aqueous solution. The donor concentration was fixed while the acceptor concentration was varied. There was no chemical reaction either aggregation between D-A pairs in solution. Absorption spectra were measured using JASCO-V570-UV-Vis-NIR spectrometer. The fluorescence spectra were recorded on a Cary Eclipse spectrofluorometer (Varian). Transmission and scanning electron microscopes (TEM, JEM 1011 and SEM, Hitachi S-480) were used to determine the shape, size of particles.

The competition of these mechanisms is depending on the spectral overlap of fluorophores and GNPs, their relative concentration, and excitation wavelength. It can be seen from Fig.2(b), the addition of GNPs in the solution of Cy5 molecules causes only the fluorescence quenching, while the interaction between GNPs and CdTe QD, RBDSNPs, or OBs causes both fluorescence enhancement and quenching, depending on the concentration of GNPs in the solution (Fig.2 (a), (b)). We concentrate on the results of localized plasmon coupled Förster energy transfer, the results on fluorescence enhancement will be discussed in next part.

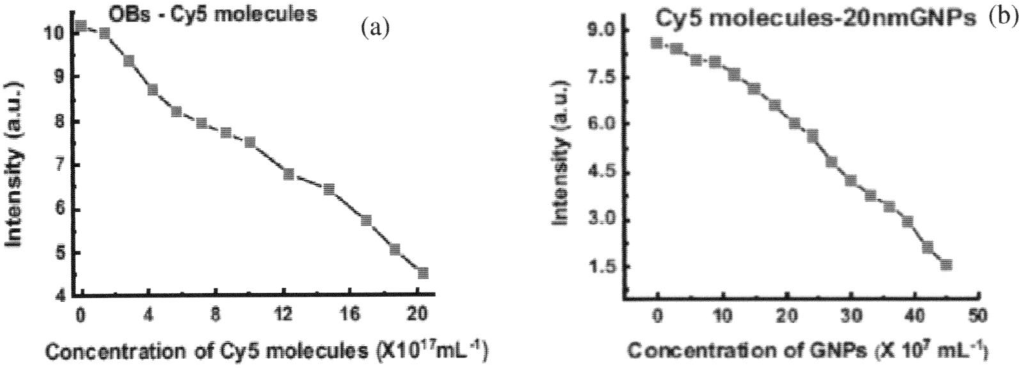


**Figure 2.** Conventional fluorescence quenching depending on the concentration of GNPs.

Unconventional fluorescence enhancement and quenching depending on the concentration of GNPs and spectrum overlap are presented in the figure 3 in the left, and right correspondingly.

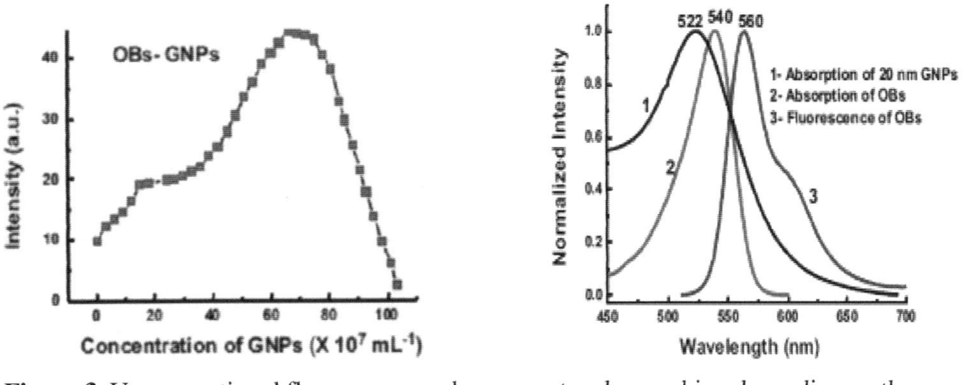


**Figure 3.** Unconventional fluorescence enhancement and quenching depending on the concentration of GNPs (left), and spectrum overlap (right).

### 3. FRET, SET and Coulomb energy transfer (CET) mechanisms

Conventional FRET (Fig.2) is a distance dependant radiationless transfer energy from an excited donor fluorophore to a suitable acceptor through long-range Coulomb dipole-dipole interactions. There are few criteria must be satisfied in order to FRET to occur: overlap between emission spectrum of D and absorption spectrum of D, approximately parallel dipole orientations of D and A, suitable fluorescence lifetime of D, and suitable D-A separation (typically up to 10 *nm*).

Forster showed that the efficiency of traditional FRET process is on the inverse sixth power of the distance between D and A pair (or distance dependence of traditional FRET mechanism). Denoting R is distance between D and A, the FRET efficiency $E_{FRET}$ is given by

$$E_{FRET} = \frac{1}{1+(R/R_{FRET})^6}, \tag{1}$$

where R is the distance between D and A, $R_{FRET}$ is the Forster radius.

Recent experiment argue that the RET between D and metallic nanoparticle A more look like the energy transfer to a metallic surface (dipole-surface energy transfer) with the inverse four power of the distance between D and A pair (or distance dependence of SET mechanism) [3-4]. The SET efficiency $E_{SET}$ is defined by the expression

$$E_{SET} = \frac{1}{1+(R/R_{SET})^4}, \tag{2}$$

where $R_{SET}$ is the SET radius.

In additional, for surface plasmon problem in the literature there are exist one more energy transfer mechanism – CET mechanism by radiate plasmon. In analogy the FRET and SET cases, the CET efficiency $E_{CET}$ is defined by the expression

$$E_{CET} = \frac{1}{1+(R/R_{CET})^2}, \tag{3}$$

where $R_{CET}$ is the CET radius.

At $R = R_{FRET}$, $R_{SET}$, and $R_{CET}$, half of excitation energy of D is transferred to nano metallic particle A.

The energy transfer rate $k_E$ approximately is a product $k_E \approx G_D G_A$ of the interaction elements of the donor $G_D$ and acceptor $G_A$ [15, 16]. These interaction elements can be simplified such that their separation distance d dependencies are some functions of their geometric arrangement. For single dipoles $G \approx \frac{1}{d^3}$ , for a 2D dipole array $G \approx \frac{1}{R}$, and for a 3D dipole array $G \approx const$, such that the power of the distance factor decreases as the dimension increases. In the case of FRET, which consists of two single dipoles, is easily derived from this rule such that

$$k_{FRET} \approx G_D G_A \approx G_{Dipole} G_{Dipole} \approx \left(1/R^3\right)\left(1/R^3\right) \approx 1/R^6, \tag{4}$$

alternatively, FRET is commonly written as $k_{FRET} \approx (1/\tau_D)(R_0/R)^6$.

Using a similar formalism, the rate of energy transfer from a dipole to a metallic surface is described interband transition, which was further extended to the conduction electrons in metals, that SET rate is

$$k_{SET} \approx G_D G_A \approx G_{Dipole} G_{Surf} \approx (1/R^3)(1/R) \approx 1/R^4. \tag{5}$$

The CET rate is

$$k_{CET} \approx G_D G_A \approx G_{Surf} G_{Surf} \approx (1/R)(1/R) \approx 1/R^2. \tag{6}$$

Thus, energy transfer to a surface follows a very different distance trend and magnitude of interaction.

## 4. General resonance energy transfer (GRET) mechanism

The RET mechanism from general donor and general acceptor in the water which investigate in modern theoretical and experimental works still quite complicate and unclear. The interaction of fluorophores with metal surfaces is different depending on the distance regime [1]. For example the interaction of fluorophores with metal surfaces at very close distances (~1nm): radiative rate enhancement is observed; at intermediate distances (~30 *nm*): energy transfer is the dominant process; and at very large distances (50 *nm*): fluorescence oscillations due to the fact that dipole-mirror effects take precedence.

The rate of EET from a fluorescent dye molecule to the surface plasmonic modes of a spherical metal nanoparticle is studied in [S. Bhowmick] where theoretical predicted that the distance dependence of the transfer rate to vary as $R^{-x}$, with $G_A$ at intermediate distances, in partial agreement with the recent experimental results, the Forster's $R^{-6}$ dependence is recovered at large separations.

In the work [7] the $R^{-2}$ dependence rule for CET from a segment of polyfluorene to tetraphenylporphyrin was showed. The RET character also depends on size of general D and general A. Based on analyze the recently obtained experimental data [15] we concluded in [17] that RET from fluorescent dipole donor to metal nanoparticle acceptor is like FRET $R^{-6}$ for small radius and like SET $R^{-4}$ for larger radius. For the GNPs the transition radius $R_C$ is quite small ($R_C \sim 1.5\ nm$), for a gold particle with radius smaller 1.5 *nm* is FRET-like, but larger that is SET-like. We supposed a new reasonable model with mixing mechanism: continue changing from FRET-mechanism for small size to SET-mechanism for large size metal nanoparticles.

Yet no single theoretical model has been explaining and cleaning which the distance dependence in actual is inverse four or six order or some other order? Some questions are raised: what mechanism observed in experiment FRET or SET or CET?, and how distinguish them?

So that a GRET mechanism is considered. The quantum efficiency of energy transfer $E_{GRET}$ can be written as

$$E_{GRET}(R,y) = \frac{1}{1+(R/R_G)^x}, \tag{7}$$

where $R_{GRET}$ is the general Forster-like radius for general mechanism, $x$ is some positive number describing the inverse order power of the general donor-acceptor pair distance.

The GRET transfer rate is

$$k_{GRET} \approx G_D G_A \approx 1/R^x, \tag{8}$$

at $R = R_{GRET}$, half of excitation energy of the general donor is transferred to the general acceptor.

Using unitless system $d = R/R_{GRET}$ and denote $\eta_i^{-1} = R_i/R_{GRET}$, where $i$ = FRET, SET, CET, GRET is the number characterized the partial mechanism, we introduce the unitless general the quantum efficiency $E_{GRET}(d,x,\eta_i)$ of energy transfer as

$$E_{GRET}(d,x,\eta_i) = \frac{1}{1+(\eta_i d)^x}, \tag{9}$$

half of excitation energy of general donor is transferred to general acceptor at $d_i$, which satisfies the condition $d_i\eta_i = 1$. It is easy to see that $E_{FRET}(d) = E_{GRET}(d,6,\eta_{FRET})$, $E_{SET}(d) = E_{GRET}(d,4,\eta_{SET})$, $E_{CET}(d) = E_{GRET}(d,2,\eta_{CET})$ for simplicity we put $\eta_i = 1$, the behavior of function $E_i(d,x,\eta_i)$, and their derivation $DE_i(d,x,\eta_i) = \partial E(d,x,\eta_i)/\partial d$ are presented in the figure 4 (a and b), for FRET- blue, SET- yellow, and CET- green, respectively.

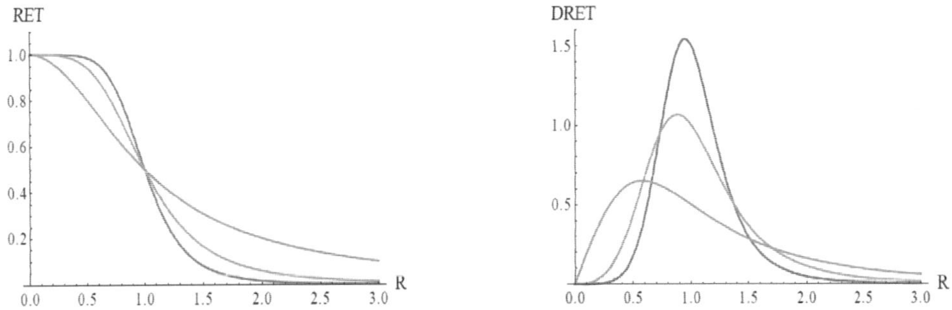


**Figure 4.** Behavior of the RET efficiency functions $E_i(d,x,\eta_i)$, and their derivation. $DE_i(d,x,\eta_i) = \partial E(d,x,\eta_i)/\partial d$ for FRET- blue, SET- yellow, and CET- green, respectively.

## 5. Unified RET theory

The normalized FRET rate $k$ can be calculated by electric Green function method [4, 16]

$$k(R) = 18\pi\varepsilon_0^2 c^4 \int_0^\infty \frac{d\omega}{\omega^4} f_D(\omega) f_A(\omega) M(\omega), \qquad (10)$$

where $\omega$ is the transfer frequency, $c$ is the speed of light, $\varepsilon_0$ is the dielectric constant of the media $M_{(\omega,R)}$ is the Green function, $f_D(\omega)$ and $f_A(\omega)$ are the normalized transfer spectrum of D and A, respectively. Using a some kind of unified theory, we have [16, 17]

$$M(\omega) = \frac{1}{(4\pi\varepsilon_0)^2} K^2 G_A(\omega)^2 G_D(\omega)^2 \frac{1}{R^6}\left[3 + (\omega R/c)^2 + (\omega R/c)^4\right], \qquad (11)$$

where $k^2$ is the orientation factor of effective dipoles and can take values from 0 (perpendicular transition effective dipoles) to 1 (parallel transition effective dipoles). We note that in the equation (11) there are 3 terms with $R^{-6}$, $R^{-4}$ , and $R^{-2}$ dependence.

## 6. Simple model for nanoparticles concentration dependence of intensity

In this part, for explaining the observed unconventional with both fluorescence intensity enhancement and quenching depending on the concentration of GNPs, we propose a simple two branches surface plasmon polariton model.

We consider a two branch model of surface plasmon polariton for GNPs. Because the obsorption spectra of GNPs are quite broaded (see the figure 3), both two branches give their contribution to the energy transfer processes.

The Green fuction $M$ can be written as

$$M(\omega,\langle R\rangle) = \frac{1}{(4\pi\varepsilon_0)^2} K^2 G_A(\omega)^2 G_D(\omega)^2 \left[\frac{3}{\langle R\rangle^6} + \frac{(\omega/c)^2}{\langle R\rangle^4} + \frac{(\omega/c)^4}{\langle R\rangle^2}\right], \qquad (12)$$

where $\langle R\rangle$ is the average D-A distance. In the equation (12) just are 3 contributions given by FRET, SET, and CET-mechanisms respectively.

Based on that fact, we assume a GFRET model with mixing 3 mechanisms: FRET, SET, and CET-mechanisms. Considering a simple relation between the average donor-acceptor distance $\langle R\rangle$ and NPs concentration $n$ as $\langle R\rangle \sim (n_A + n_D)^{1/3} \approx n^{-1/3}$ we have

$$M(\omega,\langle R\rangle) = \frac{1}{(4\pi\varepsilon_0)^2} K^2 G_A(\omega)^2 G_D(\omega)^2 \left[An^2 + Bn^{4/3} + Cn^{2/3}\right], \qquad (13)$$

where $A, B$, and $C$ are some parameters.

Taking the ussual linear optic condition $(G_A G_D)^2 \sim n$, the normalized intensity $I$ is

$$I_{(n)} = I_0 + an^3 + bn^{7/3} + cn^{5/3}, \qquad (14)$$

where $a, b$, and $c$ are the parameters, $I_0$ is the the normalized intensity at $n \approx 0$.

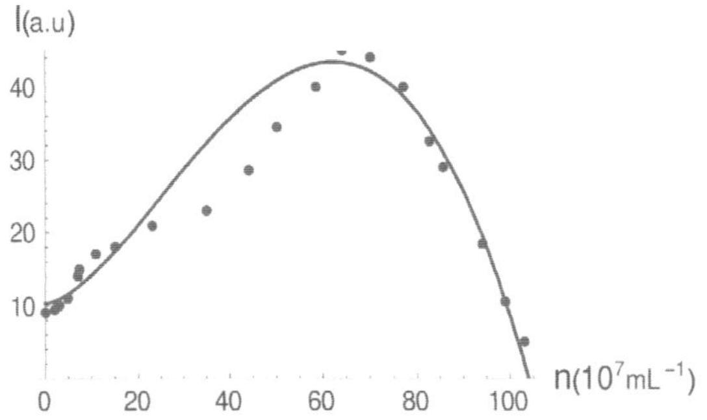


**Figure 5.** Theory-experimental comparison of unconventional fluorescence enhancement and quenching depending on the concentration of GNPs. The blue curve is obtained from equation (14) with the fitting parameters $a$ = -8.70925.10$^{-5}$, $b$ = - 2.69385.10$^{-3}$, $c$ = 9.7635.10$^{-2}$, $I_o$ = 10.32 the experiment data are in red cycles.

As an example we take the case of RET between OBs and GNPs (see the figure 3). In the figure 5 we present the curve (14) of OBs unconventional fluorescence enhancement and quenching depending on the concentration of GNPs with the fitting parameters $a$ = -8.70925.10$^{-5}$, $b$ = - 2.69385.10$^{-3}$, $c$ = 9.7635.10$^{-2}$, $I_o$ = 10.32, also the experiment data in red cycles. We obtained the quite good agreement between the calculated values of normalized intensity from our model and experimental data.

## 7. Discussion

In this work concentration of GNPs dependence of energy transfer occurs between the fluorophores and GNPs is investigated. For explaining the observed unconventional with both fluorescence intensity enhancement and quenching depending on the concentration of GNPs, a simple two branches surface plasmon polariton model is proposed. Because the absorption spectra of GNPs are quite broad both two branches give their contribution to the energy transfer processes. The concentration depending is simple with 3 fitting parameters $a$, $b$, and $c$ describing the FRET, SET and CET contributions, respectively.
As an example, we took the case of energy transfer between OBs and GNPs, and obtained quite good agreement of the theoretical results with experimental data. The sign of $a$ and $b$ are negative describe the quenching effect in conventional RET. While the sign of $c$ is positive describes the enhancement by CET mechanism observed in unconventional RET experiments. Those signs are agreed with surface plasmon polarition picture.

## Acknowledgments

This research was financially supported by the Vietnam National Foundation for Science and Technology Development (NAFOSTED) under grant number 103.01-2015.42.

## References

[1] Yun C S, Javier A, Jennings T, Fisher M, Hira S, Peterson S, Hopkins B, Reich N O, and Strouse G F 2005 Nanometal Surface Energy Transfer in Optical Rulers, Breaking the FRET Barrier *J. Am. Chem. Soc.* **127** 3115

[2] Bhowmick S, Saini S, Shenoya V B, and Bagchi B 2006 Resonance energy transfer from a fluorescent dye to a metal nanoparticle *J. Chem. Phys.* **125** 181

[3] Wong K F, Bagchi B, Rossky P J. 2004 Distance and orientation dependence of excitation transfer rates in conjugated systems: beyond the Forster theory *J. Phys. Chem. A* **108** 5752

[4] Vincent R, Carminati R 2011 Magneto-optical control of Forsterr energy transfer *Phys. Rev. B* **83** 165

[5] Gulin-Sarfraz T, Sarfraz J, Karaman D S, Zhang J, Oetken-Lindholm C, Duchanoy A 2014 FRET-reporter nanoparticles to monitor redox-induced intracellular delivery of active compounds *RSC Adv.* 4 16429

[6] Yaghini E, Giuntini F, Eggleston I M, Suhling K, Seifalian A M, MacRobert A J. 2014 Fluorescence Lifetime Imaging and FRET-Induced Intracellular Redistribution of Tat Conjugated Quantum Dot Nanoparticles through Interaction with a Phthalocyanine Photosensitiser *Small.* **10** 782

[7] Singh H, Bagchi B. 2005 Non-Forster distance and orientation dependence of energy transfer and applications of fluorescence resonance energy transfer to polymers and nanoparticles: How accurate is the spectroscopic ruler with 1/R6 rule?", *Curr. Sci.* **89** 1710

[8] Jennings T L, Singh M P, Strouse G F. 2006 Fluorescent Lifetime Quenching near d = 1.5 nm Gold Nanoparticles: Probing NSET Validity *J. Am. Chem. Soc.* **128** 5462

[9] Pons T, Medintz I L, Sapsford K E, Higashiya S, Grimes A F, English D S 2007 On the Quenching of Semiconductor Quantum Dot Photoluminescence by Proximal Gold Nanoparticles *Nano Lett.* **7** 3157

[10] West R G, Sadeghi S M 2012 Enhancement of Energy Transfer between Quantum Dots: The Impact of Metallic Nanoparticle Sizes *J. Phys. Chem. C* **116** 20496

[11] Zhang X, Marocico C A, Lunz M, Gerard V A, Gun'ko Y K, Lesnyak V 2012 Wavelength,

Concentration, and Distance Dependence of Nonradiative Energy Transfer to a Plane of Gold Nanoparticles *ACS Nano* **6** 9283

[12] Zhang X, Marocico C A, Lunz M, Gerard V A, Gun'ko Y K, Lesnyak V, Gaponik N, Susha A. S, Rogach A L, and Bradle A L 2014 Experimental and Theoretical Investigation of the Distance Dependence of Localized Surface Plasmon Coupled Forster Resonance Energy Transfer, *ACS Nano* **8** 1273

[13] Lunz M, Zhang X, Gerard V A, Gun'ko Y K, Lesnyak V, Gaponik N 2012 Effect of Metal Nanoparticle Concentration on Localized Surface Plasmon Mediated Forster Resonant Energy Transfer *J. Phys. Chem. C* **116** 26529

[14] Lakowicz J R 2005 Radiative decay engineering 5: metal-enhanced fluorescence and plasmon emission *Anal. Biochem.* **337** 171

[15] Breshike C J, Riskowski R A, and Strouse G F 2013 Leaving Forster Resonance Energy Transfer Behind: Nanometal Surface Energy Transfer Predicts the Size-Enhanced Energy Coupling between a Metal Nanoparticle and an Emitting Dipole", *J. Phys. Chem. C* **117** 23942

[16] Ha C V, Nga D T, Viet N A, Nhung T H 2015 The local field dependent effect of the critical distance of energy transfer between nanoparticles *J. Opt Comm* **353** 4955.

[17] Thien T V, Hoa N M, Quyen B T L, Xuan L V, Ha C V, Nga D T, Viet N A 2014 Forster resonance energy transfer (FRET) with a metal nanoparticle and surface plasmon energy transfer (SET) mechanism *Proc.8th Opt. Conf. (Danang) 472-477.*

[18] Andrews D L, Bradshaw D S 2004 Virtual photons, dipole fields and energy transfer: a quantum electrodynamical approach", *Eur. J. Phys.* **25** 845858

# Seesaw-deflected Anomaly Mediation and the 125 GeV Higgs Boson

**Hieu Minh Tran[1], and Nobuchika Okada[2]**

[1]Hanoi University of Science and Technology, 1 Dai Co Viet Road, Hanoi, Vietnam
[2]Department of Physics and Astronomy, University of Alabama, Tuscaloosa, Alabama 35487, USA

E-mail: hieu.tranminh@hust.edu.vn, okadan@ua.edu

**Abstract.** We investigate the phenomenology of a class of model that at the same time solves the tachyonic slepton problem of the pure anomaly mediated supersymmetry breaking (AMSB) model and generates neutrino masses. We introduce heavy fields in the seesaw mechanism that are the messengers in the deflected AMSB scenario. Various theoretical and phenomenological constraints have been taken into account, especially the Higgs mass limits. The viable parameter regions have been specified, and the properties of dark matter candidate have been studied. We point out that the type III seesaw with three generations of 24-messenger is excluded, while the type II seesaw and type III seesaw with two generations of 24-messenger are still allowed. The sparticle masses are heavy as in usual SUSY models. The spin-independent cross-section of the scattering between the lightest neutralino and proton show the possibility to see evidences of new physics from future dark matter search experiments. We find that the lepton flavor violation effects caused by the Yukawa mediation are suppressed due to the electroweak symmetry breaking condition.

## 1. Introduction

In the first running of the Large Hardron Collider (LHC), the ATLAS and CMS collaborations discovered the existence of the Higgs boson with the combined mass measurements as followed [1, 2]:

$$m_H^{\text{ATLAS}} = 125.5 \pm 0.2(\text{stat}) \ ^{+0.5}_{-0.6} \ (\text{syst}) \ \text{GeV}, \quad (1)$$

$$m_H^{\text{CMS}} = 125.3 \pm 0.4(\text{stat}) \pm 0.5(\text{syst}) \ \text{GeV}. \quad (2)$$

The existence of a scalar sector in the standard model (SM) would lead to the instability of the electro-weak scale under quantum corrections, the so-called gauge hierarchy problem. The discovery of the Higgs boson at the LHC confirms this theoretical problem of the SM. Supersymmetry (SUSY) provides a natural solution to the problem by removing quadratic divergences upto all orders of perturbative theory. The minimal supersymmetric extension of the SM (MSSM) predicts a good candidate for the dark matter. Moreover, the three gauge couplings in the MSSM naturally merge at a unification scale giving a hint for a grand unified theory (GUT) of all gauge interactions.

Here we continue our previous work [3] on the positively deflected anomaly mediated SUSY breaking (deflected AMSB) [4, 5] which was originally invented to overcome the tachyonic slepton problem in the pure AMSB [6, 7] . Along that line and following the idea of [8, 9], in this paper we study the phenomenology of more physically predictive models (the so-called seesaw-deflected anomaly mediation) in which the messengers play the roles of heavy fields in the seesaw mechanism to generate masses for neutrinos as observed from neutrino oscillation data. These models are very interesting since they address several questions at the same time: solving the hierarchy problem, preserving the unification of interaction, predicting neutrino masses and a dark matter candidate.

In order to explain neutrino masses in seesaw mechanism we introduce the Yukawa interaction between messenger fields and lepton doublets and/or Higgs doublet. These new couplings could in general cause the so-called Yukawa mediation that contributes to the renormalization group (RG) evolution in the original deflected AMSB scenario, and thus changes the SUSY breaking parameters in the MSSM. Moreover, such interaction with the lepton doublets could also affect the lepton flavor violation (LFV) and then have potential contribution to flavor changing neutral currents (FCNCs) which are strongly suppressed by experiments.

Together with the Higgs mass measurements at the LHC, we consider limits for the dark matter relic abundance, and other phenomenological constraints from collider experiments. On the theoretical side, we consider the gauge coupling blow-up constraint which imposes a lower limit upon the threshold to ensure that the gauge couplings stay within the perturbation limit.

The structure of the paper is as follows. In section 2, the deflected anomaly mediation scenario is described and incorporated with seesaw mechanism to generate neutrino masses. In section 3, the numerical analysis is performed taking into account a theoretical constraint on messenger scale and various phenomenological constraints from collider physics to cosmological observation and dark matter searches. After identifying the viable parameter regions, the sparticle mass spectrum, the structure and properties of the LSP are presented. In section 4, we discuss on the Yukawa mediation and the relevant lepton flavor violation. Section 5 is devoted for conclusion.

## 2. Seesaw-deflected Anomaly Mediation

In the AMSB scenario, once SUSY is dynamically broken in the hidden sector the $F$-component of the compensator field acquires nonzero vacuum expectation value (VEV) $F_\phi$ leading to the emergence of soft terms in the MSSM sector via supergravity interaction. This superconformal anomaly is identified as the only source of SUSY breaking in pure anomaly mediation. Nevertheless, the pure AMSB scenario is suffered from the tachyonic slepton problem.

To eliminate such problem, we introduce a messenger sector which realizes a certain representation under $SU(5)$ to preserve gauge coupling unification. We identify the messengers as heavy states in the seesaw mechanism to generate neutrino masses. Hence the messenger scale is identical to the seesaw scale. Here we shall consider type II and type III seesaw mechanism which employ $SU(2)$ triplets.

For type II seesaw case, a pair of vector-like charged $SU(2)$ triplet $\Delta = (\mathbf{1}, \mathbf{3}, 1)$ and $\bar{\Delta} = (\mathbf{1}, \mathbf{3}, -1)$ are embedded in a $\mathbf{15} + \overline{\mathbf{15}}$ representation of $SU(5)$ as in the decomposition into $SU(3)_C \times SU(2)_L \times U(1)_Y$:

$$\mathbf{15} \;=\; (\mathbf{1}, \mathbf{3}, 1) + (\mathbf{3}, \mathbf{2}, \tfrac{1}{6}) + (\mathbf{6}, \mathbf{1}, -\tfrac{2}{3}). \tag{3}$$

The superpotentials in this case for the messenger sector and for seesaw mechanism respectively read:

$$W_{\text{Mess}} \;=\; S\bar{T}T, \tag{4}$$
$$W_{\text{seesaw}} \;=\; Y_\nu^{ij}\bar{\mathbf{5}}_i\bar{\mathbf{5}}_j T + \lambda \mathbf{5}_H \mathbf{5}_H \bar{T}, \tag{5}$$

where $T$ and $\bar{T}$ represent $\mathbf{15}$ and $\overline{\mathbf{15}}$ supermultiplets. $S$ is a gauge singlet superfield that acquires a nonzero VEV when SUSY is broken. The neutrino masses are then:

$$m_\nu^{ij} = Y_\nu^{ij} \lambda \frac{v_u^2}{M_{\rm Mess}}, \tag{6}$$

where $M_{\rm Mess}$ is the messenger scale, and $v_u = \frac{\langle H_u \rangle}{\sqrt{2}} = v \sin^2\beta$ with $v = 174$ GeV to be the VEV of the SM-like Higgs field.

In type III seesaw case, we use a neutral $SU(2)$ triplet $T_R = (\mathbf{1}, \mathbf{3}, 0)$ which is contained in a **24** supermultiplet of $SU(5)$ with the branching rule:

$$\mathbf{24} = (\mathbf{1}, \mathbf{1}, 0) + (\mathbf{1}, \mathbf{3}, 0) + (\mathbf{3}, \mathbf{2}, -\tfrac{5}{6}) + (\bar{\mathbf{3}}, \mathbf{2}, \tfrac{5}{6}) + (\mathbf{8}, \mathbf{1}, 0). \tag{7}$$

The superpotentials for messenger sector and seesaw mechanism:

$$W_{\rm Mess} = S \mathrm{Tr}\left(\Sigma_j^2\right), \tag{8}$$

$$W_{\rm seesaw} = Y_\nu^{ij} \bar{\mathbf{5}}_i \Sigma_j \mathbf{5}_H, \tag{9}$$

where $\Sigma_j$ are **24** messengers. We need at least two generations of **24** multiplet to obtain a realistic neutrino mass matrix:

$$m_\nu^{ij} = (Y_\nu^\dagger Y_\nu)^{ij} \frac{v_u^2}{M_{\rm Mess}}. \tag{10}$$

In Eq. (8), we assume the degeneracy of all the **24** messengers in terms of their masses for simplicity.

When the scalar and the $F$-component of the above gauge singlet get VEVs, the seesaw messengers induce GMSB (gauge mediated SUSY breaking)-like contributions to the soft terms via gauge interactions. Since the VEVs of the singlet superfield originate from the compensator's VEV, there is a relation between them:

$$\frac{F_S}{S} = dF_\phi, \tag{11}$$

where $d$ is the deflection parameter which, in a simple model, can be calculated from the superpotential of the singlet $W(S)$ as [5]:

$$d \sim -2 \frac{\frac{\partial W}{\partial S}}{S \frac{\partial^2 W}{\partial S^2}} \tag{12}$$

Because the singlet's VEVs are just the secondary source of SUSY breaking, and in order to get rid of the tachyonic sleptons, we impose a constraint on $d$-parameter for theoretical consistency: $|d| \lesssim \mathcal{O}(1)$. The deflection parameter can be either negative [4] or positive [5]. In this paper, we restrict our study on the later case.

The soft SUSY breaking terms can be derived from the renormalized gauge couplings and the supersymmetric wave-function renormalization coefficients as follows [4, 5]:

$$\frac{M_i(\mu)}{\alpha_i(\mu)} = \frac{F_\phi}{2} \left( \frac{\partial}{\partial \ln \mu} - d \frac{\partial}{\partial \ln |S|} \right) \frac{1}{\alpha_i(\mu, S)}, \tag{13}$$

$$m_I^2(\mu) = -\frac{|F_\phi|^2}{4} \left( \frac{\partial}{\partial \ln \mu} - d \frac{\partial}{\partial \ln |S|} \right)^2 \ln Z_I(\mu, S), \tag{14}$$

$$A_I(\mu) = -\frac{F_\phi}{2} \left( \frac{\partial}{\partial \ln \mu} - d \frac{\partial}{\partial \ln |S|} \right) \ln Z_I(\mu, S), \tag{15}$$

where the gauge couplings and the wavefunction renormalization coefficients are given by

$$\alpha_i^{-1}(\mu, S) = \alpha_i^{-1}(\Lambda_{cut}) + \frac{b_i - N}{4\pi} \ln \frac{S^\dagger S}{\Lambda_{cut}^2} + \frac{b_i}{4\pi} \ln \frac{\mu^2}{S^\dagger S}, \tag{16}$$

$$Z_I(\mu, S) = \sum_{i=1,2,3} Z_I(\Lambda_{cut}) \left( \frac{\alpha(\Lambda_{cut})}{\alpha(S)} \right)^{\frac{2c_i}{b_i - N}} \left( \frac{\alpha(S)}{\alpha(\mu)} \right)^{\frac{2c_i}{b_i}}. \tag{17}$$

The index $i = 1, 2, 3$ corresponds to the MSSM gauge group $U(1)_Y \times SU(2)_L \times SU(3)_C$, and $b_i = \{-33/5, -1, 3\}$ $(i = 1, 2, 3)$ are $\beta$-function coefficients of the MSSM gauge coupling RG equations, $c_i$ are the corresponding quadratic Casimirs. Here $S = M_{\rm Mess}$ plays the role of the messenger scale which is an intermediate scale between the cutoff scale $\Lambda_{\rm cut}$ and the electroweak scale.

In Eq. (17), $N$ is the Dynkin index of the seesaw messengers. For type II seesaw, $N = 2 \times \frac{7}{2} = 7$. For type III seesaw, $N = 2 \times \frac{10}{2} = 10$ in the case with two generations of **24** messengers, and $N = 3 \times \frac{10}{2} = 15$ in the case with three generations of **24** messengers.

Substituting Eqs. (16) and (17) in Eqs. (13)-(15), we find the solutions for the RG equations of the soft terms. The gaugino masses are given at the messenger scale by

$$M_i(M_{\rm Mess}) = -\frac{\alpha_i}{4\pi} F_\phi \left( b_i + dN \right). \tag{18}$$

For the $A$-parameters of the third generation, we have

$$A_t(M_{\rm Mess}) = -\frac{F_\phi}{(4\pi)^2} \left( 6|Y_t|^2 + |Y_b|^2 - \frac{16}{3} g_3^2 - 3g_2^2 - \frac{13}{15} g_1^2 \right), \tag{19}$$

$$A_b(M_{\rm Mess}) = -\frac{F_\phi}{(4\pi)^2} \left( |Y_t|^2 + 6|Y_b|^2 + |Y_\tau|^2 - \frac{16}{3} g_3^2 - 3g_2^2 - \frac{18}{5} g_1^2 \right), \tag{20}$$

$$A_\tau(M_{\rm Mess}) = -\frac{F_\phi}{(4\pi)^2} \left( 3|Y_b|^2 + 4|Y_\tau|^2 - 3g_2^2 - \frac{9}{5} g_1^2 \right), \tag{21}$$

where $Y_{t,b,\tau}$ are the Yukawa couplings of the third generation quarks and lepton. Lastly, the sfermion squared masses are given by

$$m_{H_u}^2(M_{\rm Mess}) = m_{H_d}^2(M_{\rm Mess}) = F_\phi^2 \left[ \frac{3}{10} \left( \frac{\alpha_1}{4\pi} \right)^2 G_1 + \frac{3}{2} \left( \frac{\alpha_2}{4\pi} \right)^2 G_2 \right], \tag{22}$$

$$m_{\tilde{L}}^2(M_{\rm Mess}) = F_\phi^2 \left[ \frac{3}{10} \left( \frac{\alpha_1}{4\pi} \right)^2 G_1 + \frac{3}{2} \left( \frac{\alpha_2}{4\pi} \right)^2 G_2 \right], \tag{23}$$

$$m_{\tilde{E}}^2(M_{\rm Mess}) = F_\phi^2 \left[ \frac{6}{5} \left( \frac{\alpha_1}{4\pi} \right)^2 G_1 \right], \tag{24}$$

$$m_{\tilde{Q}}^2(M_{\rm Mess}) = F_\phi^2 \left[ \frac{1}{30} \left( \frac{\alpha_1}{4\pi} \right)^2 G_1 + \frac{3}{2} \left( \frac{\alpha_2}{4\pi} \right)^2 G_2 + \frac{8}{3} \left( \frac{\alpha_3}{4\pi} \right)^2 G_3 \right], \tag{25}$$

$$m_{\tilde{U}}^2(M_{\rm Mess}) = F_\phi^2 \left[ \frac{8}{15} \left( \frac{\alpha_1}{4\pi} \right)^2 G_1 + \frac{8}{3} \left( \frac{\alpha_3}{4\pi} \right)^2 G_3 \right], \tag{26}$$

$$m_{\tilde{D}}^2(M_{\rm Mess}) = F_\phi^2 \left[ \frac{2}{15} \left( \frac{\alpha_1}{4\pi} \right)^2 G_1 + \frac{8}{3} \left( \frac{\alpha_3}{4\pi} \right)^2 G_3 \right], \tag{27}$$

where

$$G_i = Nd^2 + 2Nd + b_i \qquad (i = 1, 2, 3). \tag{28}$$

## 3. Numerical analysis

Our analysis concentrates on a specific type of seesaw ($N$ is fixed). Therefore the free parameter set includes four independent parameters:

$$d,\ M_{\text{Mess}},\ F_{\phi},\ \tan\beta. \tag{29}$$

and the sign of $\mu$ parameter (the supersymmetric Higgs mass). Here we only consider the case $\text{sign}(\mu) = +$. By fixing the deflection parameter $d$ and $\tan\beta$ at the beginning, we are left with only two free parameters: the seesaw messenger scale $M_{\text{Mess}}$, and the VEV of the compensator's $F$-component $F_{\phi}$.

We employ the SOFTSUSY [10] and the MICROMEGAS [11] packages to compute mass spectrum and other observables. For fixed values of the deflection parameter and $\tan\beta$, we scan over the parameter space of $(M_{\text{Mess}}, F_{\phi})$, and analyze the Higgs mass together with the dark matter relic abundance. We are only interested in the region with the Higgs mass around the measured values at ATLAS and CMS (Eqs. (1) and (2)).

In Figs. 1-5, the results are shown with different types of seesaw mechanism. We plot the contours of Higgs mass on the $(M_{\text{Mess}}, F_{\phi})$ plane. While the Higgs mass and the sparticle spectrum just slightly change with respect to the messenger scale, they are substantially heavier for larger $F_{\phi}$. This can be easily seen from the boundary conditions at the messenger scale. In Eqs. (18)-(27), $F_{\phi}$ plays the role of an overall scaling factor for all the soft terms. In all the figures, the narrow red strips indicate regions that reproduce the dark matter relic density consistent with the WMAP data [12] and the Planck data [13]:

$$\Omega_{\text{CDM}}h^2 \quad = \quad 0.1138 \pm 0.0045 \tag{30}$$

As shown in these figures, the dark matter observation imposes a very severe constraint on the parameter space. This fact reveals the high sensitivity of the dark matter annihilation cross-sections with respect to any small change of free parameters.

Firstly, let us consider type II seesaw case ($N = 7$). We explore the $(M_{\text{Mess}}, F_{\phi})$ parameter space with $d = 0.7$, and $\tan\beta = 10, 15, 20, 30$. The results are plotted in Figs. 1-4 respectively. The left and right bounds of the contour region in Fig. 1 are specified by the no electroweak symmetry breaking (no-EWSB) condition. It is because the parameter $\mu^2$ become negative outside the contour region. In Fig. 2, the bounds are almost specified by the same condition as above, except for the left bound with $F_{\phi}$ below 270 GeV. This part of the left bound is defined by the $A^0$ tachyon. The conditions specify the bounds of the contour region in Fig. 3 are a bit more complicated. For $F_{\phi} < 400$ GeV, the left bound is defined by the $A^0$ tachyon condition, while above that value, the condition is $H^{\pm}$ tachyon. The region on the right hand side of the right bound is excluded because of no-EWSB condition. In Fig. 4, we can see two parts of the bound connected together not smoothly at $M_{\text{Mess}} \approx 10^{5.5}$ GeV. The left part of the bound is specified by $A^0$ tachyon condition, while the curved right part of the bound is specified by $H^{\pm}$ tachyon condition.

The red strips in Figs. 1-4 show the region with the acceptable value for dark matter relic density. Although with similar relic abundance, the properties of dark matter particle (the lightest neutralino) are different in different red strips. For those very close to the borders of the contour regions, the LSP is a good mixture of bino and Higgsino. In these cases, the LSP is annihilated via multiple processes. The largest contribution ($\sim 20\%$) comes from $\chi_1^0$ pair annihilation by transferring neutral Higgs bosons in the s-channel to create a pair of top and anti-top quark. Inside the contour regions, farther from the borders, there are red strips where the LSP structure is bino-like. Here the lightest neutralino is annihilated mostly via the s-channel with $A^0$ and $H$ resonances. Due to the physics of dark matter particle, these strips correspond to the funnel regions as in the constrained MSSM.

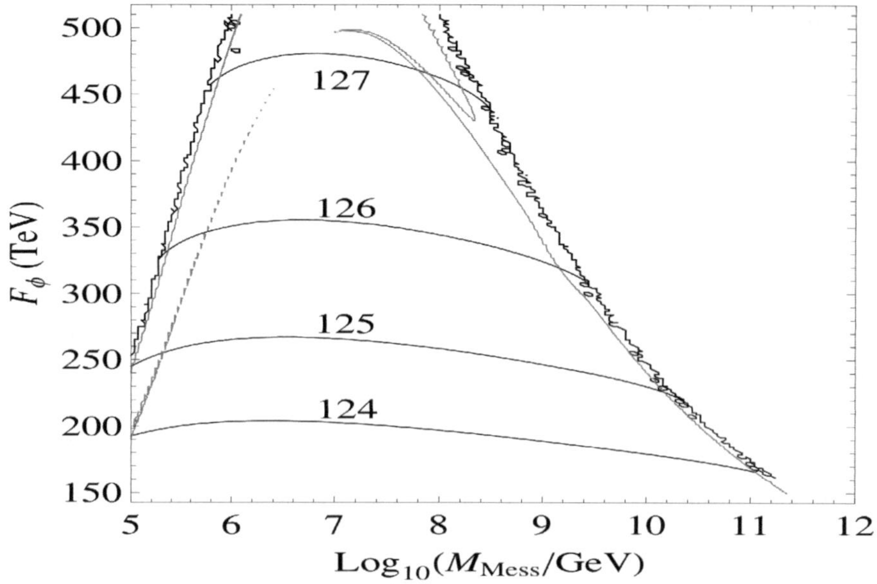


**Figure 1.** Results for type II seesaw case ($N = 7$) with $d = 0.7$, and $\tan\beta = 10$. Various values of the resultant SM-like Higgs boson mass are shown as the contours. There are two diagonal boundaries, outside of them are theoretically excluded. The red strip indicates the region where the relic abundance of neutralino dark matter is consistent with the observation.

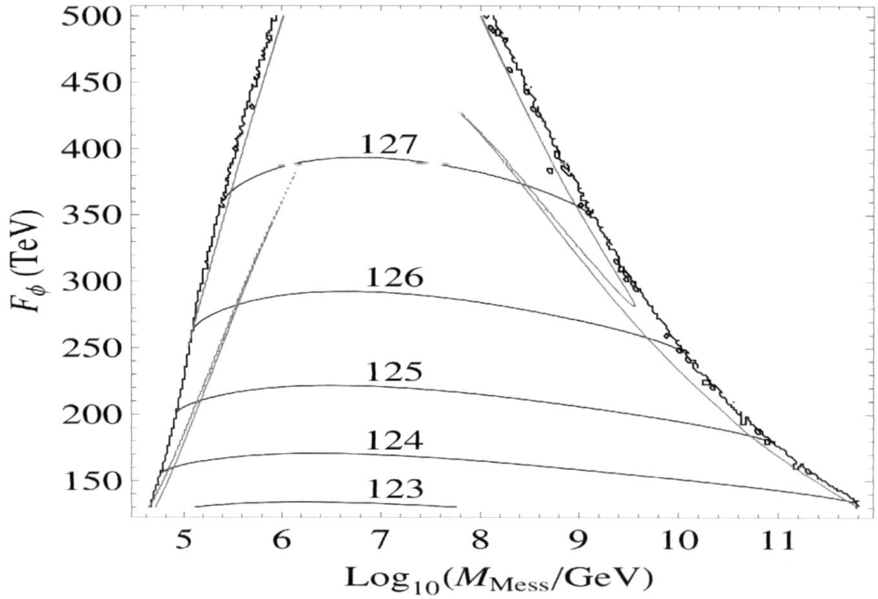


**Figure 2.** The same as Fig. 1, but for type II seesaw case ($N = 7$) with $d = 0.7$, and $\tan\beta = 15$.

Comparing the first four figures, we can see the movement of the Higgs mass contours and red strips when changing $\tan\beta$. As $\tan\beta$ increases, the Higgs mass contours move downward,

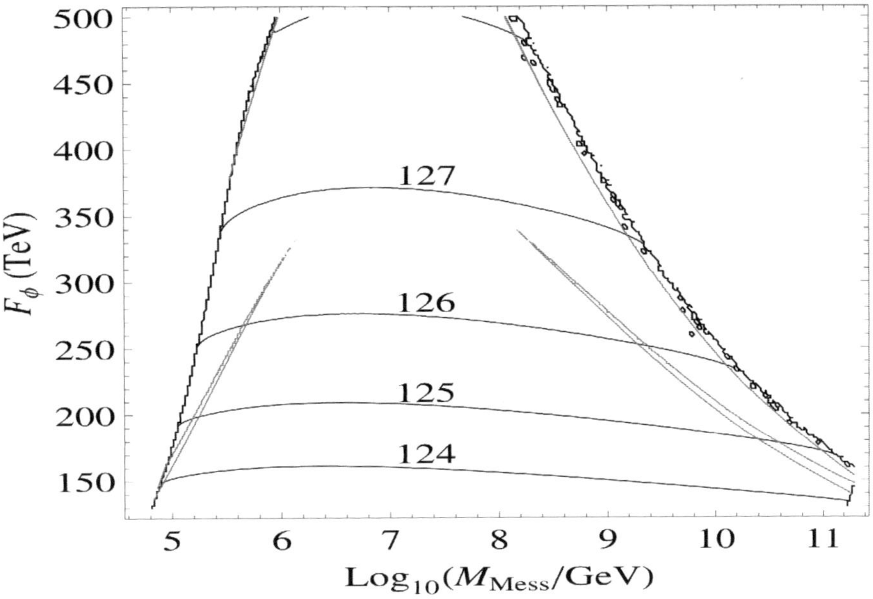


**Figure 3.** The same as Fig. 1, but for type II seesaw case ($N = 7$) with $d = 0.7$, and $\tan\beta = 20$.

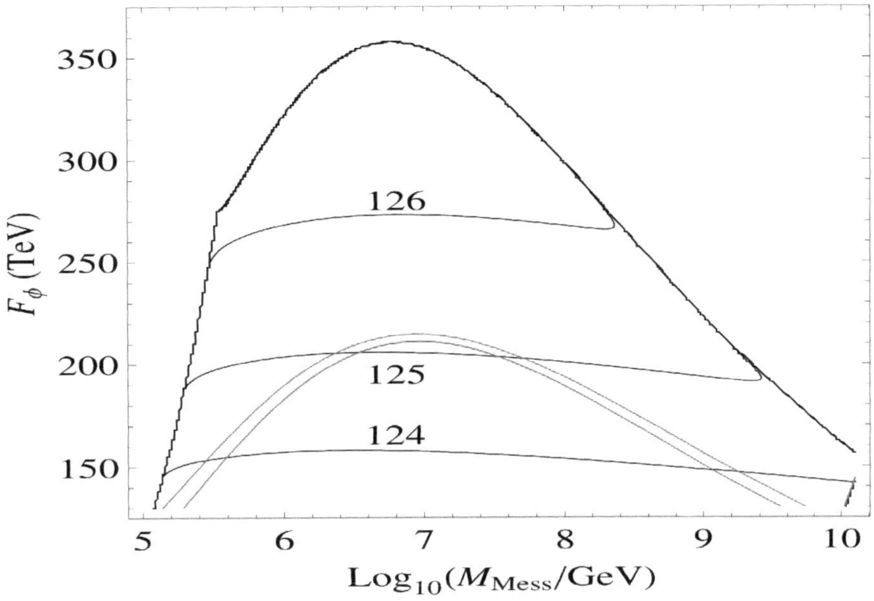


**Figure 4.** The same as Fig. 1 but for type II seesaw case ($N = 7$) with $d = 0.7$, and $\tan\beta = 30$.

so do the funnel-like red regions. Moreover, the funnel-like regions which are separated in Figs. 1-3 become connected when $\tan\beta = 30$ as in Fig. 4, while the red strips close to the border disappear due to the enlargement of the tachyonic $A^0$ and $H^\pm$ regions outside the contour area when raising $\tan\beta$ up to 30.

For larger $\tan\beta$, there is no overlapped region that satisfies both the Higgs mass and WMAP limits. With smaller $d$-parameter (for example: 0.3, 0.4, 0.5), the model has to face the tachyonic slepton problem then being excluded. For larger values of $d$ (for example: 0.8, 0.9, 1.0), the dark matter relic density is too high to meet the constraint by WMAP. The gauge coupling blow-up constraint for type II seesaw messenger with $N = 7$ is:

$$M_{\text{Mess}} \gtrsim 4.3 \times 10^6\,\text{GeV}. \tag{31}$$

This condition when applied to our scenario actually rules out a significant portion of the free parameter space. It cut away the two red strips on the left handside of Fig. 1-3, and a part of the red strips in Fig. 4. Therefore, the type II seesaw-deflected AMSB is very predictive to be tested.

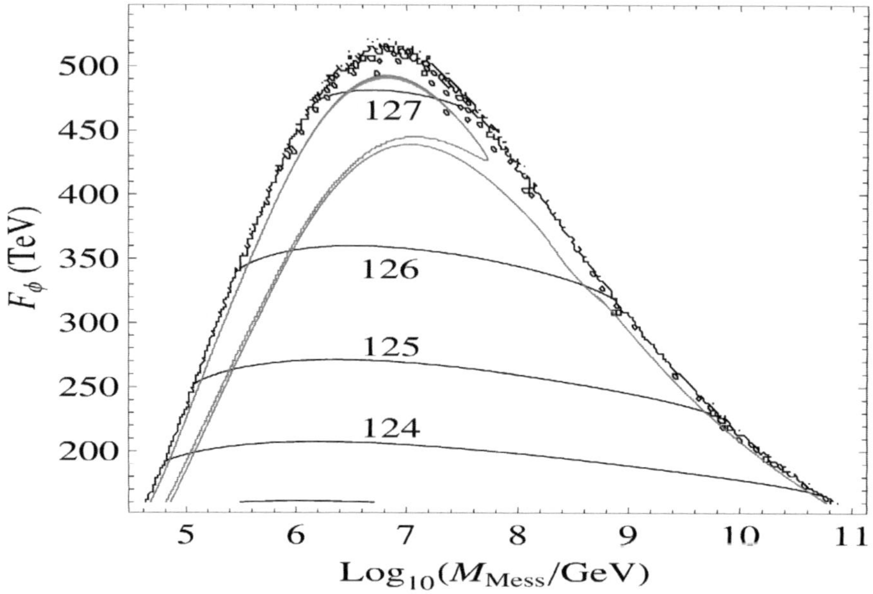


**Figure 5.** The same as Fig. 1 but for type III seesaw case with two generations of **24** messenger ($N = 10$), $d = 0.5$, and $\tan\beta = 10$.

Secondly, similar to the above case, we consider type III seesaw scenario. In type III seesaw with two generations of **24** messenger ($N = 10$), the parameter space corresponding to $d = 0.5$, and $\tan\beta = 10$ is examined. We expect that this case is not so much different from the above one since the Dynkin indices of the seesaw messengers in the two cases are roughly at the same order. The result is shown in Figs. 5. The boundary for the contour region in Fig. 5 is defined by the no-EWSB condition.

In the red strips, the LSP structures are similar to the previous case. Namely, there are red regions very close to the boundary where the lightest neutralino is mostly Higgsino or a good mixture of bino-Higgsino. The ones which locate more inside the contour region correspond to funnel-like regions with bino-like dark matter. However, the funnel-like strips on the left and right parts of the contour region are now connected. The LSP structure determines the main mechanism of neutralino annihilation in each region. The discussion for this case is the same as above.

We also find similarities between the two cases in the behavior of the contour lines and the red strips with respect to $\tan\beta$. All of them move downward when increasing $\tan\beta$. In other words, the Higgs boson becomes heavier for larger $\tan\beta$ with the same values of $M_{\rm Mess}$ and $F_\phi$. Moreover, raising $\tan\beta$ up also results in an enlargement of the excluded regions with tachyonic particles such as $A^0$, $H^\pm$ stau. When varying the deflection parameter, we observe that for smaller $d$, the tachyon stau region dominate and at some value it occupies most of the contour region. For larger $d$ (for instant: 0.6, 0.7), the dark matter relic abundance is too high and becomes larger than the WMAP's upper limit.

The constraint to keep our theory with two **24** seesaw messengers staying in a perturbative regime is as follows:

$$M_{\rm Mess} \gtrsim 3.4 \times 10^9\,\text{GeV}. \tag{32}$$

It is more severe than the previous case (Eq. (31)). Taking into account this condition, only a small part of the free parameter space is left.

For the case of type III seesaw with three generations of **24** messenger ($N = 15$), we examine the cases with $\tan\beta = 10$. When $d = 0.35$ or smaller, we find that the smaller the deflection parameter is, the larger part of the contour region is ruled out by the tachyonic slepton condition. For larger $d$ (for instant: 0.4), there is no point that satisfies both the Higgs mass limits and the WMAP constraint on dark matter relic density at the same time.

Once we introduced more messenger in the theory, the larger Dynkin index must be added to the $\beta$ function coefficients of the gauge couplings. Consequently, the gauge coupling blows up faster and reaches the perturbative limit quickly as increasing the renormalization scale. To avoid such thing to happen, the running of gauge coupling from the messenger scale to the cutoff scale must be short. In this case, the condition for a perturbative gauge coupling reads:

$$M_{\rm Mess} \gtrsim 6.2 \times 10^{11}\,\text{GeV}. \tag{33}$$

We find that all the region with acceptable Higgs mass is forbidden by the very high lower limit of the messenger scale. This is also true for different values of $\tan\beta$. Hence the type III seesaw case with three generations of **24** messenger is excluded.

To be more specific, in Table 1, we show a few examples for sparticle spectra. While the first three columns correspond to the type II seesaw case, the other ones are type III seesaw with two generations of messenger. In the first, the third and the last columns, the LSP is a good mixture of bino-Higgsino. There are many processes that the LSP goes through in its annihilation. For the remaining columns, the LSP is bino-like and the most dominant contribution to its annihilation comes from processes with $A^0$ and $H$ resonances mediated via s-channel.

We take into account the perturbative gauge coupling limits (31)-(32), the LHC measurements on Higgs mass (1)-(2), and the WMAP limits on the dark matter relic density (30). In addition, we also consider other phenomenological constraints: the branching ratios of $b \to s\gamma$ [14], $B \to \tau\nu_\tau$ [15], and the muon anomalous magnetic moment $a_\mu = (g_\mu - 2)/2$ [16],

$$2.85 \times 10^{-4} \leq \text{BR}(b \to s + \gamma) \leq 4.24 \times 10^{-4}\ (2\sigma), \tag{34}$$

$$\frac{\text{BR}^{\rm exp}(B \to \tau\nu_\tau)}{\text{BR}^{\rm SM}(B \to \tau\nu_\tau)} = 1.25 \pm 0.40, \tag{35}$$

$$\Delta a_\mu = a_\mu^{\rm exp} - a_\mu^{\rm SM} = (26.1 \pm 8.0) \times 10^{-10}\ (3.3\sigma). \tag{36}$$

Particularly, we take into consideration the current best limits of the decay process $B_s \to \mu^+\mu^-$ announced by the CMS Collaboration [17]:

$$2.1 \times 10^{-9} < \text{BR}(B_s \to \mu^+\mu^-) < 4.0 \times 10^{-9}\ (3.5\sigma), \tag{37}$$

We can see from Table 1 that all the above constraints are satisfied, except for the muon anomalous magnetic dipole moment. Since the sparticles are all heavy, the SUSY contribution to $a_\mu$ are too small. However, the constraint (36) can be loosen by the uncertainty in theoretical estimation of $a_\mu$ in the SM due to hadronic loop corrections.

Besides we also take into account the constraints from dark matter direct and indirect detection experiments. For direct dark matter search, the best upper limit on the spin-independent cross-section of the scattering between dark matter and proton is set by XENON100 [18]:

$$\sigma_{\rm SI}^{\chi-{\rm p}} \lesssim 10^{-8}{\rm pb}\ (90\%\ {\rm CL}), \quad {\rm at}\ m_{\rm WIMP} \approx \mathcal{O}(500)\,{\rm GeV}. \tag{38}$$

While for indirect dark matter search the most stringent limit on the spin-dependent WIMP-proton cross-section is set by IceCube [19]:

$$\sigma_{\rm SD}^{\chi-{\rm p}} \lesssim 10^{-4}{\rm pb}\ (90\%\ {\rm CL}), \quad {\rm at}\ m_{\rm WIMP} \approx 150 - 600\,{\rm GeV}. \tag{39}$$

The values of spin-independent and spin-dependent cross-sections for the scattering of the lightest neutralino onto proton for each benchmark points are listed in Table 1. We find that the results satisfy all the current limits set by direct and indirect dark matter searches, with the notice that the spin-independent cross-section in the first column is just marginally close to the corresponding limit. Especially, the spin-independent WIMP-proton cross-sections in the third, the fourth and the last columns are at the order of $\mathcal{O}(10^{-9})$ pb. These values can be reached by future direct dark matter searches such as XENON1T since its expected $\sigma_{\rm SI}^{\chi-{\rm p}}$ sensitivity is about $\mathcal{O}(10^{-10})$ pb for the WIMP's mass around 500 GeV [20]. Therefore, such case can be tested in the near future dark matter direct detection experiments.

## 4. Yukawa mediation and Lepton flavor violation

We would like to comment on the Yukawa mediation and the lepton flavor violation due the Yukawa interactions (Eqs. (5) and (9)) between the seesaw messengers and the MSSM particles. In order to get very small neutrino masses, Eqs. (6) and (10) show that either the seesaw messenger are extremely heavy and/or the Yukawa couplings ($Y_{ij}$ and $\lambda$ in type II seesaw, $Y_i$ in type III seesaw) are small enough. With the new Yukawa couplings at the order of $\mathcal{O}(1)$ in usual seesaw models, the seesaw scale often reads $M_{\rm Mess} \sim 10^{14}$ GeV to get the allowed neutrino masses. However in our models, as can be seen in the above contour plots, thanks to the electroweak symmetry breaking condition the seesaw messenger scale is at most $10^{11} - 10^{12}$ GeV which is about three or two order in magnitude smaller than the usual values. To obtain the right neutrino masses, the new Yukawa couplings must be roughly at the order of $\mathcal{O}(10^{-3})$ or $\mathcal{O}(10^{-2})$. Thus, the effects of Yukawa mediation are suppressed by the same factors, and so is the lepton flavor violation.

The results for the LFV branching fraction of the process $\mu \to e\gamma$ are shown in the bottom row of Table 1. We can see that the $e\bar{\nu}_e\nu_\mu$ branching ratio for all the benchmark points well satisfy the current upper limit on this process set by the MEG experiment [21]:

$$\mathrm{BR}(\mu \to e\gamma) \ \leq \ 5.7 \times 10^{-13} \quad (90\%\ \mathrm{CL}). \tag{40}$$

## 5. Conclusions

In this paper, the seesaw-deflected anomaly mediation has been presented to overcome the tachyonic slepton problem, and to provide an explanation for the neutrino masses. We have studied the phenomenology of the models by examining their free parameter space, and analyzing the sparticle mass spectra and properties of dark matter candidate. Various phenomenological constraints have been taken into account. The recent results from the LHC about the Higgs mass measurement together with the WMAP's limits on the dark matter relic density and the pertubative gauge coupling constraint have put stringent restrictions on the free parameter space

**Table 1.** Benchmark particle mass spectra for the case $N = 7$, and 10. Masses of particles are given in GeV. The values of the branching fractions of $b \to s + \gamma$, $B_s \to \mu^+\mu^-$, $B \to \tau\nu_\tau$, the muon anomalous magnetic dipole moment $\Delta a_\mu$, and the neutralino relic density are calculated for each benchmark points. The spin-independent and spin-dependent cross-sections for the neutralino-proton elastic scattering are also provided. The last row of the table shows the branching fraction of $\mu \to e + \gamma$ taking into account the Yukawa mediation.

| $N$ | 7 | 7 | 7 | 7 | 10 | 10 | 10 |
|---|---|---|---|---|---|---|---|
| $d$ | 0.7 | 0.7 | 0.7 | 0.7 | 0.5 | 0.5 | 0.5 |
| $\tan\beta$ | 10 | 10 | 20 | 20 | 15 | 20 | 20 |
| $F_\phi$ | $2.70\times10^5$ | $3.20\times10^5$ | $2.14\times10^5$ | $2.05\times10^5$ | $2.24\times10^5$ | $2.07\times10^5$ | $2.05\times10^5$ |
| $M_{\text{Mess}}$ | $4.790\times10^9$ | $1.403\times10^9$ | $8.558\times10^9$ | $3.081\times10^{10}$ | $5.648\times10^9$ | $3.770\times10^9$ | $1.215\times10^{10}$ |
| $h^0$ | 125.5 | 126.0 | 125.5 | 125.6 | 125.5 | 125.4 | 125.5 |
| $H^0$ | 1531 | 1768 | 1059 | 739.2 | 1151 | 1026 | 722.0 |
| $A^0$ | 1531 | 1768 | 1059 | 739.4 | 1151 | 1026 | 721.9 |
| $H^\pm$ | 1533 | 1770 | 1062 | 743.8 | 1153 | 1029 | 726.4 |
| $\tilde{g}$ | 12518 | 14650 | 10097 | 9702 | 10637 | 9889 | 9800 |
| $\tilde{\chi}^0_{1,2}$ | 653.7, 685.6 | 794.1, 939.5 | 534.7, 1026 | 494.3, 532.7 | 522.9, 637.7 | 487.3, 991.3 | 466.2, 507.7 |
| $\tilde{\chi}^0_{3,4}$ | 710.3, 2704 | 946.3, 3203 | 1031, 2143 | 556.3, 2052 | 646.8, 2302 | 996.1, 2128 | 529.7, 2107 |
| $\tilde{\chi}^\pm_{1,2}$ | 685.9, 2704 | 939.8, 3203 | 1027, 2143 | 533.0, 2052 | 638.0, 2302 | 991.8, 2128 | 508.1, 2107 |
| $\tilde{u}, \tilde{c}_{R,L}$ | 11367, 11787 | 13235, 13735 | 9197, 9525 | 8874, 9190 | 9562, 9892 | 8880, 9184 | 8839, 9142 |
| $\tilde{d}, \tilde{s}_{R,L}$ | 11289, 11787 | 13154, 13735 | 9132, 9525 | 8803, 9190 | 9499, 9893 | 8826, 9184 | 8778, 9142 |
| $\tilde{t}_{1,2}$ | 9728, 11019 | 11468, 12905 | 7839, 8850 | 7474, 8497 | 8184, 9229 | 7631, 8559 | 7516, 8483 |
| $\tilde{b}_{1,2}$ | 11018, 11257 | 12904, 13118 | 8848, 9028 | 8495, 8697 | 9227, 9438 | 8558, 8727 | 8481, 8676 |
| $\tilde{\nu}_L^{e,\mu}$ | 3591 | 4163 | 2875 | 2815 | 2930 | 2690 | 2718 |
| $\tilde{e}, \tilde{\mu}_L$ | 3592, 3590 | 4164, 4162 | 2876, 2875 | 2816, 2815 | 2932, 2930 | 2692, 2690 | 2719, 2718 |
| $\tilde{e}, \tilde{\mu}_R$ | 1043, 1043 | 1283, 1283 | 827.3, 827.0 | 761.9, 761.5 | 760.0, 759.7 | 722.6, 722.3 | 680.8, 680.4 |
| $\tilde{\nu}_L^\tau$ | 3586 | 4158 | 2858 | 2798 | 2921 | 2675 | 2702 |
| $\tilde{\tau}_{1,2}$ | 1008, 3587 | 1248, 4159 | 709.0, 2860 | 628.8, 2799 | 686.4, 2922 | 607.5, 2677 | 546.7, 2704 |
| $\text{BR}(b \to s + \gamma)$ | $3.42\times10^{-4}$ | $3.39\times10^{-4}$ | $3.51\times10^{-4}$ | $3.68\times10^{-4}$ | $3.49\times10^{-4}$ | $3.52\times10^{-4}$ | $3.69\times10^{-4}$ |
| $\text{BR}(B_s \to \mu^+\mu^-)$ | $3.06\times10^{-9}$ | $3.06\times10^{-9}$ | $2.98\times10^{-9}$ | $2.90\times10^{-9}$ | $3.02\times10^{-9}$ | $2.97\times10^{-9}$ | $2.89\times10^{-9}$ |
| $\frac{\text{BR}^{\text{exp}}(B\to\tau\nu_\tau)}{\text{BR}^{\text{SM}}(B\to\tau\nu_\tau)}$ | 0.998 | 0.998 | 0.981 | 0.960 | 0.991 | 0.979 | 0.958 |
| $\Delta a_\mu$ | $3.10\times10^{-11}$ | $2.15\times10^{-11}$ | $8.65\times10^{-11}$ | $1.10\times10^{-10}$ | $7.60\times10^{-11}$ | $9.98\times10^{-11}$ | $1.24\times10^{-10}$ |
| $\Omega h^2$ | 0.1121 | 0.1121 | 0.1121 | 0.1121 | 0.1121 | 0.1121 | 0.1121 |
| $\sigma_{\text{SI}}^{\chi-\text{p}}$(pb) | $1.550\times10^{-8}$ | $1.820\times10^{-9}$ | $3.605\times10^{-11}$ | $5.234\times10^{-9}$ | $2.448\times10^{-9}$ | $2.392\times10^{-11}$ | $4.462\times10^{-9}$ |
| $\sigma_{\text{SD}}^{\chi-\text{p}}$(pb) | $2.460\times10^{-5}$ | $2.106\times10^{-6}$ | $2.503\times10^{-7}$ | $3.538\times10^{-5}$ | $7.147\times10^{-6}$ | $2.647\times10^{-7}$ | $3.760\times10^{-5}$ |
| $\text{BR}(\mu \to e + \gamma)$ | $1.70\times10^{-16}$ | $4.48\times10^{-17}$ | $1.33\times10^{-15}$ | $1.88\times10^{-14}$ | $5.35\times10^{-16}$ | $3.37\times10^{-16}$ | $3.36\times10^{-15}$ |

of those models. The possibility of type III seesaw with three generations of **24** messenger is excluded. For type II seesaw and type III seesaw with two generations of **24** messenger, although the sparticle masses are too heavy to be directly detected at the LHC there is a good chance to test a part of their free parameter spaces in future dark matter direct detection experiments like XENON1T. We have also argued that in our scenarios the effects of Yukawa mediation caused by the new Yukawa couplings between the MSSM and the seesaw messenger sector are negligible. As a result, the branching ratio of the lepton flavor violating processes such as $\mu \to e + \gamma$ are shown to well satisfy the current limit from the MEG experiment data.

## Acknowledgments

The work of H. M. T. is supported in part by Vietnam National Foundation for Science and Technology Development (NAFOSTED) under the grant No. 103.01-2014.22.

## References

[1] ATLAS Collaboration 2013 *ATLAS-CONF*-2013-014

[2] CMS Collaboration 2013 *JHEP* **06** 081 (*Preprint* arXiv:1303.4571 [hep-ex])
[3] Okada N and Tran M H 2013 *Phys. Rev.* D **87** 035024 (*Preprint* arXiv:1212.1866 [hep-ph])
[4] Pomarol A and Rattazzi R 1999 *JHEP* **9905** 013 (*Preprint* hep-ph/9903448)
[5] Okada N 2002 *Phys. Rev.* D **65** 115009 (*Preprint* hep-ph/0202219)
[6] Randall L and Sundrum R 1999 *Nucl. Phys.* B **557** 79 (*Preprint* hep-th/9810155)
[7] Giudice G F, Luty M A, Murayama H and Rattazzi R 1998 *JHEP* **9812** 027 (*Preprint* hep-ph/9810442)
[8] Mohapatra R N, Okada N and Yu H B 2008 *Phys. Rev.* D **77** 115017 (*Preprint* arXiv:0711.0956 [hep-ph])
[9] Mohapatra R N, Okada N and Yu H B 2008 *Phys. Rev.* D **78** 075011 (*Preprint* arXiv:0807.4524 [hep-ph])
[10] Allanach B C 2002 *Comput. Phys. Commun.* **143** 305 (*Preprint* arXiv:hep-ph/0104145)
[11] Belanger G, Boudjema F, Pukhov A and Semenov A 2006 *Comput. Phys. Commun.* **174** 577 (*Preprint* arXiv:hep-ph/0405253); Belanger G, Boudjema F, Pukhov A and Semenov A 2002 *Comput. Phys. Commun.* **149** 103 (*Preprint* arXiv:hep-ph/0112278); Belanger G *et al.* 2010 (*Preprint* arXiv:hep-ph/1004.1092)
[12] Hinshaw G *et al.* [WMAP Collaboration] 2012 (*Preprint* arXiv:1212.5226 [astro-ph.CO])
[13] Ade P A R *et al.* [Planck Collaboration] 2013 (*Preprint* arXiv:1303.5076 [astro-ph.CO])
[14] Barberio E *et al.* (Heavy Flavor Averaging Group (HFAG) Collaboration) 2007 (*Preprint* arXiv:hep-ex/0704.3575)
[15] Buchmueller O, Cavanaugh R, De Roeck A, Ellis J R, Flacher H, Heinemeyer S, Isidori G and Olive K A *et al.* 2009 *Eur. Phys. J.* C **64** 391 (*Preprint* arXiv:0907.5568 [hep-ph])
[16] Hagiwara K, Liao R, Martin A D, Nomura D and Teubner T 2011 *J. Phys.* G **38** 085003 (*Preprint* arXiv:1105.3149 [hep-ph])
[17] Chatrchyan S *et al.* [CMS Collaboration] 2013 (*Preprint* arXiv:1307.5025 [hep-ex])
[18] Aprile E *et al.* [XENON100 Collaboration] 2012 *Phys. Rev. Lett.* **109** 181301 (*Preprint* arXiv:1207.5988 [astro-ph.CO])
[19] Aartsen M G *et al.* [IceCube Collaboration] 2013 *Phys. Rev. Lett.* **110** 131302 (*Preprint* arXiv:1212.4097 [astro-ph.HE])
[20] Aprile E [XENON1T Collaboration], (*Preprint* arXiv:1206.6288 [astro-ph.IM])
[21] Adam J *et al.* [MEG Collaboration], *Phys. Rev. Lett.* **110** 201801 (2013) (*Preprint* arXiv:1303.0754 [hep-ex])

Journal of Physics: Conference Series **726** (2016) 012011 doi:10.1088/1742-6596/726/1/012011

# Effects of pairing correlations on the inverse level density parameter of hot rotating nuclei

**Le Thi Quynh Huong**[1,2], **Nguyen Quang Hung**[3], **Le Thi Quynh Trang**[4]

[1] Department of Natural Science and Technology , Khanh Hoa University, Nha Trang City, Khanh Hoa Province, Vietnam
[2] Faculty of Physics and Engineering Physics, Ho Chi Minh University of Science, Ho Chi Minh City, Vietnam
[3] Institute of Research and Development, Duy Tan University, K7/25 Quang Trung, Danang city, Vietnam
[4] School of engineering, Tan Tao University, Tan Tao University Avenue, Tan Duc E.City, Duc Hoa, Long An Province, Vietnam

E-mail: lethiquynhhuong@yahoo.com (L.T.Q.H)

**Abstract.** Angular momentum dependence of the inverse level density parameter $K$ in the excitation-energy region of $\sim 30 - 40$ MeV is studied within the finite-temperature Bardeen-Cooper-Schrieffer (FTBCS) theory and the FTBCS theory that includes the effect due to quasiparticle-number fluctuations (FTBCS1). The two theories take into account the noncollective rotation of the nucleus at nonzero values of z-projection $M$ of the total angular momentum. The comparison between the results obtained within the FTBCS and FTBCS1 as well as the case without pairing correlations and the experimental data for two medium-mass even-even nuclei $^{108}$Cd and $^{122}$Te shows that by including the pairing corrections the FTBCS and FTBCS1 reproduces quite well all the experimental data, whereas the non-pairing case always overestimates the data. Due to the effect of quasiparticle-number fluctuations, the FTBCS1 gaps at different $M$ values do not collapse at critical temperature $T_C$ as in the FTBCS ones but monotonously decrease with increasing $T$ and being finite even at high $T$. As the result, the values of $K$ obtained within the FTBCS1 are always closer to the experimental data than those obtained within the FTBCS.

## 1. Introduction

Pairing correlation is known to play a major role in strongly interacting many-body systems ranging from large ones such as superconductors to very small ones such as atomic nuclei [1]. In nuclear systems, pairing correlation has significant effects on various nuclear properties such as the binding and excitation energies, collective motions, rotations, level densities, etc. Dependence of pairing correlations on temperature and angular momentum has been widely studied within the finite-temperature Bardeen-Cooper-Schrieffer (FTBCS) theory taking into account the $z$-projection $M$ of total angular momentum [2, 3]. It has been shown in Ref. [2] that the pairing gap obtained within the FTBCS theory decreases with increasing both temperature $T$ and/or angular momentum $M$ and vanishes at a given critical temperature $T_C$ or angular momentum $M_C$. However, by taking into account the quasiparticle-number fluctuations (QNF), which are neglected within the FTBCS, the pairing gap obtained within the so-called FTBCS1 theory, which is the FTBCS plus QNF, for light and medium mass nuclei does not colapse at

$T_C$ or $M_C$ but monotously decreases with increasing $T$ or $M$ and keeps finite even at high $T$ or $M$ [4, 5, 6]. The goal of present work is to apply the FTBCS and FTBCS1 theories to study the effect of pairing correlation on the inverse level density parameter of some medium-mass hot rotating nuclei. The latter is an important input in the calculations of cross sections of fission or fusion reactions as well as astrophysical reaction rate of the nucleonsynthesis processes [7]. The results obtained are compared with the recent data extracted from the measured $\gamma$-ray multiplicity fold-gated $\alpha$-particle energy spectra of two medium-mass even-even nuclei $^{108}$Cd and $^{122}$Te at the excitation energy of $\sim 30 - 40$ MeV and angular momentum $\sim 0 - 20\hbar$ [8].

## 2. FORMALISM

### *2.1. Model Hamiltonian*

In this work, we consider a system, which consists of $N$ particles (protons or neutrons) interacting via a constant pairing interaction $G$ and rotating about the symmetry axis (noncollective rotation) with a fixed projection $M$ of the total angular momentum along the $z$-axis. The Hamiltonian of this type has the form as [2, 5]

$$H = H_p - \lambda\hat{N} - \gamma\hat{M} \ , \tag{1}$$

where $\lambda$ and $\gamma$ are the chemical potential and angular velocity, respectively and $H_p$ is the standard pairing Hamiltonian, which is given as

$$H_p = \sum_k \epsilon_k(a_k^\dagger a_k + a_{-k}^\dagger a_{-k}) - G\sum_{k,k'} a_k^\dagger a_{-k}^\dagger a_{-k'}a_{k'} \ , \tag{2}$$

with $a_{\pm k}^\dagger(a_{\pm k})$ denoting the operator that creates (annihilates) a particle with angular momentum $k$, spin projection $\pm m_k$, and energy $\epsilon_k$. The subscripts $k$ here label the single-particle states $|k,\, m_k>$ in deformed basis, whereas -$k$ denotes the time-reversal ones $|k,\, -m_k>$ namely the states, which have the same spin projection with the states $k$ but having the opposite direction. In Eq. (1), the particle-number operator $\hat{N}$ and total angular momentum operator $\hat{M}$, which coincides with its $z$-projection, are given as

$$\hat{N} = \sum_k (a_k^\dagger a_k + a_{-k}^\dagger a_{-k}), \quad \hat{M} = \sum_k (a_k^\dagger a_k - a_{-k}^\dagger a_{-k}) \ . \tag{3}$$

### *2.2. FTBCS1 theory*

The FTBCS1 equations at finite temperature and angular momentum are derived in the same way as the conventional FTBCS ones, namely based on the variational procedure to minimize the expectation value of the pairing Hamiltonian (1) in the grand-canonical ensembles. The only difference between the FTBCS and FTBCS1 is that the FTBCS1 equations contain the corrections due to the quasiparticle-number fluctuations (QNF) on the pairing gap, average particle number $N$, and angular momentum $M$, whereas the QNF are zero within the FTBCS. The derivation of the FTBCS1 equations was mentioned in details in Refs. [5, 6]. The final FTBCS1 equations for the pairing gap, particle number and total angular momentum are then given as

$$\begin{aligned}
\Delta_k &= \Delta + \delta\Delta_k \ , \\
N &= 2\sum_k \left[(1 - n_k^+ - n_k^-)v_k^2 + \frac{1}{2}(n_k^+ + n_k^-)\right] , \\
M &= \sum_k m_k(n_k^+ - n_k^-) \ ,
\end{aligned} \tag{4}$$

Journal of Physics: Conference Series **726** (2016) 012011 doi:10.1088/1742-6596/726/1/012011

where

$$\begin{aligned}
\Delta &= G\sum_k u_k v_k(1-n_k^+-n_k^-); \quad \delta\Delta_k = G\frac{\delta N_k^2}{1-n_k^+-n_k^-}u_k v_k\ , \\
u_k^2 &= \frac{1}{2}\left(1+\frac{\epsilon_k-\lambda}{E_k}\right); \quad v_k^2 = 1-u_k^2\ , \\
E_k &= \sqrt{(\epsilon_k-\lambda)^2+\Delta_k^2}\ , \\
n_k^{\pm} &= \frac{1}{1+e^{\beta(E_k\mp\gamma m_k)}}\ ; \quad \beta = 1/T\ ,
\end{aligned} \tag{5}$$

with the quasiparticle-number fluctuations $\delta N_k^2$ at nonzero angular momentum being defined as

$$\delta N_k^2 = (\delta N_k^+)^2 + (\delta N_k^-)^2 = n_k^+(1-n_k^+) + n_k^-(1-n_k^-)\ . \tag{6}$$

The conventional FTBCS equations are obtained from the FTBCS1 by setting $\delta N_k^2 = 0$. The FTBCS1 (FTBCS) total (internal) energy and entropy are calculated as

$$E(T,M) = < H >\ , \tag{7}$$

$$S(T,M) = -\sum_k [n_k^+ \ln n_k^+ + (1-n_k^+)\ln(1-n_k^+) + n_k^- \ln n_k^- + (1-n_k^-)\ln(1-n_k^-)]\ . \tag{8}$$

### *2.3. Inverse level density parameter*

The level density parameter a as a function of angular momentum $M$ and temperature $T$ is given as [9]

$$a(T,M) = \frac{U(T,M)}{T^2}\ , \tag{9}$$

where *U(T,M)* is the corrected total excitation energy of the system

$$U(T,M) = E^*(T,M) - E_{pair} - E_{rot}\ , \tag{10}$$

with $E_{pair}$ and $E_{rot}$ being the corrections due to the pairing and rotational vibration of the total excitation energy, respectively [10]

$$E_{pair} = 12/\sqrt{A}\ , \qquad E_{rot} = \frac{M(M+1)}{\frac{4}{5}mR^2\left(1+\sqrt{\frac{5}{16\pi}}\beta_2\right)} \tag{11}$$

In Eq. (11), $\beta_2$ is the quadrupole deformation parameter ($\beta_2 = 0$ for a spherical nucleus) and $R = 1.2A^{1/3}$ ($A$ is mass number) is the nuclear radius. The total excitation energy $E^*(T,M)$ is calculated based on the FTBCS (FTBCS1) total energy as

$$E^*(T,M) = E(T,M) - E(0,0)\ . \tag{12}$$

The inverse level density parameter $K$ is obtained as the mass number ($A$) divided by the level density parameter $a$

$$K(T,M) = \frac{A}{a(T,M)}\ . \tag{13}$$

Journal of Physics: Conference Series **726** (2016) 012011 doi:10.1088/1742-6596/726/1/012011

## 3. NUMERICAL RESULTS AND DISCUSSION

### *3.1. Ingredients of numerical calculations*

The numerical calculations are carried out for two medium-mass even-even nuclei, namely $^{108}$Cd and $^{122}$Te. The single-particle spectra are obtained within the axially deformed Woods-Saxon (WS) potential including the spin-orbit and Coulomb interactions. The parameters of the WS potential are taken from Ref. [11], in which the quadrupole deformation parameters $\beta_2$ are fixed and equal to 0.135 (prolate shape) for $^{108}$Cd and -0.139 (oblate shape) for $^{122}$Te, respectively. The pairing interaction parameter $G$ is adjusted so that the pairing gap at $T = 0$ and $M = 0$ obtained within the FTBCS (FTBCS1) fits the experimental odd-even mass difference [12]. In order to compare the results of $K$ obtained within the FTBCS (FTBCS1) with the experimental values reported in Ref. [7], we perform the calculations at the same angular momenta and in the same interval of excitation energy ( 30 − 40)MeV as those of the measured data.

### *3.2. Inverse level density parameter*

Shown in Fig. 1 are the neutron and proton pairing gaps $\Delta_N$ [Fig. 1 (a)] and $\Delta_Z$ [Fig. 1 (b)], excitation energy $E^*$ [Fig. 1 (c)], and inverse level density parameter $K$ [Fig. 1 (d)] obtained within the FTBCS and FTBCS1 calculations for prolately deformed $^{108}$Cd nucleus. In Fig. 1 (d), the filled circles with error bars denote the experimental data of $K$, whereas the triangles depict the results obtained within the case of zero pairing ($\Delta = 0$). Similar results obtained for an oblately deformed $^{122}$Te nucleus are displayed in Figs. 2 (a) − (d).

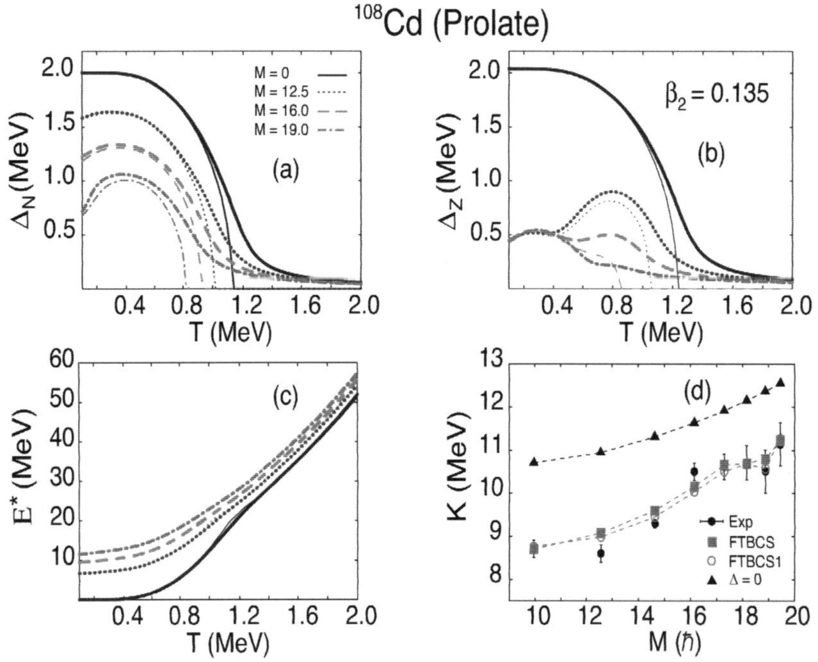


**Figure 1.** *(a) - (c): neutron and proton pairing gaps $\Delta_N$ (a) and $\Delta_Z$ (b) and total excitation energy $E^*$ (c) as functions of temperature $T$ obtained within the FTBCS (thin lines) and FTBCS1 (thick lines) at different values of total angular momentum $M$ for $^{108}$Cd nucleus. (d): inverse level density parameter $K$ as functions of $M$ obtained within the FTBCS (open circles), FTBCS1 (squares), and the case without pairing correlation (triangles) versus the experimental data (full circles with error bars) taken from Ref. [8].*

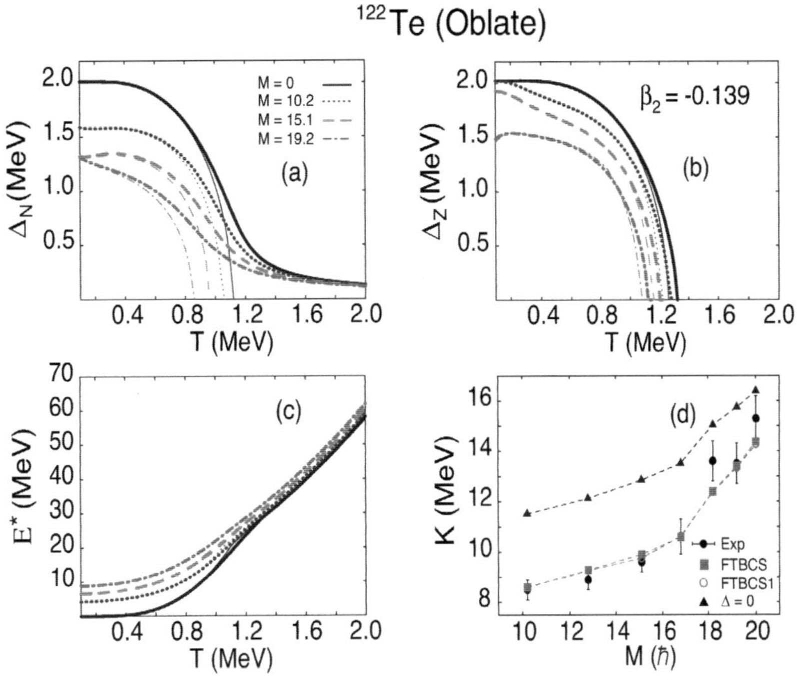


**Figure 2.** *The same as in Fig. 1 but for $^{122}Te$ nucleus*

In Figs. 1 (a)–(b) and Fig. 2 (a), one can see that the FTBCS and FTBCS1 gaps both decrease with increasing either temperature $T$ or total angular momentum $M$. However, the FTBCS1 gaps at different $M$ do not collapse at critical temperature $T_C$ as in the case of the FTBCS ones. In stead of that the FTBCS1 gaps monotonously decrease with increasing $T$ and being finite even at $T$ as high as 2 MeV. This feature is well-known due to the quasiparticle-number fluctuations within the FTBCS1 [4, 5]. Due to the shell structure in an oblately deformed $^{122}$Te nucleus, the proton gap obtained within the FTBCS1 is nearly similar as that of the FTBCS one as seen in Fig. 2 (b). As a result, the excitation energy $E^*$ obtained within the FTBCS1 is lower than that obtained within the FTBCS only in the low and intermediate regions of temperature and total angular momentum [See e.g., the thin and thick lines in the Figs. 1 (c) and 2 (c)], whereas in the high $T$ and $M$ regions both FTBCS and FTBCS1 predict nearly the same excitation energies. Consequently, the corresponding inverse level density parameter $K$ obtained within the FTBCS1 in the region of intermediate $M \sim 10\hbar - 18\hbar$ for boh nuclei [See Fig. 1 and 2 (d)] is closer to the experimental data than that obtained within the FTBCS. It is aslo clear to see from Figs. 1 (d) and 2 (d) that the $K$ values obtained within the zero-pairing case always deviate significantly from the measured data in both nuclei. This results show that pairing correlation has significant effects on the inverse level density parameter of light and medium-mass nuclei at low and intermediate temperature and angular momentum. In addition, when the total angular momentum $M$ is high enough (but not too high), one can see clearly the reentrance effect [5, 6, 13] of the pairing gap of $^{108}$Cd nucleus [See e.g., red dash dotted lines of Fig. 1 (a) for the neutron gap or blue dotted lines in Fig. 1 (b) for the proton gap], in which the pairing gap at low $T$ increases to reach a maximum and decreases with further increasing of $T$, whereas this effect is not seen in $^{122}$Te nucleus due to the shell structure. However, this pairing reentrance does not affect the $K$ value considered here as it appears only in the low temperature region, whereas the experimental data are measured in the intermediate region of temperature and/or excitation energy $\sim 30 - 40$ MeV.

## 4. CONCLUSIONS

Present work studies the temperature and angular momentum dependence of the inverse level density parameter within the conventional FTBCS theory and the FTBCS theory taking into account the quasiparticle-number fluctuations (FTBCS1) as well as the case without pairing. The numerical calculations are carried out for two medium-mass even-even $^{108}$Cd and $^{122}$Te nuclei with the use of single-particle spectra obtained within the axially deformed Woods-Saxon potential at the fixed values of the quadrupole deformation parameters. The results obtained show that due to the effects of quasiparticle-number fluctuations, the pairing gaps obtained within the FTBCS1 for the neutrons and protons in $^{108}$Cd and for the neutrons in $^{122}$Te at the given total angular momenta $M$ do not collapse at critical temperature as those obtained within the FTBCS but monotonously decrease with increasing $T$ and keep finite even at high $T$. The FTBCS1 and FTBCS gaps for the protons in $^{122}$Te are nearly the same due to the shell structure of this nucleus. This feature of the FTBCS1 pairing gaps leads to the lowering of the FTBCS1 excitation energies around the regions of intermediate temperature and angular momentum as compared with the FTBCS ones. As a result, the inverse level density parameters obtained within the FTBCS1 are always closer to the experimental data than that of the FTBCS, whereas the results obtained within the zero-pairing case are quite far from the data. Based on these results, we conclude that pairing correlations have significant effects on the inverse level density parameters of hot rotating nuclei.

## ACKNOWLEDGEMENT

We acknowledge Mamta Aggarwal for providing us the experimental data. This work is supported by the National Foundation for Science and Technology Development (NAFOSTED) through Grant No. 103.04-2013.08.

### References

[1] Dean D J and Hjorth-Jensen H 2003 *Rev. Mod. Phys.* **75** 607.
[2] Moretto L G 1972 *Phys. Lett.* **B40** 1; Moretto L G 1972 *Nucl. Phys.* **A185** 145.
[3] Goodman A L 1981 *Nucl. Phys.* **A352** 30; Goodman A L 1984 *Phys. Rev.* **C29** 1887.
[4] Dang N D and Hung N Q 2008 *Phys. Rev.* **C77** 064315.
[5] Hung N Q and Dang N D 2008 *Phys. Rev.* **C78** 064315; Hung N Q and Dang N D 2011 *Phys. Rev.* **C84** 054324.
[6] Hung N Q *et. al.* 2015 *J. Phys: Conf. Ser.* **627** 012006; Dey B *et. al.* 2015 *Phys. Rev.* **C91** 044326.
[7] Wallace R K and Woosely S. E 1981 *Astrophys. J. Suppl.* **45** 389.
[8] Gupta Y K *et. al.* 2008 *Phys. Rev.* **C78** 054609.
[9] Agrawal B K and Ansari A 1994 *Nucl. Phys.* **A567** 189.
[10] Hilaire S and Goriely S 2006 *Nucl. Phys.* **A779** 63-81.
[11] Cwiok S *et al.* 1987 *Comput. Phys. Commun.* **46** 379.
[12] Ring P and Schuck P 1980 *The nuclear many-body problem* (Springer-Verlag, New York)
[13] Dean D J, Langanke K, Nam H, and W. Nazarewicz 2010 *Phys. Rev. Lett.* **105** 212504.

# Influence of phonon confinement on the optically detected electron-phonon resonance linewidth in quantum wells

**Nguyen Dinh Hien[1], Le Dinh[1], Vo Thanh Lam[2], Tran Cong Phong[3]**

[1] Hue University's College of Education, 34 Le Loi Str., Hue City, Viet Nam.

[2] Department of Physics, Sai Gon University, 273 An Duong Vuong Str., District 5, Ho Chi Minh City, Viet Nam.

[3] Vietnam National Institute of Educational Sciences, 101 Tran Hung Dao Str., District Hoan Kiem, Ha Noi, Viet Nam.

E-mail: dinhhienphysics@gmail.com

**Abstract.** We investigate the influence of phonon confinement on the optically detected electrophonon resonance (ODEPR) effect and ODEPR line-width in quantum wells. The obtained numerical result for the GaAs/AlAs quantum well shows that the ODEPR line-widths depend on the well's width and temperature. Besides, in the two cases of confined and bulk phonons, the linewidth (LW) decreases with the increase of well's width and increases with the increase of temperature. Furthermore, in the small range of the well's width, the influence of phonon confinement plays an important role and cannot be neglected in considering the ODEPR line-width.

## 1. Introduction

Electrophonon resonance (EPR) effects in low-dimensional electron gas systems have generated considerable interest in recent year [1]. EPR effects can only be observed for the resonant scattering of electrons confined in electric subband levels by longitudinal optical (LO) phonon whenever the LO phonon energy is equal to the energy separation between two electric subband levels. Scattering process with LO phonons is dominant in limiting the mobility of electrons in the polar semiconductors for temperature $T > 50$ K [2]. EPR was introduced by Bryskin and Firsov [2] who predicted EPR for non-degenerate semiconductors in a very strong electric fields.

The absorption line-width (LW) is well-known as a good tool for investigating the scattering mechanisms of carriers. Hence, it can be used to probe electron-phonon scattering processes. The absorption LW was investigated based on the interaction of electrons and bulk phonons, but the absorption LW in quantum wells (QWs) due to confined electrons and confined LO-phonons interaction is still open for study.

Phonon confinement is an essential part of the description of electrophonon interactions [3]. It causes the increase of electrophonon scattering rates [4] and significant nonlinearities in the dispersion relations of acoustic phonon modes, and modifies the phonon density of states [5]. There have been many models dealing theoretically with phonon modes, such as the Huang-Zhu (HZ) model, the slab mode model and guided mode model [6]. Phonon confinement is shown to be important whenever the transverse dimensions of a quantum well are smaller than the phonon coherence length [3] and should be taken into account in order to obtain realistic estimates for electron-phonon scattering in low-dimensional structures [7]. Phonon confinement affects ODEPR mainly through changes in the selection rules for transitions involving subband of electrons, phonon modes. Phonon confinement also affects the ODEPR linewidth (ODEPRLW) through changes in the probability of the electron-phonon scattering. The linewidth is defined as the profile of curves describing the dependence of the absorption power on the phonon energy or frequency [8]. The linewidth has been measured in various kinds of semiconductors, such as quantum wells [9], quantum wires [10], and quantum dots [11]. These results show that the absorption linewidth has a weak dependence on temperature and has a strong dependence on the sample size. However, in these studies, the absorption linewidth was investigated based on the interaction of electrons and bulk phonons, so that the absorption linewidth in quantum wells (QWs) due to confined optical phonon-electron interaction is still open for study. Recently, our group has proposed a method, called the profile method. This method can be used to computationally obtain the linewidth from graphs of the absorption power [12], and we used this method to determine the cyclotron resonance linewidth in cylindrical quantum wires [13] and the influence of phonon confinement on the optically detected electrophonon resonance linewidth in GaAs/AlAs quantum wires [14, 15].

In the present work, we investigate the ODEPRLW in QWs, the dependence of the ODEPRLW on the well's width and the temperature of system is obtained. The results of the present work are fairly different from the previous theoretical results because the phonon confinement is considered. The paper is organized as follows. In Section 2, we introduce several models of the phonon confinement in QWs. Calculations of analytical expression of the absorption power in QWs are presented in Section 3. The graphical dependence of the absorption power on the photon energy in the GsAs/AlAs QW is shown in Section 4. From this dependence, we obtain the linewidth and examine not only the location of resonance peaks, but also the dependence of the linewidth on temperature and well's width. Finally, remarks and conclusions are shown in Section 5.

## 2. Confined phonon models

For a square well of infinite height, the one electron eigenfunctions, $\psi_{k_x,k_y,n}(\vec{r})$, and the eigenvalues, $E_n(k_x,k_y)$, in the conduction band are, respectively, given by [2]

$$\psi_{k_x,k_y,n}(\vec{r}) \equiv \langle \vec{r}|k_x,k_y,n\rangle = \frac{1}{\sqrt{L_xL_y}}e^{ik_xx}e^{ik_yy}\varphi_n(z) = \sqrt{\frac{2}{L_xL_yL_z}}.e^{ik_xx}e^{ik_yy}.\sin\left(\frac{n\pi z}{L_z}+\frac{n\pi}{2}\right), \tag{1}$$

$$E_n(k_x,k_y) = \frac{\hbar^2(k_x^2+k_y^2)}{2m^*}+n^2\varepsilon_0, \tag{2}$$

where $\vec{r}$ is the position vector of electron; $k_x$ and $k_y$ are, respectively, the wave-vector component of the electron in the $x$ and $y$ direction; $m^*$ being the effective mass of an electron and $\varepsilon_0 = \pi^2\hbar^2/(2m^*L_z^2)$ is the energy of the lowest electric subband, $n$ is the electric subband quantum number ($n = 1, 2, 3, \cdots$); $L_x$, $L_y$, and $L_z$ are, respectively, the $x$-, $y$,- and $z$-directional normalization lengths. The electron wave function $\varphi_n(z)$ in $z$-direction (well centered at $z = 0$) as determined by the barrier potential $V(z)$ ($V(z) = 0$ for $|z| < L_z/2$ and $V(z) = \infty$ for $|z| > L_z/2$) [16].

The matrix element for electron-confined phonon interaction in quantum well in the extreme quantum limits can be written as [17]

$$|\langle i|H_{e-ph}|f\rangle|^2 = |C_{i,f}^m(\vec{q})|^2 = |C_m(q_\perp)|^2|G_{n_i,n_f}^{m\alpha}(q_z)|^2\delta_{k_\perp^i,k_\perp^f\pm q_\perp}, \tag{3}$$

where

$$|C_m(q_\perp)|^2 = \frac{2\pi e^2\hbar\omega_{m,\vec{q}_\perp}}{\epsilon_0\Omega}\Big(\frac{1}{\chi_\infty} - \frac{1}{\chi_0}\Big)\frac{1}{q_\perp^2 + q_m^2}, \tag{4}$$

where $\chi_0, \chi_\infty, \epsilon_0$ and $\Omega$ are the static, high-frequency, vacuum dielectric constants and the normalization volume of specimen, respectively; $q_\perp$ is a two dimensional vector in the $(x, y)$ plane of phonon; $q_m = m\pi/L_z$. The term $G^{m\alpha}_{n_i,n_f}(q_z)$ is given by

$$G^{m\alpha}_{n_i,n_f}(q_z) = \int_{L_z/2}^{L_z/2} \varphi^*_{n_f}(z)u_{m\alpha}(z)\varphi_{n_i}(z)dz, \tag{5}$$

where $\alpha$ distinguishes the even (-) and odd (+) confined phonon mode, $u_{m\alpha}(z)$ being the parallel component of displacement vector of $m$-th phonon mode in the direction of spatial confinement. It is different for different models and has been calculated for some confined models such as the HZ model, slab mode model, and guided mode model [17]. In the next section, we will use the HZ model to calculate the optical absorption power in quantum wells because among these models, the HZ model has received wide acceptance and best describes the electron-phonon interaction in quasi-two-dimensional systems [18]. For example, calculations of electron intra- and inter-subband scattering rates in GaAs quantum wells due to confined LO phonons using the HZ model [19] have been found to be in good agreement with experimental results [20]. For this model [17]

$$u_{m+}(z) = \sin\Big(\frac{\mu_m\pi z}{L_z}\Big) + \frac{C_m z}{L_z}, \qquad m = 3, 5, 7, \cdots, \tag{6}$$

where, $\mu_m$ and $C_m$ are constants: $\mu_3 = 2.8606$, $\mu_5 = 4.918$, $\mu_7 = 6.95, ...,$ and $C_3 = 1.9523$, $C_5 = -1.983$, $C_7 = 1.992, ...,$

$$u_{m-}(z) = \cos\Big(\frac{m\pi z}{L_z}\Big) - (-1)^{m/2}, \qquad m = 2, 4, 6, \cdots. \tag{7}$$

The overlap integral (5) can be evaluated for intra-subband transition $(1 \to 1)$, and is obtained for the HZ model as

$$G^{m+}_{11} = 0, \qquad m = 3, 5, 7, \cdots, \tag{8}$$

$$G^{m-}_{11} = \frac{3}{2}\delta_{m,2} - (-1)^{m/2}(1 - \delta_{m,2}), \qquad m = 2, 4, 6, \cdots. \tag{9}$$

## 3. Analytical results

In this section, we utilize the HZ model for confined phonon to calculate the absorption power in above mentioned quantum wells, subjected to an electric field with amplitude $E_0$ and frequency $\omega$. The absorption power is obtained by using the operator projection technique to calculate the conductivity tensor $(\sigma_{zz}(\omega))$. Finally, absorption power with confined phonons can be written as

$$P(\omega) = \frac{E_0^2}{2}\mathrm{Re}[\sigma_{zz}(\omega)] = \frac{E_0^2}{2\omega}\sum_{\alpha,\beta}(j_z)_{\alpha,\beta}(j_z)_{\beta,\alpha}\frac{(f_\alpha - f_\beta)\hbar B_{\alpha,\beta}(\omega)}{(\hbar\omega - E_{\alpha,\beta})^2 + \hbar^2 B^2_{\alpha,\beta}(\omega)}, \tag{10}$$

$$(j_z)_{\alpha,\beta} = i\frac{2e\pi\hbar}{\Omega m^*}n_\beta\delta(k_{\perp\beta} - k_{\perp\alpha})\Bigg[\frac{1 - (-1)^{n_\alpha+n_\beta}}{n_\alpha + n_\beta} + \frac{1 - (-1)^{n_\alpha-n_\beta}}{n_\alpha - n_\beta}\Bigg], \tag{11}$$

$$(j_z)_{\beta,\alpha} = i\frac{2e\pi\hbar}{\Omega m^*}n_\alpha\delta(k_{\perp\alpha} - k_{\perp\beta})\Bigg[\frac{1 - (-1)^{n_\alpha+n_\beta}}{n_\alpha + n_\beta} + \frac{1 - (-1)^{n_\beta-n_\alpha}}{n_\beta - n_\alpha}\Bigg], \tag{12}$$

$f_\alpha$ and $f_\beta$ are the Fermi-Dirac distribution function of electron at state $|\alpha\rangle$ and $|\beta\rangle$,

$$\begin{aligned} B_{\alpha,\beta}(\omega) &= C_0^m \sum_m |G^{m\alpha}_{n_\beta,n_\eta}(q_z)|^2 \sum_{n_\eta}\sum_{\pm} \frac{m^* k_1^{\pm}(k_{\perp\beta} - k_1^{\pm})}{\hbar^2 |k_1^{\pm}|[3(k_{\perp\beta} - k_1^{\pm})^2 + (\frac{m\pi}{L_z})^2]} \\ &\times \{[f_\alpha(1+N_q) - (f_\alpha + N_q) f_\eta(k_1^+)] - [N_q f_\alpha - (1+N_q - f_\alpha) f_\eta(k_1^-)]\} \\ &+ C_0^m \sum_m |G^{m\alpha}_{n_\alpha,n_\eta}(q_z)|^2 \sum_{n_\eta}\sum_{\pm} \frac{m^* k_2^{\pm}(k_{\perp\alpha} - k_2^{\pm})}{\hbar^2 |k_2^{\pm}|[3(k_{\perp\alpha} - k_2^{\pm})^2 + (\frac{m\pi}{L_z})^2]} \\ &\times \{(1+N_q - f_\beta) f_\eta(k_2^+) - N_q f_\beta] - [(f_\beta + N_q) f_\eta(k_2^-) - \times(1+N_q) f_\beta]\}, \end{aligned} \tag{13}$$

where $N_q$ is the Planck distribution function for confined phonon at the state $|q\rangle = |m, q_\perp\rangle$,

$$C_0^m = \frac{2\pi e^2 \hbar\omega_{m,\vec{q}_\perp}}{\epsilon_0 \Omega}\left(\frac{1}{\chi_\infty} - \frac{1}{\chi_0}\right), \tag{14}$$

$$k_1^{\pm} = \left\{k_{\perp\beta}^2 + \frac{2m^*}{\hbar^2}[(n_\eta^2 - n_\beta^2)\varepsilon_0 + \hbar(\omega \pm \omega_{LO_{m,q_\perp}})]\right\}^{1/2}, \tag{15}$$

$$k_2^{\pm} = \left\{k_{\perp\alpha}^2 - \frac{2m^*}{\hbar^2}[(n_\alpha^2 - n_\eta^2)\varepsilon_0 + \hbar(\omega \pm \omega_{LO_{m,q_\perp}})]\right\}^{1/2}. \tag{16}$$

We can see that these analytical results appear very involved. However, physical conclusion can be drawn from graphical representations and numerical results, obtained from adequate computational methods.

## 4. Numerical results and Discussion

To clarify the obtained results we numerically evaluate the absorption power, $P(\omega)$, for a specific GaAs/AlAs QW. The absorption power is considered to be a function of the photon energy. The parameters used in our computational evaluation are as follows [21]: $\chi_\infty = 10.9, \chi_0 = 12.9, m^* = 0.067 \times m_0$ ($m_0$ being the mass of free electron), $\hbar\omega_0 = 36.25$ meV, $E_0 = 5.0 \times 10^6$ Vm$^{-1}$, and $\beta = 4.73 \times 10^3$ ms$^{-1}$. The following conclusions are obtained in the extreme quantum limit, where the low subbands of confined phonon for Huang-Zhu model $m = 2$ and $m = 4$; $n_\alpha = 1$, $n_\beta = 2$ for confined electron.

Dependence of the absorption power in QW on the photon energy for two different models of phonons is shown in figure 1. The solid curve corresponds to confined phonon, i.e., the energy of confined phonon for $m = 2$, $\hbar\omega_{m,q_\perp} = \hbar\sqrt{\omega_0^2 - \beta^2(m\pi/L_z)^2} = 36.22$ meV, $\varepsilon_0 = 28.67$ meV (for $L_z = 14$ nm). From the graph we can see that each peak describes a specific resonance. By using the computational method, we easily determine from the left to the right (of solid curve).

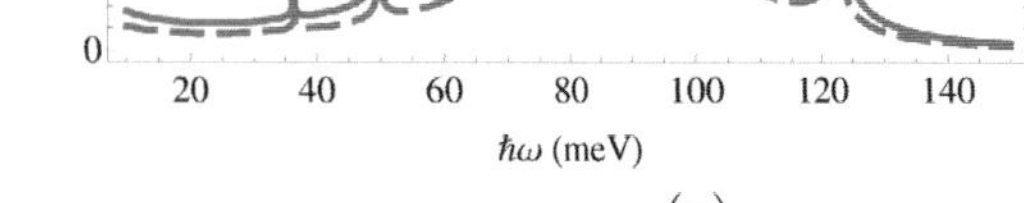


(a)

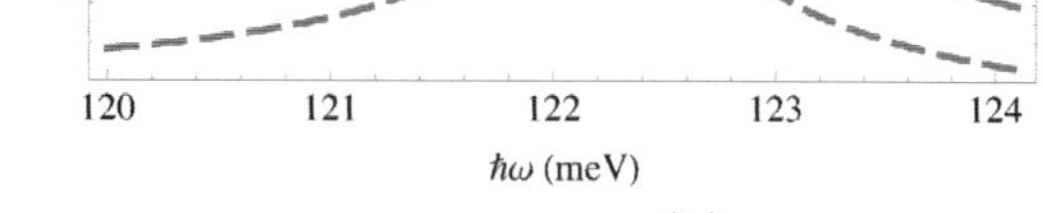


(b)

Journal of Physics: Conference Series **726** (2016) 012012 doi:10.1088/1742-6596/726/1/012012

**Figure 1:** Dependence of the absorption power in QW on the photon energy for two different models of phonon. Here, T=200 K, $L_z$=14 nm. The solid and the dashed curves correspond to confined phonons and bulk phonons, respectively. (a) All peaks, (b) At the fourth peak in (a).

The dependence of the absorption power on the photon energy in the case of confined phonon is presented in figure 1a. The graph exhibits four peaks which can be explained as follows:

- The first peak corresponds to the photon energy $\hbar\omega$ = 36.22 meV, which satisfies the condition $\hbar\omega = \hbar\omega_{m,\vec{q}_\perp}$. This is the condition for intrasubband transitions.
- The second peak corresponds to the photon energy of $\hbar\omega$ = 49.79 meV, which satisfies the condition $\hbar\omega = (n_\beta^2 - n_\alpha^2)\varepsilon_0 - \hbar\omega_{m,\vec{q}_\perp}$, i.e., 49.79 meV=$(2^2 - 1^2)\times$ 28.67 meV - 36.22 meV. This condition implies that an electron in the $n_\alpha = 1$ can move to $n_\beta = 2$ by absorbing of a photon with energy $\hbar\omega$ along with the absorption of a phonon with energy $\hbar\omega_{m,q_\perp}$. This is the condition for optically detected electrophonon resonance (ODEPR).
- The third peak corresponds to the photon energy $\hbar\omega$ = 86.01 meV, which satisfies the condition $\hbar\omega = (n_\beta^2 - n_\alpha^2)\varepsilon_0$, i.e., 86.01 meV= $(2^2 - 1^2)\times$ 28.67 meV. This condition implies that an electron in the $n_\alpha = 1$ can move to $n_\beta = 2$ by absorbing a photon with energy $\hbar\omega$. This is the condition for direct transitions.
- The fourth peak corresponds to the energy $\hbar\omega$ = 122.23 meV, which satisfies the condition $\hbar\omega = (n_\beta^2 - n_\alpha^2)\varepsilon_0 + \hbar\omega_{m,\vec{q}_\perp}$, i.e., 122.23 meV=$(2^2 - 1^2)\times$ 28.67 meV + 36.22 meV. This condition implies that an electron in the $n_\alpha = 1$ can move to $n_\beta = 2$ by absorbing a photon with energy $\hbar\omega$ along with the emission of a phonon with energy $\hbar\omega_{m,q_\perp}$. This is the condition for optically detected electrophonon resonance (ODEPR).

The dependence of the absorption power in QW on the photon energy for two different models of phonon at the fourth peak on the figure 1a is showed in figure 1b. In the case of bulk phonon this peak is located at the photon energy of 122.27 meV, while it is at 122.23 meV in the case of confined phonon. The difference between these two values is insignificant and can be neglected.

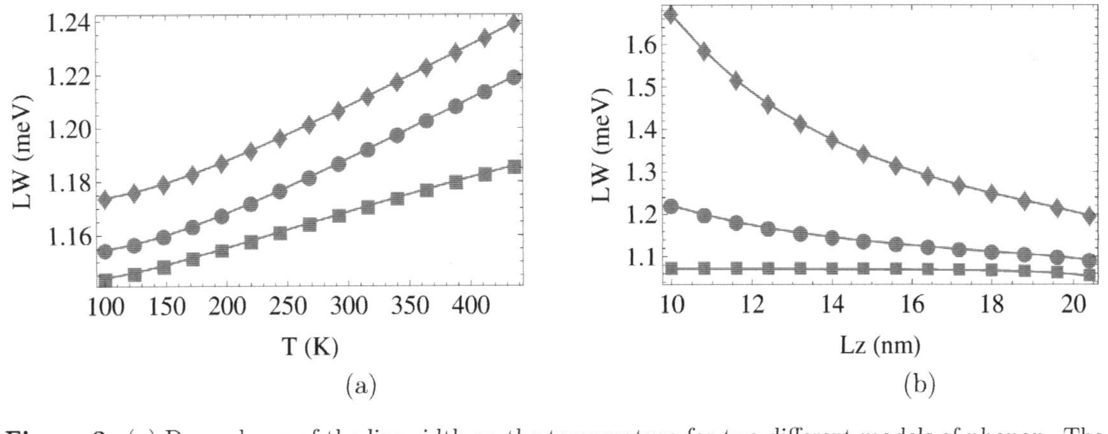


**Figure 2:** (a) Dependence of the linewidth on the temperature for two different models of phonon. The closed circles and the closed squares curves correspond to the case of confined phonons and bulk phonons, respectively. Here, $L_z$=14nm. (b) Dependence of the linewidth on the width of QW for two different models of phonon. Here, T=200 K. In the two cases (a) and (b), the closed diamond and the closed circles curves correspond to the case of confined phonons with $m$ = 4 and $m$ = 2, the closed squares curves bulk phonons, respectively.

The dependence of the linewidth on temperature in both models of phonon is showed in figure 2a. From the figure we can see that the linewidth increases with the temperature. This is reasonable because the possibility of the electron-phonon scattering rises as the temperature increases. It is also seen that linewidths of the ODEPR peak in the cases of confined phonons are larger than the bulk one. This is because when phonons are confined the probability electron-phonon scattering is increased.

The dependence of the linewidth on the well's width in both models of phonon is showed in figure 2b. It can be seen from the figure that the linewidths decrease with the well's width. This can be explained physically by the fact that the possibility of electron-phonon scattering decreases when the well's width increases. Furthermore, the LW in the confined phonon cases vary faster and have greater values than it does in the bulk phonon case. In addition, the LW for the confined phonon with $m = 4$ varies faster and has a greater value than it does for the confined phonon with $m = 2$ and the smaller the well's width is, the more pronounced the difference is. Thus, as the well's width decreases, the phonon confinement becomes more important and cannot be neglected. For well with width so large, the influence of phonon confinement on the LW is very small and can be ignored.

## 5. Conclusions

In the present paper, we have obtained analytical expressions for the absorption power in quantum wells due to electrons-confined optical phonons interaction. Carrying out numerical computations and graphically plotting the dependence of the absorption power on the photon energy, we can find the resonant peaks. Besides, using the profile method, we have determined the linewidth based on graphs of the absorption power versus the photon energy with the help of the computational program. From the numerical results, we can write out some important remarks as follows: The effect of the phonon confinement on the peak location is very weak, so for theoretical studies on determining the peak locations that do not require a high precision, we can ignore the phonon confinement effect. In both bulk and confined phonon models, the linewidth increases with the increase of temperature and decreases with the increase of well's width. However, the linewidth in the case of confined phonon model varies faster and has a greater value than it does in the case of bulk phonon model. The linewidth in the case of strong confined phonon (with $m = 4$) varies faster and has a greater value than it does in the case of weak confined phonon (with $m = 2$). The effect of the phonon confinement on the linewidth is very weak and can ignore when the well's width is larger 20 nm, but the phonon confinement cannot be neglected, especially at small well's width.

## References

[1] Se Gi Yu, Pevzner V B, Kim K W, Stroscio M A 1998 *Phys. Rev.* B **58** 3580
[2] Sang Chil Lee et al 2005 *Physica* E **28** 402
[3] Bennett C R, Guven K, Tanatar B 1998 *Phys. Rev.* B **57** 3994
[4] Nishiguchi N 1995 *Phys. Rev.* B **52** 5279
[5] Svizhenko A, Balandin A, Bandyopadhyay S, Stroscio M A 1998 *Phys. Rev.* B **57** 4687
[6] Zheng, Matsura M 2000 *Phys. Rev.* B **61** 12624
[7] Ridley B K 1989 *Phys. Rev.* B **39** 5282
[8] Cho Y J, Choi S D 1994 *Phys. Rev.* B **49** 14301
[9] Spector H N, Lee J, Melman P 1986 *Phys. Rev.* B **34** 2554
[10] Weman H, Sirigu L, Karlsson K F, Leifer K, Rudra A, Kapon E 2002 *Appl. Phys. Lett.* **81** 2839
[11] Matthiesen C, Vamivakas A N, Atatre M 2012 *Phys. Rev. Lett.* **108** 093602
[12] Phong T C, Phuc H V 2011 *Mod. Phys. Lett.* B **25** 1003
[13] Phong T C, Phuong L T T, Phuc H V 2012 *Superlattices Microstruct.* **52** 16
[14] Phuong L T T, Phuc H V, Phong T C 2014 *Physica* E **56** 102
[15] Phong T C, Phuong L T T, Phuc H V, Vinh P T 2013 *J. Korean Phys. Soc.* **62** 305
[16] Bhat J S, Kapatkar S B, Kubakaddi S S, Mulimani B G *Phys. StatusSolidi* B **209** 37
[17] Bhat J S, Mulimani B G, Kubakaddi S S 1994 *Phys. Rev.* B **49** 16459
[18] Tsen K T, Wald K R, Ruf T, Yu P Y, Morkoc H 1991 *Phys. Rev. Lett.* **67** 2557
[19] Rudin S, Reinecke T L 1990 *Phys. Rev.* B **41** 7713
[20] Thatham M C, Ryan J F, Foxon C T 1989 *Phys. Rev. Lett.* **63** 1637
[21] Masale M, Constantious N C 1993 *Phys. Rev.* B **48** 11128

# Lattice distortion inducing electron-hole pair condensation in two dimensional systems

**Do Thi Hong Hai[1], Nguyen Thi Giang[2], Nguyen Huu Nha[3] and Phan Van Nham[4]**

[1] Department of Physics, Hanoi University of Mining and Geology, Duc Thang, Bac Tu Liem, Hanoi, Vietnam
[2] Institute of Physics, Vietnam Academy of Science and Technology, 10 Dao Tan, Hanoi, Vietnam
[3] Department of Physics, HCMC University of Science, 227 Nguyen Van Cu, Ho Chi Minh city, Vietnam
[4] Institute of Research and Development, Duy Tan University, K7/25 Quang Trung, Danang, Vietnam

E-mail: dothihonghai@humg.edu.vn

**Abstract.** A stability of the electron-hole pair condensation at low temperature has been addressed by involving the coupling of exciton to vibrational degrees of freedom in two-dimensional two-band $f-c$ electron system. By mean of the unrestricted Hartree-Fock approximation, we find a formation of the insulating state typifying an excitonic condensate accompanied by a finite lattice distortion if the exciton-phonon coupling is large enough. As functions of temperature both excitonic condensation order parameter and lattice distortion behave in a same way which manifests the continuous transition in analogy to the superconductivity in the BCS theory. Inspecting to the microscopic properties in momentum space we strongly specify the BCS type of the excitonic condensation driven by the exciton-phonon interaction at low temperature.

## 1. Introduction

Exciton is an electrically neutral quasiparticles formed by coupling of electron and hole. It exists in insulators, semiconductors and in some liquids. The first theory of exciton was proposed by Yakov Frenkel in 1931 [1] when he was describing excitation of the atoms in the crystal lattice of insulators. Normally, quasiparticle excitons do not form the ground state because electrons and holes can reunite easily and radiate photons, or excitons can decay due to lattice defects. However, at the semimetal-semiconductor transition, the ground state of the crystals may become unstable with the spontaneous formation of excitons, provided the very small overlap or band gap between the valence and conduction bands. At low temperature, these composite quasiparticle bosons might condense and a macroscopic phase-coherent quantum state can be created. The semimetallic or semiconducting conguration in this situation would transform to an insulating state. This new state is called an insulator excitonic - EI phase, which had been theoretically proposed since the 1960s [2, 3]. The EI phase received below the critical temperature $T_{EI}$ may be similar to BCS condensate in the semimetal region or Bose Einstein condensate in the semiconductor region [4].

Journal of Physics: Conference Series **726** (2016) 012013 doi:10.1088/1742-6596/726/1/012013

The EI phase is very rare in nature. However, with the increasing of experimental data on the materials, that are candidates for discernible EI phase which strongly supports the theoretical predictions. For example, in the pressure-sensitive mixed-valence rare-earth chalgogenide $TmSe_{0.45}Te_{0.55}$, electrical and thermal transport measurements indicates exciton condensation at temperature below 20K in the pressure range 5-11 kbar [5]. One other EI state in quasi one-dimensional (1D) $Ta_2NiSe_5$ are found recently by X-ray photoemission spectroscopy data [6]. Besides, the quasi-2D transition-metal dichalgogenide $1T - TiSe_2$, is the semimetal material, the temperature dependence of electron spectroscopy confirmed the existence of EI phase [7].

In this paper, we study the phase transition in two-dimensional two-band $f - c$ electron system. Our main purpose is considering the electron-hole pairs condensed prosess due to temperature and intensity electron-phonon interaction by the mean-field theory but only being interested in the effects of electron-phonon interaction and disregarding the Coulomb interaction. This model has been examined meticulously in one-dimensional by the projector-based renormalization method at the ground state temperature $T = 0$ and only noticing the effects of electron-phonon interaction, too [8]. Or investigating the influence of the Coulomb interaction strength, excitation energy of $f$ electrons,... while ignoring electron-phonon interaction, in the expanded Falicov-Kimball model in both one-dimensional [9] and two-dimensional [10]. Although ignoring the contribution of fluctuations, but the mean-field theory has helped to see the correlation picture of electron system adequately.

This paper is organized as follows. Section 2 presents the model. In Section 3 we use the mean-field theory to calculate directly the energies and the average particle densities. Section 4 presents our numerical results: temperature dependent of both order parameter and lattice distortion in the EI phase. The conclusion is given in Section 5.

## 2. Model

We start to consider the Hamiltonian of the system contains two types of spinless electrons $(c, f)$ carrying momentum $\mathbf{k}$ and dispersionless phonons $(b)$

$$\mathcal{H} = \sum_{\mathbf{k}} \varepsilon^f_{\mathbf{k}} f^\dagger_{\mathbf{k}} f_{\mathbf{k}} + \sum_{\mathbf{k}} \varepsilon^c_{\mathbf{k}} c^\dagger_{\mathbf{k}} c_{\mathbf{k}} + \omega_0 \sum_{\mathbf{q}} b^\dagger_{\mathbf{q}} b_{\mathbf{q}} + \frac{g}{\sqrt{N}} \sum_{\mathbf{kq}} \left[ c^\dagger_{\mathbf{k+q}} f_{\mathbf{k}} \left( b^\dagger_{-\mathbf{q}} + b_{\mathbf{q}} \right) + H.c. \right], \quad (1)$$

where $\omega_0$ is the dispersionless phonon energy and the electronic excitation energies are given by

$$\varepsilon^{f,c}_{\mathbf{k}} = \varepsilon^{f,c} - t^{f,c}\gamma_{\mathbf{k}} - \mu, \quad (2)$$

with $\varepsilon^{f,c}$ are the on-site energies, $t^{f,c}$ are the nearest-neighbor particle transfer amplitudes, $\gamma_{\mathbf{k}} = 2\left(cosk_x + cosk_y\right)$ indicates a nearest-neighbor hopping in a two-dimensional lattice and $\mu$ denotes the chemical potential. The last term in Eq.(1) expresses a local electron-phonon interaction with coupling constant $g$, written in $\mathbf{k}$ space, between local $f - c$ particle-hole excitations and lattice displacements. It can be regarded as an effective "exciton"-phonon interaction when considering to annihilate an f electron correspondingly create a hole in the valence band.

In Eq.(1) we have brought in Fourier-transformed quantities $f^\dagger_{\mathbf{k}} = \left(\frac{1}{\sqrt{N}}\right)\sum_i f^\dagger_i e^{i\mathbf{kR_i}}, c^\dagger_{\mathbf{k}} = \left(\frac{1}{\sqrt{N}}\right)\sum_i c^\dagger_i e^{i\mathbf{kR_i}}$ and $b^\dagger_{\mathbf{q}} = \left(\frac{1}{\sqrt{N}}\right)\sum_i b^\dagger_i e^{i\mathbf{qR_i}}$, where $f^\dagger_i, c^\dagger_i$, and $b^\dagger_i$ are the local quantities. N counts the number of lattice sites $i$.

In what folows, we study the half - filled band case, with total electron density

$$\langle n \rangle = \left\langle n^f_i \right\rangle + \langle n^c_i \rangle = 1, \quad (3)$$

Journal of Physics: Conference Series **726** (2016) 012013 doi:10.1088/1742-6596/726/1/012013

with $n_i^f = \left(\frac{1}{N}\right)\sum_{\mathbf{k}} f_{\mathbf{k}}^\dagger f_{\mathbf{k}}$ and $n_i^c = \left(\frac{1}{N}\right)\sum_{\mathbf{k}} c_{\mathbf{k}}^\dagger c_{\mathbf{k}}$. The chemical potential $\mu$ has to satisfy Eq.(3). In this paper, we choose $t^c = 1$ as the unit of energies and calculating with $\varepsilon^c = 0, \varepsilon^f = -1, t^f = -0.3$. Therefore, the $c$ electrons will be considered as "light" while the $f$ electrons (respectively holes) are "heavy", i.e., $|t_f| < 1$. For $t_c t_f < 0$, leading to a picture of direct $c - f$ hopping, which proposes a possible condensation of bound $c - f$-electron-hole pairs with finite momentum

$$d_{\mathbf{k}} = \left\langle c_{\mathbf{k}+\mathbf{Q}}^\dagger f_{\mathbf{k}} \right\rangle \neq 0, \tag{4}$$

where $\mathbf{Q} = (\pi, \pi)$ in two-dimensional. Nonvanishing $d_{\mathbf{k}}$ leads to the spontaneous symmetry breaking, so small infinitesimal fields must be taken into Eq.(1). We write

$$\begin{aligned}\mathcal{H} = & \sum_{\mathbf{k}} \varepsilon_{\mathbf{k}}^f f_{\mathbf{k}}^\dagger f_{\mathbf{k}} + \sum_{\mathbf{k}} \varepsilon_{\mathbf{k}}^c c_{\mathbf{k}}^\dagger c_{\mathbf{k}} + \omega_0 \sum_{\mathbf{q}} b_{\mathbf{q}}^\dagger b_{\mathbf{q}} \\ & + \Delta_0 \sum_{\mathbf{k}} \left( c_{\mathbf{k}+\mathbf{Q}}^\dagger f_{\mathbf{k}} + f_{\mathbf{k}}^\dagger c_{\mathbf{k}+\mathbf{Q}} \right) + \sqrt{N} h_0 \left( b_{-\mathbf{Q}}^\dagger + b_{\mathbf{Q}} \right) \\ & + \frac{g}{\sqrt{N}} \sum_{\mathbf{kq}} \left[ c_{\mathbf{k}+\mathbf{q}}^\dagger f_{\mathbf{k}} \left( b_{-\mathbf{q}}^\dagger + b_{\mathbf{q}} \right) + H.c. \right],\end{aligned} \tag{5}$$

where $\Delta_0 = 0^+$ and $h_0 = 0^+$ while the fields $h_0$ and $\Delta_0$ are mutually dependent. Furthermore, from $b_{\mathbf{Q}} = b_{-\mathbf{Q}}$, we can substitute $\sqrt{N} h_0 \left( b_{-\mathbf{Q}}^\dagger + b_{-\mathbf{Q}} \right)$ for the field contribution $\sqrt{N} h_0 \left( b_{-\mathbf{Q}}^\dagger + b_{\mathbf{Q}} \right)$.

## 3. Mean-field theory

In the mean-field limit, we can perform approximately interaction operators in Hamiltonian (5) by introducing fluctuation operators $\delta\mathcal{A} = \mathcal{A} - \langle \mathcal{A} \rangle$ in the electron (exciton)-phonon interaction. We get

$$\begin{aligned}\delta \left( c_{\mathbf{k}+\mathbf{q}}^\dagger f_{\mathbf{k}} \right) \delta \left( b_{-\mathbf{q}}^\dagger + b_{\mathbf{q}} \right) = & c_{\mathbf{k}+\mathbf{q}}^\dagger f_{\mathbf{k}} \left( b_{-\mathbf{q}}^\dagger + b_{\mathbf{q}} \right) - [\left\langle c_{\mathbf{k}+\mathbf{q}}^\dagger f_{\mathbf{k}} \right\rangle \left( b_{-\mathbf{q}}^\dagger + b_{\mathbf{q}} \right) \\ & + c_{\mathbf{k}+\mathbf{q}}^\dagger f_{\mathbf{k}} \left\langle (b_{-\mathbf{q}}^\dagger + b_{\mathbf{q}} \right\rangle \delta_{\mathbf{q},\mathbf{Q}}] + \left\langle c_{\mathbf{k}+\mathbf{q}}^\dagger f_{\mathbf{k}} \right\rangle \left\langle (b_{-\mathbf{q}}^\dagger + b_{\mathbf{q}} \right\rangle \delta_{\mathbf{q},\mathbf{Q}},\end{aligned} \tag{6}$$

Then, the Hamiltonian $\mathcal{H}$ is rewritten

$$\mathcal{H} = \mathcal{H}_0 + \mathcal{H}_1 \tag{7}$$

with

$$\begin{aligned}\mathcal{H}_0 = & \sum_{\mathbf{k}} \varepsilon_{\mathbf{k}}^f f_{\mathbf{k}}^\dagger f_{\mathbf{k}} + \sum_{\mathbf{k}} \varepsilon_{\mathbf{k}}^c c_{\mathbf{k}}^\dagger c_{\mathbf{k}} + \omega_0 \sum_{\mathbf{q}} b_{\mathbf{q}}^\dagger b_{\mathbf{q}} \\ & + \Delta \sum_{\mathbf{k}} \left( c_{\mathbf{k}+\mathbf{Q}}^\dagger f_{\mathbf{k}} + f_{\mathbf{k}}^\dagger c_{\mathbf{k}+\mathbf{Q}} \right) + \sqrt{N} h \left( b_{-\mathbf{Q}}^\dagger + b_{-\mathbf{Q}} \right),\end{aligned} \tag{8}$$

$$\mathcal{H}_1 = \frac{g}{\sqrt{N}} \sum_{\mathbf{kq}} \left[ \delta \left( c_{\mathbf{k}+\mathbf{q}}^\dagger f_{\mathbf{k}} \right) \delta \left( b_{-\mathbf{q}}^\dagger + b_{\mathbf{q}} \right) + H.c. \right]. \tag{9}$$

Here the fields have gotten additional shifts, which will play as order parameters in the following

$$\Delta = \Delta_0 + \frac{g}{\sqrt{N}} \left\langle b_{-\mathbf{Q}}^\dagger + b_{-\mathbf{Q}} \right\rangle, \tag{10}$$

$$h = h_0 + \frac{g}{N} \sum_{\mathbf{k}} \left\langle c_{\mathbf{k}+\mathbf{Q}}^\dagger f_{\mathbf{k}} + f_{\mathbf{k}}^\dagger c_{\mathbf{k}+\mathbf{Q}} \right\rangle, \tag{11}$$

Note that the infinitesimal $\Delta_0 = 0^+$ and $h_0 = 0^+$ can be neglected.

Finally, we exclude the term $\propto \left(b^\dagger_{-\mathbf{Q}} + b_{-\mathbf{Q}}\right)$ in Eq.(8) by defining new phonon operators

$$B^\dagger_{\mathbf{q}} = b^\dagger_{\mathbf{q}} + \sqrt{N}\,(h/\omega_0)\,\delta_{\mathbf{q},\mathbf{Q}}. \tag{12}$$

Using $\delta B^\dagger_{-\mathbf{q}} = \delta b^\dagger_{-\mathbf{q}}$ and $\delta B_{\mathbf{q}} = \delta b_{\mathbf{q}}$ we get

$$\mathcal{H}_0 = \sum_{\mathbf{k}} \varepsilon^f_{\mathbf{k}} f^\dagger_{\mathbf{k}} f_{\mathbf{k}} + \sum_{\mathbf{k}} \varepsilon^c_{\mathbf{k}} c^\dagger_{\mathbf{k}} c_{\mathbf{k}} + \omega_0 \sum_{\mathbf{q}} B^\dagger_{\mathbf{q}} B_{\mathbf{q}}$$
$$+\Delta \sum_{\mathbf{k}} \left(c^\dagger_{\mathbf{k}+\mathbf{Q}} f_{\mathbf{k}} + f^\dagger_{\mathbf{k}} c_{\mathbf{k}+\mathbf{Q}}\right) + const, \tag{13}$$

$$\mathcal{H}_1 = \frac{g}{\sqrt{N}} \sum_{\mathbf{kq}} \left[\delta\left(c^\dagger_{\mathbf{k}+\mathbf{q}} f_{\mathbf{k}}\right) \delta\left(B^\dagger_{-\mathbf{q}} + B_{\mathbf{q}}\right) + H.c.\right]. \tag{14}$$

Neglecting the fluctuation part $\mathcal{H}_1$ and annuling constant in the Eq.(13), the Hamiltonian in mean-field approximation is given

$$\mathcal{H}_{MF} = \sum_{\mathbf{k}} \varepsilon^f_{\mathbf{k}} f^\dagger_{\mathbf{k}} f_{\mathbf{k}} + \sum_{\mathbf{k}} \varepsilon^c_{\mathbf{k}} c^\dagger_{\mathbf{k}} c_{\mathbf{k}} + \omega_0 \sum_{\mathbf{q}} B^\dagger_{\mathbf{q}} B_{\mathbf{q}}$$
$$+\Delta \sum_{\mathbf{k}} \left(c^\dagger_{\mathbf{k}+\mathbf{Q}} f_{\mathbf{k}} + f^\dagger_{\mathbf{k}} c_{\mathbf{k}+\mathbf{Q}}\right). \tag{15}$$

We have defined quasiparticle operators

$$C^\dagger_{1,\mathbf{k}} = \xi_{\mathbf{k}} c^\dagger_{\mathbf{k}+\mathbf{Q}} + \eta_{\mathbf{k}} f^\dagger_{\mathbf{k}}, \tag{16}$$
$$C^\dagger_{2,\mathbf{k}} = -\eta_{\mathbf{k}} c^\dagger_{\mathbf{k}+\mathbf{Q}} + \xi_{\mathbf{k}} f^\dagger_{\mathbf{k}}, \tag{17}$$

with

$$\xi^2_{\mathbf{k}} = \frac{1}{2}\left[1 + sgn\left(\varepsilon^f_{\mathbf{k}} - \varepsilon^c_{\mathbf{k}+\mathbf{Q}}\right)\frac{\varepsilon^f_{\mathbf{k}} - \varepsilon^c_{\mathbf{k}+\mathbf{Q}}}{W_{\mathbf{k}}}\right], \tag{18}$$

$$\eta^2_{\mathbf{k}} = \frac{1}{2}\left[1 - sgn\left(\varepsilon^f_{\mathbf{k}} - \varepsilon^c_{\mathbf{k}+\mathbf{Q}}\right)\frac{\varepsilon^f_{\mathbf{k}} - \varepsilon^c_{\mathbf{k}+\mathbf{Q}}}{W_{\mathbf{k}}}\right], \tag{19}$$

and

$$W_{\mathbf{k}} = \sqrt{\left(\varepsilon^c_{\mathbf{k}+\mathbf{Q}} - \varepsilon^f_{\mathbf{k}}\right)^2 + 4|\Delta|^2}. \tag{20}$$

Using the Bogoliubov transformation, the electronic part of $\mathcal{H}_{MF}$ is diagonalized and then the Hamiltonian $\mathcal{H}_{MF}$ reads

$$\mathcal{H}_{MF} = \sum_{\mathbf{k}} E^c_{\mathbf{k}} C^\dagger_{1,\mathbf{k}} C_{1,\mathbf{k}} + E^f_{\mathbf{k}} C^\dagger_{2,\mathbf{k}} C_{2,\mathbf{k}} + \omega_0 \sum_{\mathbf{q}} B^\dagger_{\mathbf{q}} B_{\mathbf{q}}, \tag{21}$$

here the electronic quasiparticle energies are given

$$E^c_{\mathbf{k}} = \frac{\varepsilon^f_{\mathbf{k}} + \varepsilon^c_{\mathbf{k}+\mathbf{Q}}}{2} - \frac{sgn\left(\varepsilon^f_{\mathbf{k}} - \varepsilon^c_{\mathbf{k}+\mathbf{Q}}\right)}{2} W_{\mathbf{k}}, \tag{22}$$

$$E^f_{\mathbf{k}} = \frac{\varepsilon^f_{\mathbf{k}} + \varepsilon^c_{\mathbf{k}+\mathbf{Q}}}{2} + \frac{sgn\left(\varepsilon^f_{\mathbf{k}} - \varepsilon^c_{\mathbf{k}+\mathbf{Q}}\right)}{2} W_{\mathbf{k}}. \tag{23}$$

The quadratic form of Eq.(21) allows us to calculate all expectation values. From Eqs.(10)-(12) we easily take the following implicit equation for the order parameters $\Delta = -\frac{2g}{\omega_0}h$

$$1 = \frac{4g^2}{\omega_0}\frac{1}{N}\sum_{\mathbf{k}} sgn\left(\varepsilon^f_{\mathbf{k}} - \varepsilon^c_{\mathbf{k}+\mathbf{Q}}\right)\frac{f^F\left(E^c_{\mathbf{k}}\right) - f^F\left(E^f_{\mathbf{k}}\right)}{W_{\mathbf{k}}}, \tag{24}$$

where $f^F\left(E^{c,f}_{\mathbf{k}}\right)$ are the Fermi-Dirac functions which satisfy the equation

$$f^F(\epsilon) = \frac{1}{e^{\beta\epsilon} - 1} \tag{25}$$

with $\beta = 1/T$ is the inverse of temperature.
Note that Eq. (24) stands for a BCS-like equation for $\Delta$. A nonzero $\Delta$ explains a exciton condensation phase as was explained above.

Now we consider forming of the EI state, here $d_{\mathbf{k}} = \left\langle c^\dagger_{\mathbf{k}+\mathbf{Q}} f_{\mathbf{k}}\right\rangle \neq 0$ [Eq.(4)] plays as EI order parameter. From Eqs.(16)-(20) we get the mean particle number densities of $c$ and $f$ electrons for a system with N lattice sites

$$n^c_{\mathbf{k}+\mathbf{Q}} = \left\langle c^\dagger_{\mathbf{k}+\mathbf{Q}} c_{\mathbf{k}+\mathbf{Q}}\right\rangle = \xi^2_{\mathbf{k}} f^F\left(E^c_{\mathbf{k}}\right) + \eta^2_{\mathbf{k}} f^F\left(E^f_{\mathbf{k}}\right), \tag{26}$$

$$n^f_{\mathbf{k}} = \left\langle f^\dagger_{\mathbf{k}} f_{\mathbf{k}}\right\rangle = \eta^2_{\mathbf{k}} f^F\left(E^c_{\mathbf{k}}\right) + \xi^2_{\mathbf{k}} f^F\left(E^f_{\mathbf{k}}\right), \tag{27}$$

$$n^{cf}_{\mathbf{k}} = \left\langle c^\dagger_{\mathbf{k}+\mathbf{Q}} f_{\mathbf{k}}\right\rangle = -\left[f^F\left(E^c_{\mathbf{k}}\right) - f^F\left(E^f_{\mathbf{k}}\right)\right] sgn\left(\varepsilon^f_{\mathbf{k}} - \varepsilon^c_{\mathbf{k}+\mathbf{Q}}\right)\frac{\Delta}{W_{\mathbf{k}}}. \tag{28}$$

Finally, using Eq.(12), the lattice displacement in the EI state at momentum $\mathbf{Q}$ is given by

$$x_{\mathbf{Q}} = \frac{1}{\sqrt{N}}\frac{1}{\sqrt{2\omega_0}}\left\langle b^\dagger_{-\mathbf{Q}} + b_{\mathbf{Q}}\right\rangle = -\frac{h}{\omega_0}\sqrt{\frac{2}{\omega_0}}. \tag{29}$$

## 4. Numerical results and discussion

This section presents the numerical results which are obtained by solving self-consistently equations above to consider the picture of electron-hole pairs condensation due to exciton-phonon interaction. The calculation results are implemented in the unit of energy for $\hbar = k_B = 1$.

### *4.1. The electronic quasiparticle energies*

The first, let us describe momentum dependent of the quasiparticle energies of $c$ and $f$ electrons.

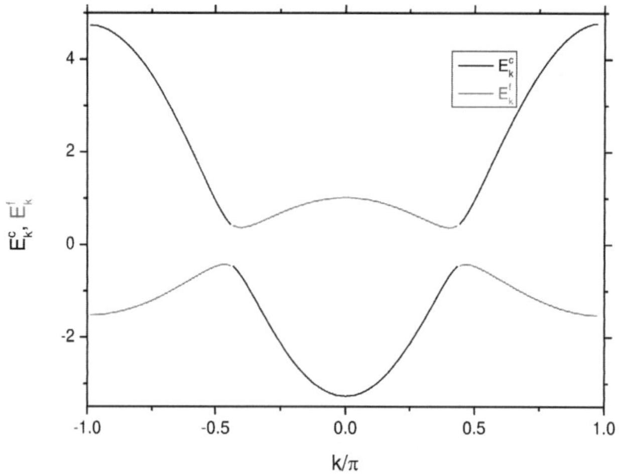


**Figure 1.** Quasiparticle dispersions $E^c_{\mathbf{k}}$ (black lines) and $E^f_{\mathbf{k}}$ (red lines) along the major axes of the **k**.

Figure 1 shows the quasiparticle dispersions $E^c_{\mathbf{k}}$ of $c$ electrons (black lines) and $E^f_{\mathbf{k}}$ of $f$ electrons (red lines) along the major axes of the momentum **k** at the temperature $T = 0$ with the coupling constant of electron-phonon interaction $g = 1.1$ and the phonon frequency $\omega_0 = 2.5$. The excitonic condensate state is formed indicating by opening a gap at the Fermi level. This is similar to the condensation of electron-electron pairs with their opposite spin in the BCS theory. These condensate states are broken if there exists a sufficient external energy which decouples the bound states, then the gap vanishes.

The following are detailed examination of the picture of phase transition with variation of the temperature and the coupling constant of electron-phonon interaction.

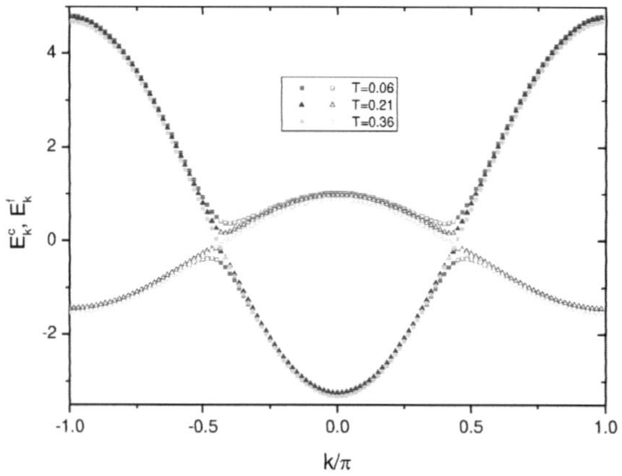


**Figure 2.** Quasiparticle dispersions $E^c_{\mathbf{k}}$ (filled symbols) and $E^f_{\mathbf{k}}$(open symbols) along the major axes of the **k** at various temperatures.

Figure 2 presents the renormalized quasiparticle bands $E^c_{\mathbf{k}}$ (filled symbols) and $E^f_{\mathbf{k}}$ (open symbols) along the major axes of the momentum **k** at various temperatures with the coupling constant of electron-phonon interaction $g = 1.1$. At high temperature, such as $T = 0.36$ the two quasiparticle-energy bands overlap, thus no gap exists. Decreasing the temperature, an energy

IOP Publishing
doi:10.1088/1742-6596/726/1/012013

gap opens at the Fermi level indicating the formation of an excitonic insulating state. This gap expands when the temperature decreases.

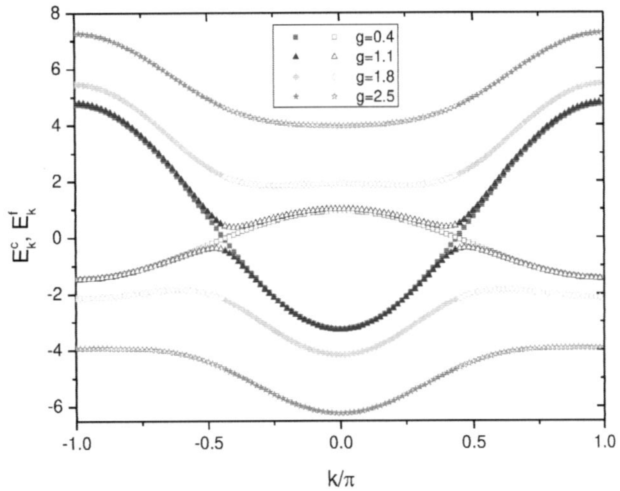


**Figure 3.** Quasiparticle dispersions $E^c_{\mathbf{k}}$ (filled symbols) and $E^f_{\mathbf{k}}$ (open symbols) along the major axes of the **k** at various coupling constants of electron-phonon interaction.

Figure 3 describes the renormalized quasiparticle energies $E^c_{\mathbf{k}}$ (filled symbols) and $E^f_{\mathbf{k}}$ (open symbols) along the major axes of the momentum **k** at various coupling constants of electron-phonon interaction $g$ for $T = 0$. We see that in a case of small electron-phonon interaction such as $g = 0.4$, the two quasiparticle-energy bands of $c$ and $f$ electron overlap, the gap does not exist. Increasing the coupling constant, an energy gap opens at the Fermi surface. This shows that the EI state is formed. This gap expands when the coupling constant increases.

The above results show that the temperature and also the coupling constant of electron-phonon interaction affect to the formation of the excitonic condensate. Clearly, the EI phase is only formed at low enough temperature and large enough coupling constant.

*4.2. The mean particle number densities*

In order to consider more meticulously the picture of the excitonic condensation, firstly we survey temperature and coupling constant of electron-phonon interaction dependent of the mean particle number densities.

Journal of Physics: Conference Series **726** (2016) 012013
doi:10.1088/1742-6596/726/1/012013

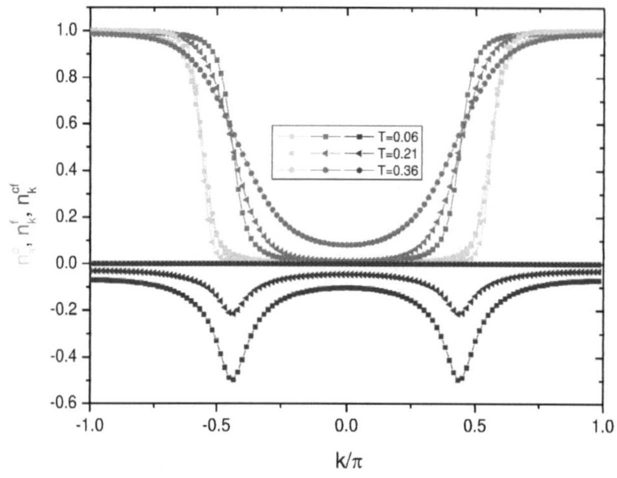


**Figure 4.** The mean particle number densities of $c$ and $f$ electrons along the major axes of the $\mathbf{k}$ at various temperatures.

Figure 4 displays the mean particle number densities of $c$ and $f$ electrons as a function of the momentum $\mathbf{k}$ when the temperature was changed at coupling constant of electron-phonon interaction $g = 1.1$. For low temperatures, when existing a small energy gap at the Fermi level, $n^{cf}_{\mathbf{k}}$ is strongly peaked at $\mathbf{k}$ which closes to the Fermi surface. Increasing the temperature above some critical temperature $T_{EI}$, the mean particle number density $n^{cf}_{\mathbf{k}}$ is more weakly peaked. As a matter of course, at higher temperature, when no gap exists, $n^{cf}_{\mathbf{k}}$ also vanishes, resulting in a state of Fermi liquid. While $n^{c}_{\mathbf{k}}$ and $n^{f}_{\mathbf{k}}$ monotonously increases (decreases) as $\mathbf{k}$ varies from 0 to $\pi$ (or from $-\pi$ to 0).

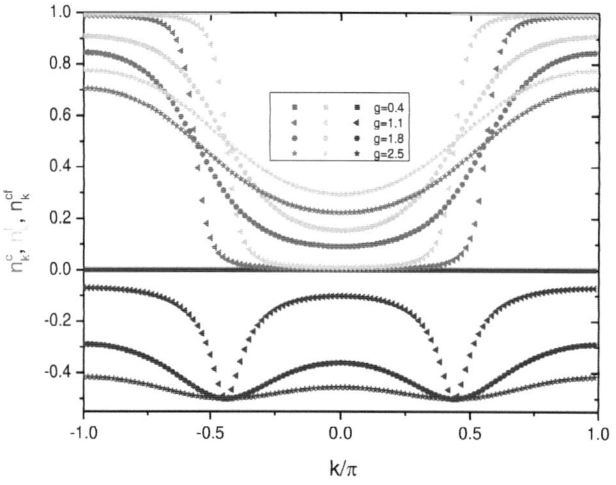


**Figure 5.** The mean particle number densities of $c$ and $f$ electrons along the major axes of the $\mathbf{k}$ at various coupling constant of electron-phonon interaction.

Figure 5 shows the mean particle number densities of $c$ and $f$ electrons as a function of the momentum $\mathbf{k}$ at various coupling constant of electron-phonon interaction, for temperature

$T = 0$. We see that for high coupling constant, a large gap opens at the Fermi surface, $n_{\mathbf{k}}^{cf}$ is peaked. Decreasing the coupling constant close to critical coupling constant $g_{EI}$, $n_{\mathbf{k}}^{cf}$ is strongly peaked. However, decreasing the coupling constant below some $g_{EI}$, when both quasiparticle bands $c$ and $f$ electrons overlap leads to the gap vanishes, the mean particle number density of $c$ and $f$ electrons also vanishes. While $n_{\mathbf{k}}^{c}$ and $n_{\mathbf{k}}^{f}$ monotonously increases (decreases) as $\mathbf{k}$ varies from 0 to $\pi$ (or from $-\pi$ to 0).

These results showed that the mean particle number density of $c$ and $f$ electrons only is nonzero when system is existing in EI phase, i.e. at low enough temperature and large enough coupling constant.

### *4.3. The EI order parameter*

This section examines in detail the influence of the electron-phonon interaction intensity and the temperature on the formation of exciton condensate by performing dependence the EI order parameter as a function of the coupling constant of electron-phonon interaction and the temperature.

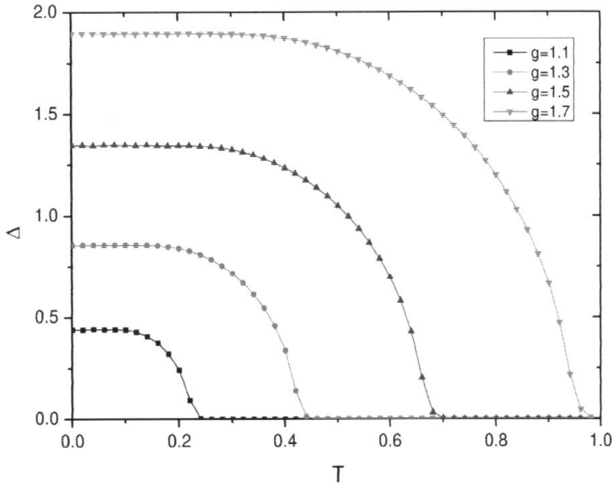


**Figure 6.** The EI order parameter along the major axes of the temperature T at various coupling constant $g$.

Figure 6 describes the EI order parameter along the major axes of the temperature T at various coupling constant $g$ for the phonon frequency $\omega_0 = 2.5$. We see that, when the temperature is smaller than a critical value $T_{EI}$, the order parameter is nonzero. Whereas, above that temperature, the EI order parameter completely equal zero. That temperature is so called an EI transition temperature. Below the critical temperature, the order parameter increases rapidly and then saturates at low temperature. Increasing the coupling constant the order parameter and $T_{EI}$ also increases.

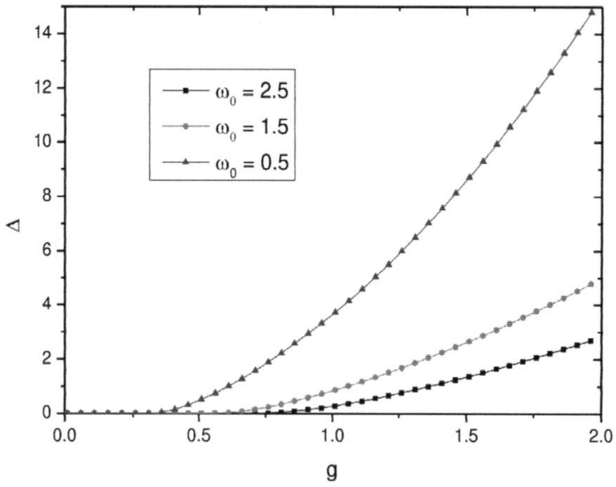


**Figure 7.** The EI order parameter along the major axes of the coupling constant at various phonon frequency $\omega_0$, for $T = 0$.

Figure 7 shows that when the coupling constant reaches a critical value, the EI order parameter $\Delta$ is nonzero, where the EI phase transitions occurs. Critical value of the coupling constant of electron-phonon interaction increases when the phonon frequency increased. The order parameter decreases, this means that quasiparticles energy gap also decreases with increasing of the phonon frequency. Clearly, largering the phonon frequency, phonon excitation is harder, so strong enough interaction intensity needs to form the bound state of the electron-hole pairs, corresponding to exciton condensation.

### *4.4. The lattice displacement in the EI state*

The following considering the relationship between the lattice displacement at the momentum **Q** with the exciton condensate state. Like above, we express the dependence of the EI lattice distortion in the electron-phonon interaction intensity and the temperature.

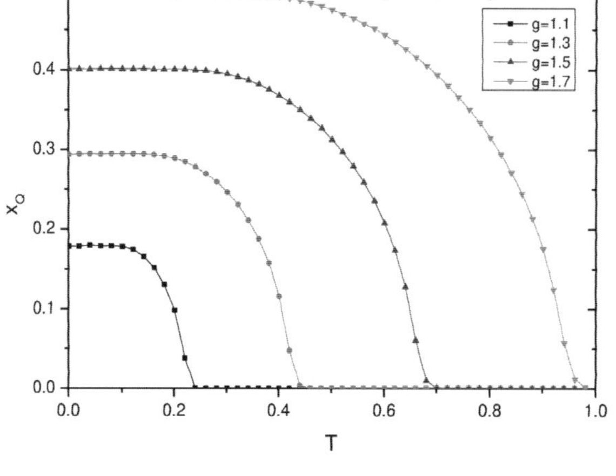


**Figure 8.** The EI lattice distortion along the major axes of the temperature T at various coupling constant $g$.

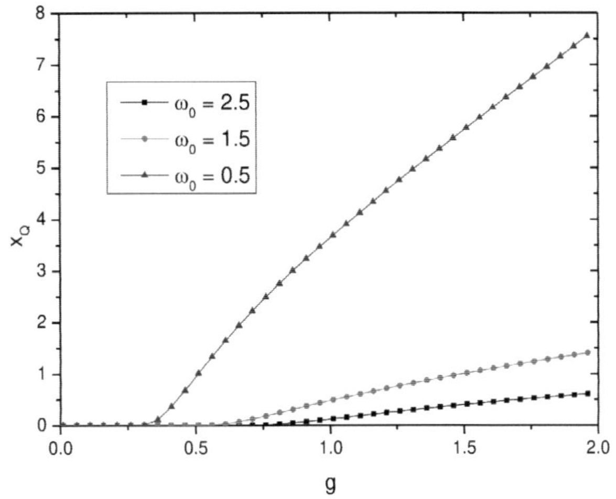


**Figure 9.** The EI lattice distortion along the major axes of the coupling constant at various phonon frequency $\omega_0$, for $T = 0$.

Figure 8 and 9 indicate that at low temperature and large coupling corresponding to forming the EI state, a finite lattice distortion has been found. The lattice distortion enhances nearly linearly if the interaction increases. The critical value of the coupling constant of electron-phonon interaction increases if the phonon is harder.

We can affirm that the condensate state of the electron-hole pairs leads to an finite crystal lattice distortion. This only occurs in the situation that the electron-phonon interaction is large and the temperature is small enough.

*4.5. The dependence on the momentum of the order parameter*

Let us examine more meticulously the nature of the electron-hole pairs condensation state due to the electron-phonon interaction. This section presents momentum dependent of the order parameter at various temperature and the coupling constant of electron-phonon interaction.

Figure 10 displays the dependence on the momentum of the EI order parameter at various coupling constant of electron-phonon interaction $g$, for $T = 0$ and $\omega_0 = 2.5$. With small coupling constant $g = 0.7$, the order parameter $\Delta_k$ also is very small and the system in the uniform state. However, increasing $g$ to $g = 1.0$, $\Delta_k$ increases, particularly at the momentum closes the Fermi momentum. This asserts that the Fermi level plays a role in the formation of the excitonic condensate. This is similar to the superconducting picture which is described by BCS theory. Therefore, in this case, the excitonic condensate is called the BCS-type excitonic condensation state. Increasing strength of interaction with $g = 1.3$, we see that the order parameter also has the maximum value at momentum around the Fermi momentum. Although, there is still significant contribution of electron-hole pairs at the momenta which different from the Fermi momentum, even in the center or the bound Brillouin zone. Nevertheless, we still affirm that the exciton condensate in the BCS-type state.

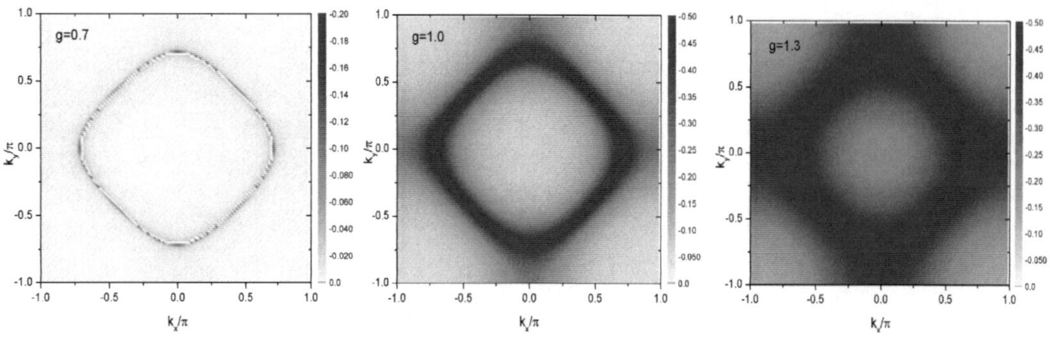


**Figure 10.** Momentum dependent of the order parameter at various coupling constant of electron-phonon interaction, for $T = 0$.

The next, we consider the dependence on temperature of the order parameter for $g = 1.3$ and $\omega_0 = 2.5$ (see Fig. 11). When fixing the coupling constant of electron-phonon interaction, the maximum value of the order parameter which has momentum is around the Fermi momentum sharply reduces if the temperature increased. However, only the electron-hole pairs which close the Fermi level play important role to form bound state. So we still have the picture of BCS-type electron-hole condensed. If the temperature is higher than the transition temperature, the order parameter becomes very small, even at the Fermi momentum. System exists in the non-order state.

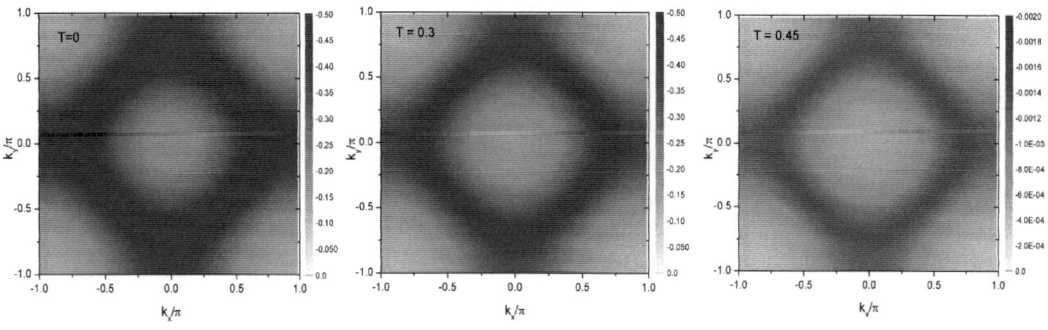


**Figure 11.** Momentum dependent of the order parameter at various temperature, for $g = 1.3$.

Lastly, let us describe magnitude of the EI order parameter depending on momentum and the temperature or the coupling constant of electron-phonon interaction.

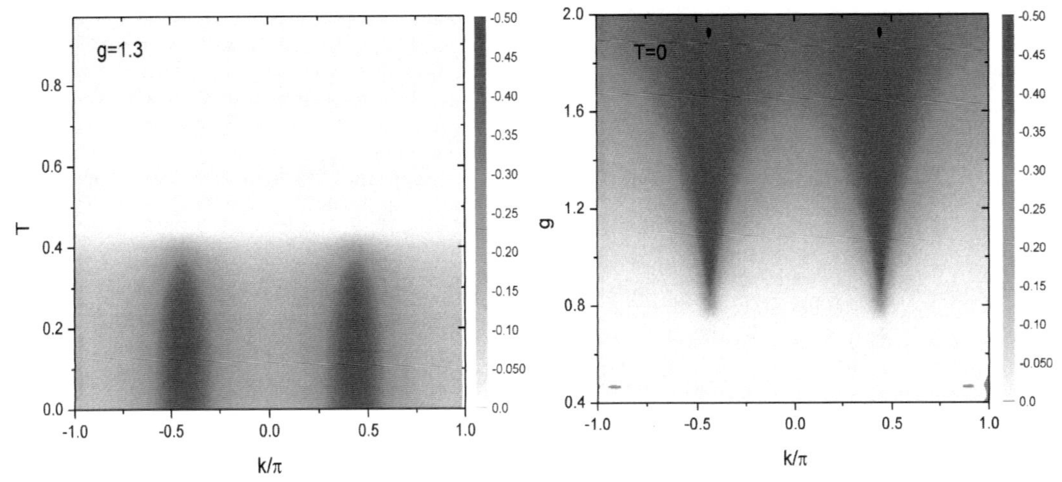


**Figure 12.** Magnitude of the EI order parameter depending on momentum **k** and the temperature $T$ (left-hand panel); momentum **k** and the electron-phonon coupling strength $g$ (right-hand panel)

Figure 12 shows the profile of the EI order parameter $\Delta_k$ dependents on the temperature $T$ for $g = 1.3$ (left-hand panel) and the electron-phonon coupling $g$ for $T = 0$ (right-hand panel) at phonon frequency $\omega_0 = 2.5$. This quantity points out the range in momentum space where $c$ electrons and $f$ holes are perceptibly involved in the electron-hole pair formation and exciton condensation process. Obviously $\Delta_k$ vanishes for all $\mathbf{k}$ above a critical temperature $T \simeq 0.45$ and below a critical coupling strength $g \simeq 0.7$. At and just below (above) the critical temperature (coupling strength) $\Delta_k$ is solely finite at and near the Fermi momentum $\mathbf{k}_F$, respectively, indicating a BCS-type electron-hole pairing instability. However, a further increasing electron-phonon coupling implicates more and more electron and hole states in the pairing process up to the point where Fermi-surface (nesting) effects are inefficient. Thus we expect that local, closely bound excitons will form in the strong interaction limit and, as a result, Bose-Einstein rather than BCS-like condensation occurs.

## 5. Conclusion

This paper presents the results of theoretical research about the effects on the electron-hole pairs condensation state (exciton) by the mean-field theory in two-dimensional two-band $f - c$ electron system that only caring about the effects of electron-phonon interaction. The results confirm the excitonic condensate state is formed only at low temperature and large electron-phonon interaction. Particularly, we have found that in the EI state, the EI order parameter is non-zero and a finite lattice distortion can be detected. And by reviewing the microscopic properties in momentum space we strongly specify the BCS type of the excitonic condensation driven by the exciton-phonon interaction at low temperature.

## References

[1] Frenkel J 1931 On the Transformation of light into Heat in Solids I *Phys. Rev.* **37**.
[2] Mott N F 1961 The transition to the metallic state *Philos. Mag.* **6**.
[3] Kohn W 1968 *Metals and Insulators in Many Body Physics* edited by de Witt C and Balian R (Gordon & Breach, New York).
[4] Bronold F X and Fehske H 2006 *Phys. Rev.* B **74** 165107.
[5] Neuenschwander J and Wachter P 1990 *Phys. Rev.* B **41** 12693; Bucher B, Steiner P and Wachter P 1991

*Phys. Rev. Lett* **67** 2717; Wachter P, Bucher B and Malar J 2004 *Phys. Rev.* B **69** 094502; Bucher B, Park T, Thompson J D, and Wachter P 2008 URL arXiv:0802.3354[cond-mat.str-el].

[6] Wakisaka Y, Sudayama T, Takubo K, Mizokawa T, Arita M, Namatame H, Taniguchi M, Katayama N, Nohara M, and Takagi H 2009 *Phys. Rev. Lett.* **103** 026402.

[7] Monney C *et al* 2009 *Phys. Rev.* B **79** 045116.

[8] Phan V N, Becker K W and Fehske H 2013 Exciton condensation due to electron-phonon interaction *Phys. Rev.* B **88** 205123.

[9] Phan V N, Becker K W and Fehske H 2010 *Phys. Rev.* B **81** 205117.

[10] Phan V N, Fehske H, and Becker K W 2011 Excitonic resonances in the 2D extended Falicov-Kimball model *Europhys. Lett.* **95** 17006.

# Dependence of the Hall Coefficient on a length of rectangular quantum wires with infinitely high potential under the influence of a Laser Radiation

**Nguyen Thu Huong[1], Nguyen Quang Bau[1], Le Thai Hung[2], Dao Manh Hung[3]**

[1]Faculty of Physics, Hanoi University of Science, Vietnam National University, 334 Nguyen Trai, Thanh Xuan, Hanoi, Vietnam
[2]University of Education, Vietnam National University, Hanoi, Vietnam
[3]Academy of Air Defense and Air Force, Son Tay, Hanoi, Vietnam

E-mail: huong146314@yahoo.com

**Abstract.** The Hall coefficient (HC) of a strong electromagnetic wave (EMW) caused by confined electrons in a rectangular quantum wire (RQW) is theoretically studied by using the quantum kinetic equation for electrons. The problem is considered in the case of electrons - acoustic phonons scattering. Wave function and energy spectrum in a RQW are different from those in a cylindrical quantum wire (CQW) or two dimensional systems (2D). Therefore analytical expressions for the HC in a RQW is obtained, different from CQW or 2D. Numerical calculations are carried out with a specific GaAs/GaAsAl RQW to show clearly the dependence of HC on a length Lx (Ly) RQW with different low temperature values. We can see that the length Lx (Ly) increases in value within the domain that HC increases. The HC reaches a peak before slightly decreases when the length Lx (Ly) continues going up. However, the HC depends on the radius and the length of CQW and wire size of RQW Lx and Ly at different values of temperatures; this is the fundamental difference between CQW and RQW. If the length Lx (Ly) continues to increase, the HC remains constant. It means that HC is no longer dependent on the length of quantum wires (This behavior is similar to the case of the independence of the HC on the length in bulk semiconductor).

## 1. Introduction

It is well-known that in one-dimensional systems (1D), the motion of electrons is restricted in two dimensions, so that they can flow freely in one dimension. The confinement of electron in these systems has changed the electron mobility remarkably. The Hall Effect in bulk semiconductor in the presence of an EMW has been studied in much detail [1, 2, 3, 4, 5], two of which odd magnetophotoresistance effect have been observed [1, 2]. These two works only considered the case when the EMW was absent. However, the magnetoresistance was derived in the presence of a strong EMW for two cases of the magnetic field vector and the electric field vector of the EMW: perpendicular [3], and parallel [4]. The quantum kinetic equation was used to calculate the nonlinear absorption coefficients of an intense EMW in quantum wells [6]. The authors of [7] showed the influence of a strong electromagnetic wave on the HC in Doped Semiconductor Superlattices with an in-plane magnetic field in quantum wires, the Hall Effect has been studied in many aspects (see Refs [8, 9, 10]). In recent works, we have used the quantum kinetic equation

Journal of Physics: Conference Series **726** (2016) 012014 doi:10.1088/1742-6596/726/1/012014

method to study the influence of a high frequency EMW on the HC in Cylindrical quantum wire (CQW) [11]. However, wave function and energy spectrum in a CQW are different from those in a RQW. Therefore analytical expressions for the HC in a RQW is obtained, different from CQW [11]. In this work, we study HC in a RQW with infinitely high potential and in the presence of a laser radiation, subjected to a crossed dc electric field and magnetic field in the presence of a strong EMW characterized by electric field. Main tool is the quantum kinetic equation for distribution function of electrons interacting with acoustic phonons. In CQW, the HC also depends on the length, but the dependence does not have wire size Lx and Ly as in RQW. Because the expression for the HC in RQW varies from that in CQW [11], the dependence of the HC on in RQW is not similar to that in CQW, numerical calculations are the difference between CQW and RQW. The main purpose of this work is to make a comparison between our calculation and other in bulk semiconductor and in 2D and in CQW. The dependence of the HC on a length of wire does not have in bulk semiconductor and in 2D. Nevertheless, the length of wire is large enough (The quantum wire becomes bulk semiconductor), the HC is independent on the length of quantum wire anymore which is true in bulk semiconductor. In CQW, the HC depends on the radius and the length of quantum wires. However, the HC depending on wire size of RQW Lx and Ly at different values of temperatures has not been studied. The work is organized as follows. Sec. 2 indicates the dependence of the HC on a length of RQW with infinitely high potential under the influence of a Laser Radiation. Numerical results and discussion are given in Sec. 3. Conclusions are shown in Sec. 4.

## 2. Dependence of the Hall Coefficient on a length of rectangular quantum wires with infinitely high potential under the influence of a Laser Radiation

In a model, we consider a wire of GaAs with rectangular cross section ($L_x \times L_y$) and the length $L_z$ embedded in GaAsAl. Wave function and energy spectrum in a RQW with electric field $\vec{E}_1 = (0, 0, E_1)$ and MF $\vec{B} = (0, B, 0)$ in the presence of a strong EMW are characterized by electric field $\vec{E} = (0, 0, E_0 \sin \Omega t)$, which are different in CQW (see Ref [11]):

$$\psi_{n,N(x,y,z)} = \begin{cases} 0 & 0 \le x \le L_x, 0 \le y \le L_y \\ \sqrt{\frac{1}{L_z}} e^{i\vec{k}z} \sqrt{\frac{2}{L_x}} \sin\left(\frac{n\pi x}{L_x}\right) \sqrt{\frac{2}{L_y}} \sin\left(\frac{N\pi y}{L_y}\right) & x < 0 \cup x > L_x, y < 0 \cup y > L_y \end{cases} \quad (1)$$

$$\varepsilon_{n,l}(k) = \frac{\hbar^2 k_x^2}{2m} + \frac{\pi^2\hbar^2}{2m}\left(\frac{n^2}{L_x^2} + \frac{l^2}{L_y^2}\right) + \omega_c(N + \frac{1}{2}) - \frac{1}{2m}\left(\frac{eE_1}{\omega_c}\right)^2 \quad (2)$$

where $k_x$ is the electron wave momentum (along the wire's axis: $x$ axis); $\omega_c = eB/m$ is the cyclotron frequency; $n, l, n', l'$ are the quantum numbers; $N, N'$ are the Landau level ($N = 0, 1, 2, \ldots$); $L_x$ ($L_y$) is the size of RQW in the $x$ ($y$) - direction; $m$ and $e$ are the effective masses and the charge of a conduction electron, respectively.

Hamiltonian of the electron - acoustic phonon system in a RQW can be written as:

$$\begin{aligned} H = &\sum_{n,l,\vec{k}} \varepsilon_{n,l}(\vec{k} - \frac{e}{c}\vec{A}_{(t)}) a^+_{n,l,\vec{k}} a_{n,l,\vec{k}} + \sum_{\vec{q}} \omega_{\vec{q}} b^+_{\vec{q}} b_{\vec{q}} + \\ &+ \sum_{n,l,n',l',\vec{k},\vec{q}} |C_{\vec{q}}|^2 |I_{n,l,n',l'}(\vec{q})|^2 a^+_{n,l,\vec{k}+\vec{q}} a_{n',l',\vec{k}} (b_{\vec{q}} + b^+_{-\vec{q}}) + \sum_{\vec{q}} \varphi(\vec{q}) a^+_{n,l,\vec{k}+\vec{q}} a_{n',l',\vec{k}} \end{aligned} \quad (3)$$

where $a^+_{n,l,\vec{k}}$ and $a_{n,l,\vec{k}}$ ($b^+_{\vec{q}}$ and $b_{\vec{q}}$) are the creation and annihilation operators of electron (optical phonon); $\vec{k}$ is the electron wave momentum; $\vec{q}$ is the phonon wave vector; $\omega_{\vec{q}}$ are acoustic phonon frequency; $I_{n,l,n',l'}(\vec{q})$ is the electron form factor, which is determined by [12], different from that in cylindrical quantum wire. $\varphi(\vec{q})$ is the potential undirected:

$$\varphi(\vec{q}) = (2\pi i)^3(e\vec{E} + \omega_c[\vec{q}\vec{h}])\frac{\partial}{\partial\vec{q}}\delta(\vec{q}).$$

From above wave function Eq. (1), energy spectrum Eq. (2), using Hamiltonian of electrons and acoustic phonons system in Eq. (3) and realizing calculations, we obtain the quantum kinetic equation for electrons in the presence of an EMW utilizing the same procedures as in Ref [11]. Then by considering the electron - acoustic phonon interaction with the assumption that $|C_{\vec{q}}|^2 = \frac{\hbar q\xi^2}{2\rho v_s V}$, which is different from the electron - optical phonon interaction constants in CQW (see Ref [11]) ($\vec{q}$ is the phonon wave vector; are the sound velocity, the acoustic deformation potential, the mass density and the normalization volume of specimen, respectively), we obtain the expression for the HC in RQW for case of electron - acoustic phonon scattering:

$$R_H = \frac{1}{B}\frac{\frac{\omega_c\tau}{1+\omega_c^2\tau^2}\left[\frac{ea}{\omega_c} + \frac{b}{m}\frac{\tau^2}{1+\omega_c^2\tau^2}\right]}{\left\{\frac{-\omega_c\tau}{1+\omega_c^2\tau^2}\left[\frac{ea}{\omega_c} + \frac{b}{m}\frac{\tau^2}{1+\omega_c^2\tau^2}\right]\right\}^2 + \left\{\frac{\tau}{1+\omega_c^2\tau^2}\left[ea + \frac{b}{m}\frac{\tau}{1+\omega_c^2\tau^2}(1-\omega_c^2\tau^2)\right]\right\}^2} \tag{4}$$

in which the quantities in the equation (4) vary from those in CQW [11]. Therefore the expression for the HC in RQW is different from that in CQW or 2D.

$$a = \frac{e\beta\hbar}{m^2}\frac{\tau}{1+\omega_c^2\tau^2}\exp\left\{\beta\left[\varepsilon_F + \frac{1}{2m}\left(\frac{eE_1}{\omega_c}\right)^2 - \frac{\pi^2\hbar^2}{2m}\left(\frac{n^2}{L_x^2} + \frac{l^2}{L_y^2}\right) - \omega_c\left(N + \frac{1}{2}\right)\right]\right\}\left(\frac{2m}{\beta\hbar^2}\right)^{3/2}\frac{\sqrt{\pi}}{2}, \tag{5}$$

$$b = \frac{\pi e k_B T\xi^3}{m\rho v_s^2 V}I_{n,l,n',l'}(s_1 + s_2 + s_3 + s_4 + s_5 + s_6 + s_7 + s_8), \tag{6}$$

$$s_1 = \frac{\beta m^2 V}{32\sqrt{2}\pi^4}e^{-\beta\frac{S_{12}}{4}}\left(S_{12}K_o\left(\frac{S_{12}}{\sqrt{2}}\right) + \frac{1}{2m}K_1\left(\frac{S_{12}}{\sqrt{2}}\right)\right),$$

$$s_2 = \frac{\beta m^3 S_{12}V}{16\pi^2}e^{-\beta\frac{S_{12}}{4}}\left(\sqrt{2}K_1\left(\frac{\beta S_{12}}{\sqrt{2}}\right) + \frac{1}{\sqrt{2}}K_2\left(\frac{\beta S_{12}}{\sqrt{2}}\right)\right),$$

$$s_3 = \frac{\beta m^3 S_{13}V}{32\pi^2}e^{-\beta\frac{S_{13}}{4}}\left(\sqrt{2}K_1\left(\frac{\beta S_{13}}{\sqrt{2}}\right) + \frac{1}{\sqrt{2}}K_2\left(\frac{\beta S_{13}}{\sqrt{2}}\right)\right),$$

$$s_4 = \frac{\beta m^3 S_{14}V}{32\pi^2}e^{-\beta\frac{S_{14}}{4}}\left(\sqrt{2}K_1\left(\frac{\beta S_{14}}{\sqrt{2}}\right) + \frac{1}{\sqrt{2}}K_2\left(\frac{\beta S_{14}}{\sqrt{2}}\right)\right),$$

$$s_5 = \frac{\beta m^2 V}{32\sqrt{2}\pi^4}e^{-\beta\frac{S_{15}}{4}}\left(S_{15}K_o\left(\frac{S_{15}}{\sqrt{2}}\right) + \frac{1}{2m}K_1\left(\frac{S_{15}}{\sqrt{2}}\right)\right),$$

$$s_6 = \frac{\beta m^3 S_{15}V}{16\pi^2}e^{-\beta\frac{S_{15}}{4}}\left(\sqrt{2}K_1\left(\frac{\beta S_{15}}{\sqrt{2}}\right) + \frac{1}{\sqrt{2}}K_2\left(\frac{\beta S_{15}}{\sqrt{2}}\right)\right),$$

$$s_7 = \frac{\beta m^3 S_{17}V}{32\pi^2}e^{-\beta\frac{S_{17}}{4}}\left(\sqrt{2}K_1\left(\frac{\beta S_{17}}{\sqrt{2}}\right) + \frac{1}{\sqrt{2}}K_2\left(\frac{\beta S_{17}}{\sqrt{2}}\right)\right),$$

$$s_8 = \frac{\beta m^3 S_{17}V}{32\pi^2}e^{-\beta\frac{S_{17}}{4}}\left(\sqrt{2}K_1\left(\frac{\beta S_{17}}{\sqrt{2}}\right) + \frac{1}{\sqrt{2}}K_2\left(\frac{\beta S_{17}}{\sqrt{2}}\right)\right),$$

$$S_{12} = \frac{\pi^2\hbar^2}{2m}(\frac{n'^2 - n^2}{L_x^2} + \frac{l'^2 - l^2}{L_y^2}) - \omega_c(N' - N) - \omega_{\vec{q}}; S_{13} = S_{12} + \Omega,$$

$$S_{14} = S_{12} - \Omega;\ \beta = 1/(k_B T); S_{15} = \frac{\pi^2\hbar^2}{2m}(\frac{n'^2 - n^2}{L_x^2} + \frac{l'^2 - l^2}{L_y^2}) - \omega_c(N' - N) + \omega_{\vec{q}},$$

$$S_{17} = S_{15} + \Omega,\ S_{18} = S_{15} - \Omega,\ I_{n,l,n',l'} = \int_{-\infty}^{+\infty} \left|I_{n,l,n',l'}(\vec{q})\right|^2 d\vec{q},$$

$\varepsilon_F$ is the Fermi level, $\tau$ is the momentum relaxation time, $k_B$ is Boltzmann constant, $T$ is temperature, $K_i(x)$ are modified Bessel functions.

## 3. Numerical results and discussions

In this section, we present detailed numerical calculations dependence of the HC on the length of quantum wires of the system when the temperature changes with parameters [9, 10, 11]:

$$\varepsilon = 12.5, m = 6.006 \times 10^{-24}, \varepsilon_F = 50meV, \hbar\omega_0 = 36.25meV, \Omega = 3 \times 10^{13}s^{-1},$$

$$k_B = 1.38 \times 10^{-23}kg/m^3, n = 1, n' = 1, l = 1, l' = 1, \tau = 10^{-12}s,$$

$$\rho = 5320kgm^{-3}, q = 2 \times 10^5 m^{-1}, v_s = 5220m/s, \xi = 2.2 \times 10^{-18}J.$$

Figure 1 (Figure 2) shows the dependence of the HC on $L_x$ ($L_y$) of RQW for case of presence of an EMW at different values of the low temperature. When the $L_x$ ($L_y$) increases for the region of small values ($L_x < 1.5 \times 10^{-9}m$), the HC is nonlinear dependent on size limits $L_x$ ($L_y$) of RQW. The HC reaches the maximum value and then decrease as the length of wire keeps increasing. This differs from one observed in CQW (see Ref [11]), whose HC is only dependent on $L$. When the $L_x$ ($L_y$) increases continuously, the HC slightly changes and reaches saturation. This behavior is similar to the results obtained previously in some works for case of absence of an EMW (see Ref [9, 10]).

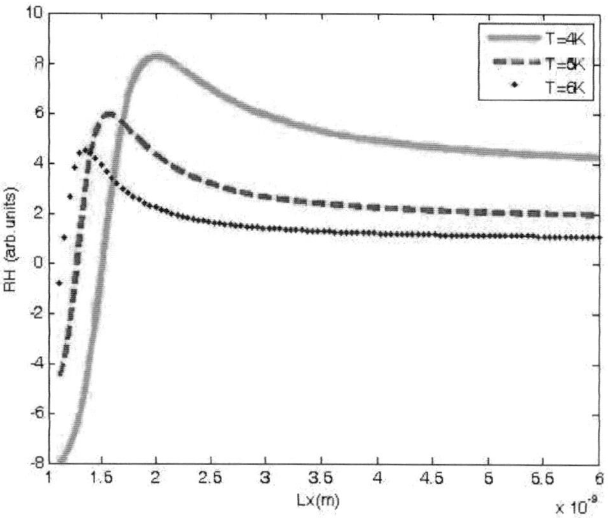


**Figure 1.** The dependence of the HC on $L_x$ of RQW for case of presence of an EMW at different values of the temperature

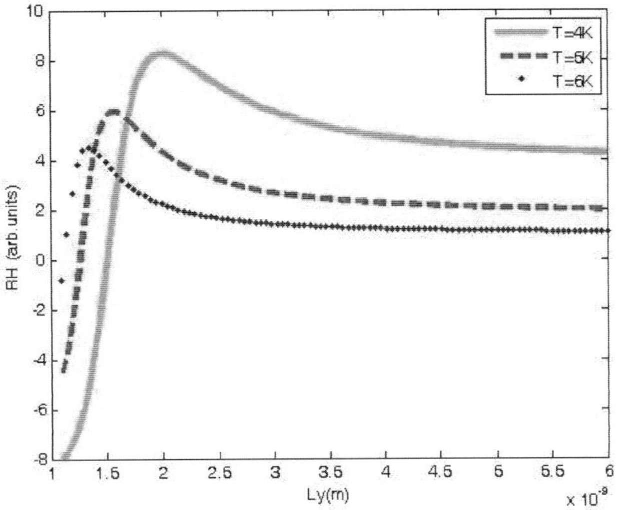


**Figure 2.** The dependence of the HC on $L_y$ of RQW for case presence of an EMW at different values of the temperature

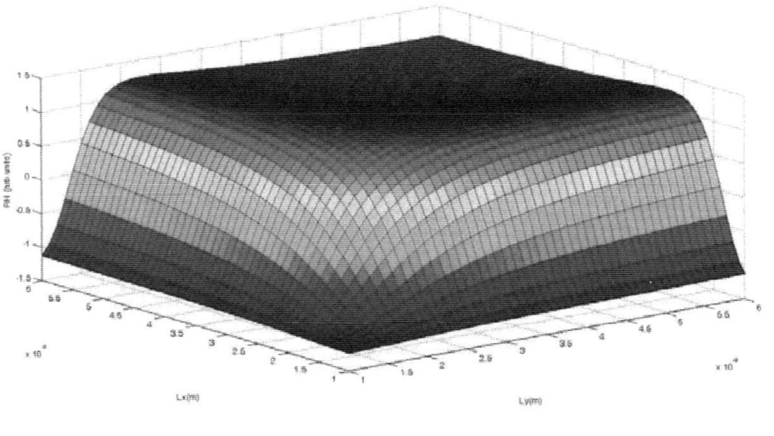


**Figure 3.** The dependence of the HC under wire size of RQW $L_x$ and $L_y$ at different values of temperatures

It is also seen that the value of the length at which the saturation reaches depends on the characteristic of quantum wires. The length $L_x$ ($L_y$) keeps rising at this time the properties of the wire are similar to those of bulk semiconductor. It means that HC is no longer dependent on the length quantum wires at all different temperatures, which is the characteristic of bulk semiconductor. The HC reaches saturation (see Ref [8, 9, 10]). This behavior is similar to the case of the dependence of the HC on the length of CQW. However in CQW, the HC depends on the radius and the length of quantum wires [11]. The HC depends on wire size of RQW $L_x$ and $L_y$ at different values of temperatures. This is the difference between CQW and RQW.

## 4. Conclusion

In this work, we have done some research into the dependence of the HC on a length of RQW when an EMW is present. The electron - acoustic phonon interaction is taken into account. We obtained the expression of HC, which is different from that in CQW [11]. The analytical results are numerically evaluated and plotted for a specific quantum wires GaAs/AlGaAs to show clearly the dependence of HC on the length Lx and Ly of the wire, which is also different from that in CQW: if $L_x$ and $L_y$ are small, the HC will increase. The HC hits a peak and then declines as the size of wire continues to increase. When $L - x$ and $L_y$ go on increasing, the HC has the saturation value. While quantum wires become bulk semiconductor, the HC does not depend on $L_x$ and $L_y$, which is the characteristic of bulk semiconductor. The smaller temperature, the larger HC. In CQW, the HC depends on the radius and the length of quantum wires [11] whereas the HC depends on wire size of RQW $L_x$ and $L_y$ at different values of temperatures. The fact was not seen in bulk semiconductor [3] as well as in quantum wells [5] and CQW [11].

## Acknowledgments

This work was completed with financial support from the National Foundation of Science and Technology Development of Vietnam (NAFOSTED) (Grant No. 103. 01 - 2015. 22).

## References

[1] Epshtein E M 1976 *Sov. Phys. Semicond.* **10** 1414
[2] Manlevich V L and Epshtein E M 1976 *J. Sov. Phys.* **19** 230
[3] Pavlovich V V and Epshtein E M 1977 *Fiz. Tekh. Poluprovodn.* **11** 809
[4] Bau N Q and Hoi B D 2014 *Int. J. Mod. Phys.* B **28** 1450001
[5] Bau N Q and Hoi B D 2012 *J. Korean Phys. Soc.* **60** 59
[6] Bau N Q *et al* 2013 *PIERS Proceedings* **March** 25
[7] Bau N Q and Hoi B D *Integrated Ferroelectrics: An International Journal.* **155** 39
[8] Akera H and Ando T 1990 *Int. J. Mod. Phys.* B **41** 11967
[9] Akera H and Ando T 1989 *Int. J. Mod. Phys.* B **39** 5508
[10] Akera H and Ando T 1991 *Int. J. Mod. Phys.* B **43** 11676
[11] Bau N Q and Huong N T 2015 *Journal of physics: Conference Series.* **627** 012023
[12] Trien H D and Bau N Q 2010 *VNU Journal of Science, Mathematics* **26** 115-120

# The Radioelectric effect in doped superlattices under the influence of confined phonon

**Nguyen Quang Bau***, **Dao Thu Hang and Do Tuan Long**
Faculty of Physics, Hanoi University of Science, Vietnam National University,
334 Nguyen Trai, Thanh Xuan, Hanoi, Vietnam

E-mail: nguyenquangbau54@gmail.com

**Abstract.** The Radioelectric effect in doped superlattices under the influence of confined phonon has been theoretically studied. The analytical expression for the Radioelectric field is obtained by quantum kinetic equation method. The theoretical expression shows that the Radioelectric field in doped superlattices depends on the frequencies and amplitudes of the laser and the linearly polarized electromagnetic wave, the period of the superlattices and especially the quantum number *m* characterizing the phonon confinement. Numerical calculation is also applied for GaAs:Si/GaAs:Be doped superlattices. It is found that the Radioelectric field is different from that in the normal bulk semiconductor as well as two-dimensional systems in case of unconfined phonon and in case of confined phonon when the contribution of confined potential of doped superlattices and confined phonon is remarkable. The Radioelectric field has multiple resonance peaks and increases as the increasing of quantum number *m*.

## 1. Introduction

In recent years, it has been revealed that not only the confined electron but also the phonon confinement can lead to unusual kinetic properties of low-dimensional semiconductors (quantum well, compositional superlattices, doped superlattices, quantum wire, ...). In these systems, the motion of both electrons and phonon are restricted and their energy levels become discrete [1,2,3]. Thus, under certain conditions, the effect of electron and phonon confinement enhanced the nonlinear kinetic properties of semiconductors, for example, the nonlinear absorption coefficient of electromagnetic waves, the Acoustoelectric effect, the Hall effect, the Radioelectric effect [2,4,5], ...

The Radioelectric effect, the effect of drag of charge carriers by electromagnetic waves, has been extensively studied: the photostimulated Radioelectrical transverse effect in semiconductors [6], the Radioelectric effect in a superlattice [7], the photostimulated kinetic effects in semiconductors [8], ... However, almost those work only considered the electron – unconfined phonon interaction while recent works show the remarkable contribution of confined phonon in these kinetic properties [4,9,10]. So, in this work, we study the Radioelectric effect in doped superlattices under the influence of confined phonon.

In literature, there are some methods to solve this problem such as: the Boltzmann equation, the Kubo-Mori method, the quantum kinetic equation method, ... Throughout [2,4] the quantum kinetic equation method shows its power, therefore, this method is chosen to use in the present work.

The article is organized as follows: in section 2 we present basic formulae for the calculation and compute the expression of the Radioelectric field. Numerical results and discussion for the

GaAs:Si/GaAs:Be doped superlattices are given in section 3. And the final section shows remarks and conclusions.

## 2. The Radioelectric field in doped superlattices under the influence of confined optical phonon

In doped superlattices (DSL) the motion of carriers is restricted, the electron wave function and its discrete energy levels are given by [2]:

$$\Psi_{n,\vec{p}_\perp}(\vec{r},z)=\sqrt{\frac{1}{V}}\exp(i\vec{p}_\perp\vec{r})\sum_{s=0}^{N_d}\phi_n(z-sd)\exp(ip_z sd), \quad (1)$$

$$\varepsilon_n(\vec{p}_\perp)=\frac{\hbar^2}{2m^*}\vec{p}_\perp^2+\hbar\omega_p\left(n+\frac{1}{2}\right), \quad (2)$$

here $n = 0,1,2,3, \ldots$ being the quantum number characterizing the electron confinement; $\vec{p}=(\vec{p}_\perp, p_z)$ is the total wave vector of electron, with $\vec{p}_\perp$ and $p_z=\frac{n\pi}{L}$ are components in the (x, y) plane and z direction; $m^*$ is the effective mass of an electron; $L$ is the well width in the z direction; $d$, $N_d$, $n_D$ are the period, the number of periods and the concentration of the doped superlattice; $\omega_p=\left(\frac{4\pi e^2 n_D}{\varepsilon_0 m^*}\right)^{1/2}$ is the plasma frequency; $\varepsilon_0$ is the electric constant; $\phi_n(z)$ being the electron subband wave functions $\phi_n(z)=\sqrt{\frac{1}{2^n n!\sqrt{\pi}\ell_z}}\exp\left(-\frac{z^2}{2\ell_z^2}\right)H_n\left(\frac{z}{\ell_z}\right)$ with $\ell_z=\sqrt{\hbar/(m^*\omega_p)}$ and $H_n(z)$ is the Hermite polynomial of $n^{th}$ order.

The wave vector of confined optical phonon and its frequency are also quantized [3,10]:

$$\vec{q}=(\vec{q}_\perp, q_m),\ q_m=\frac{m\pi}{L}, \quad (3)$$

$$\omega_{m,\vec{q}_\perp}=\left[\omega_0^2-v^2\left(q_\perp^2+q_m^2\right)\right]^{1/2}, \quad (4)$$

here $v$ is veclocity parameter and $m$ being the quantum number characterizing the phonon confinement.

Also, the confined electron – confined optical phonon interaction constant $C_{m,\vec{q}_\perp}$ as well as the electron form factor $I_{nn'}^m$ now become [1,9,10]:

$$\left|C_{m,\vec{q}_\perp}\right|^2=\frac{2\pi e^2\hbar\omega_{m,\vec{q}_\perp}}{\varepsilon_0 V_0}\left(\frac{1}{\chi_\infty}-\frac{1}{\chi_0}\right)\frac{1}{q_\perp^2+q_m^2},$$

$$I_{nn'}^m=\sum_{s=0}^{N_d}\int_{-L/2}^{L/2}\left(\eta(m)\cos\frac{m\pi z}{L}+\eta(m+1)\sin\frac{m\pi z}{L}\right)\phi_{n'}^*(z-sd)\phi_n(z-sd)dz,$$

here $\eta(m)=1$ if $m$ is even and $\eta(m)=0$ if $m$ is odd; $V_0$ is the normalization volume of specimen; $\chi_0$ and $\chi_\infty$ are the static and the high frequency dielectric constants; $\phi_{n'}^*(z)$, $\phi_n(z)$ are the electron subband wave functions in the initial and final states, respectively.

Let us consider doped superlattices subjected to a linearly polarized electromagnetic wave propagating along $Oy$ axis: $\vec{n}\parallel Oy$, $\vec{E}(t)=\vec{E}\left(e^{-i\omega t}+e^{i\omega t}\right)$; $\vec{H}(t)=\frac{1}{c}\left[\vec{n},\vec{E}(t)\right]$ and a laser radiation $\vec{F}(t)=\vec{F}\sin\Omega t$.

Hence, electrons absorb energy of electromagnetic wave (EMW), being generated with directed motion and the Radioelectric field arises under open circuit condition. In addition, the effect of confined phonon and the intensive EMW (the laser radiation) change the probability of electron scattering by phonon [6,10]. Thus, the quantum kinetic equation for electron distribution function $f_{n,\vec{p}_\perp}(t)=\left\langle a^+_{n,\vec{p}_\perp}a_{n,\vec{p}_\perp}\right\rangle_t$ now becomes:

$$\frac{\partial f_{n,\vec{p}_\perp}(t)}{\partial t}+\left(e\vec{E}(t)+e\vec{E}_0+\hbar\omega_c\left[\vec{p}_\perp,\vec{h}(t)\right],\frac{\partial f_{n,\vec{p}_\perp}(t)}{\hbar\partial\vec{p}_\perp}\right)=\sum_{\substack{n,n',m\\ \vec{p}_\perp,\vec{q}_\perp}}\frac{2\pi}{\hbar^2}\left|C_{m,\vec{q}_\perp}\right|^2\left|I^m_{nn'}\right|^2N_{m,\vec{q}_\perp}\sum_{l=-\infty}^{+\infty}J_l^2\left(\frac{e\vec{F}}{m^*\Omega^2},\vec{q}_\perp\right)\times$$

$$\times\left\{f_{n',\vec{p}_\perp+\vec{q}_\perp}(t)-f_{n,\vec{p}_\perp}(t)\right\}\left\{\delta\left(\varepsilon_{n',\vec{p}_\perp+\vec{q}_\perp}-\varepsilon_{n,\vec{p}_\perp}+\hbar\omega_{m,\vec{q}_\perp}-l\hbar\Omega\right)+\delta\left(\varepsilon_{n',\vec{p}_\perp+\vec{q}_\perp}-\varepsilon_{n,\vec{p}_\perp}-\hbar\omega_{m,\vec{q}_\perp}-l\hbar\Omega\right)\right\},\quad(5)$$

here $\vec{h}(t)=\frac{\vec{H}}{H}$, $\omega_c=\frac{eH}{m^*}$ is the cyclotron frequency; $J_l(x)$ is the $l^{th}$ order Bessel function of argument $x$; $\delta(...)$ being the Dirac's delta function.

The electron distribution function is now non-equilibrium and the total current density are given by:

$$\vec{j}_{tot}=\int_0^\infty\vec{R}(\varepsilon)d\varepsilon,\quad(6)$$

$$\vec{R}(\varepsilon)=\sum_{n,\vec{p}_\perp}\frac{e\hbar}{m}\vec{p}_\perp f_{n,\vec{p}_\perp}\delta\left(\varepsilon-\varepsilon_{n,\vec{p}_\perp}\right).\quad(7)$$

Let suppose the electron gas is non-degenerate, consider the confined electron – confined optical phonon interaction and assume that the sample in all direction is opened $\left\langle\vec{j}_{tot}\right\rangle=0$. We obtain the expression for the longitudinal Radioelectric field:

$$E_{0y}=\frac{u+v}{r+s}\left(E_xh_z-E_zh_x\right),\quad(8)$$

where:

$$r=\sum_n\frac{e^2\tau kT}{\pi\hbar^2\varepsilon_F}e^{\frac{\varepsilon_F}{kT}}\left(2kT-\varepsilon_n\right),\quad(9)$$

$$u=\sum_n\frac{2e^2\omega_c}{\pi\hbar^2\omega^2}e^{\frac{\varepsilon_F}{kT}}\left\{kT-\varepsilon_n+\frac{\gamma^2}{kT}\left(\varepsilon_nI_1-I_2\right)\right\},\quad(10)$$

$$I_1=\frac{1}{\gamma}\left[ci(\gamma\mu)\sin\gamma\mu-si(\gamma\mu)\cos\gamma\mu+\frac{1}{2}\pi\cos\gamma\mu\right],$$

$$I_2=-ci(\gamma\mu)\cos\gamma\mu-si(\gamma\mu)\sin\gamma\mu+\frac{1}{2}\pi\sin\gamma\mu,$$

$$\varepsilon_n=\hbar\omega_p\left(n+\frac{1}{2}\right),\varepsilon_{n'}=\hbar\omega_p\left(n'+\frac{1}{2}\right),\varepsilon_m=\frac{\hbar^2\pi^2}{2m^*L^2}m^2,\gamma=\frac{\varepsilon_F}{\omega\tau},\mu=\frac{1}{kT},$$

$$s=\sum_{nn'm}\sum_{\alpha=\pm1}\sum_{\beta=0,\pm1}a_0\left|I^m_{nn'}\right|^2\frac{\tau^2}{\varepsilon_F^2}(-1)^\beta\left(2-|\beta|\right)(s1-s2)\left(1-\frac{\varepsilon_m}{\left|\varepsilon_{n'}-\varepsilon_n-a_{\alpha\beta}\right|}\right)e^{\frac{\varepsilon_F}{kT}},\quad(11)$$

$$s1=2k^3T^3\left(3kT-\varepsilon_n\right),$$

$$s2=e^{\frac{a_{\alpha\beta}}{kT}}k^2T^2\left[6k^2T^2-2\left(2a_{\alpha\beta}+\varepsilon_n\right)kT+a_{\alpha\beta}\left(a_{\alpha\beta}+\varepsilon_n\right)\right],$$

$$a_0=\frac{e^6\omega_0F^2}{4\varepsilon_0m^*\hbar^4\Omega^4kT}\left(\frac{1}{\chi_\infty}-\frac{1}{\chi_0}\right)\left(e^{\frac{\hbar\omega_0}{kT}}-1\right)^{-1},\ a_{\alpha\beta}=\alpha\hbar\left(\omega_0^2-\beta^2\frac{\pi^2m^2}{L^2}\right)^{1/2}+\beta\hbar\Omega,$$

$$v=\sum_{nn'm}\sum_{\alpha=\pm1}\sum_{\beta=0,\pm1}a_0\left|I^m_{nn'}\right|^2\frac{2\omega_c\tau}{\omega^2\varepsilon_F}(-1)^\beta\left(2-|\beta|\right)(v1-v2)\left(1-\frac{\varepsilon_m}{\left|\varepsilon_{n'}-\varepsilon_n-a_{\alpha\beta}\right|}\right)e^{\frac{\varepsilon_F}{kT}},\quad(12)$$

$$v1=-2k^3T^3+\varepsilon_n k^2T^2+3\gamma^2 kT+\gamma^2\varepsilon_n-\gamma^4\left(4+\mu\varepsilon_n\right)I_1+\gamma^2\left(\gamma^2\mu-3\varepsilon_n\right)I_2,$$

$$v2=e^{\frac{a_{\alpha\beta}}{kT}}\left[-2k^3T^3+\left(a_{\alpha\beta}+\varepsilon_n\right)k^2T^2+3\gamma^2kT+AI_2-BI_1\right]+MI_2-NI_1,$$

$$A=-\left(a_{\alpha\beta}+\varepsilon_n\right)\gamma^2+\frac{\gamma^4}{a_{\alpha\beta}},\qquad B=\left(2+\frac{\varepsilon_n}{a_{\alpha\beta}}\right)\gamma^4,$$

$$M=\left(a_{\alpha\beta}-2\varepsilon_n\right)\gamma^2-\frac{\gamma^4}{a_{\alpha\beta}},\qquad N=\varepsilon_n a_{\alpha\beta}\gamma^2+\left(2-\frac{\varepsilon_n}{a_{\alpha\beta}}\right)\gamma^4.$$

Formula (8) shows the dependence of the Radioelectric field on parameters of the system, include: the frequency $\Omega$ and amplitude $F$ of the laser radiation, the frequency $\omega$ of the linearly polarized electromagnetic wave, the period $d$ and the doped concentration $n_D$ of DSL, the temperature $T$ and especially the quantum number $m$ characterizing the phonon confinement. Hence, the confined potential of doped superlattices and the influence of confined optical phonon lead to new properties of the Radioelectric field comparing with that in the normal bulk semiconductor [6] as well as two-dimensional systems in case of unconfined phonon [7] and in case of confined phonon [11].

## 3. Numerical results and discussions

In order to clarify the effect of phonon confinement, we estimate numerical values of the Radioelectric field and graph for the GaAs:Si /GaAs:Be doped superlattices. The parameters used in this calculation are as follows: $m^*=0.067m_0$ ($m_0$ is the mass of a free electron); $\chi_\infty=10.9;\chi_0=12.9;\varepsilon_F=50meV;$ $\tau=10^{-12}s;\hbar\omega_0=36.6meV;v=4730m.s^{-1};n=0\rightarrow 1;n'=0\rightarrow 2;\mathrm{N}_d=2;n_D=10^{20};L=15nm;d=20nm.$

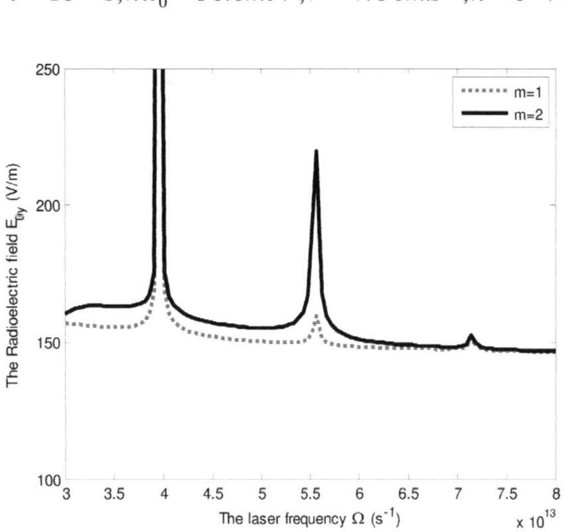


**Figure 1.** *The dependence of the Radioelectric field on the frequency of the laser radiation for* $m=1$ *(the dot line) and* $m=2$ *(the solid line), here* $F=3.10^6V/m$, $\omega=10^{11}\mathrm{s}^{-1},T=300K$.

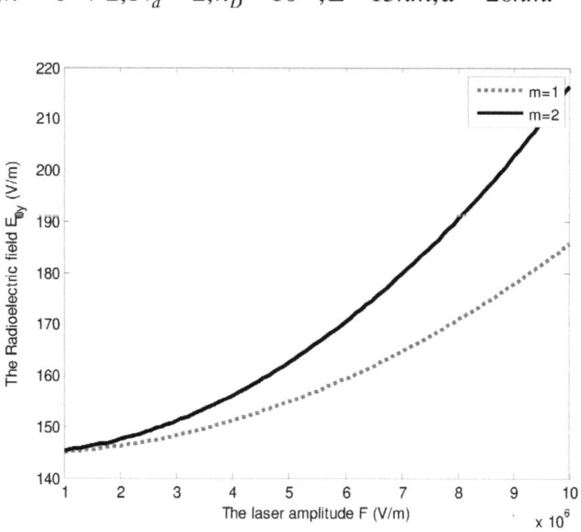


**Figure 2.** *The dependence of the Radioelectric field on the laser amplitude for* $m=1$ *(the dot line) and* $m=2$ *(the solid line), here* $\Omega=6.10^3s^{-1},\omega=10^{11}\mathrm{s}^{-1},T=300K$.

Figure 1 and figure 2 show the dependence of the Radioelectric field on the frequency and amplitude of the laser radiation at different value of quantum number $m$. The Radioelectric field depends nonlinearly on the laser amplitude and increases as the increasing of the quantum number $m$ and amplitude $F$ of the laser radiation. Beside, the Radioelectric field has multiple resonant peaks as the condition $\varepsilon_{n'} - \varepsilon_n \pm \hbar\left(\omega_0^2 - v^2\frac{\pi^2 m^2}{L^2}\right)^{1/2} \pm \hbar\Omega = 0$ is satisfied. These peaks are corresponding to the transition between mini-bands $n \to n'$ and the effect of phonon confinement. These results are different from that in the normal bulk semiconductor [6] as well as the superlattices in case of unconfined phonon [7] and the quantum well in case of confined phonon [11] because of the remarkable contribution of the confined potential of doped superlattices and the quantum number $m$ characterizing the phonon confinement.

## 4. Conclusions

In this paper, the effect of phonon confinement on the Radioelectric field in doped superlattices has theoretically studied. We have established the quantum kinetic equation for electrons distribution function and computed the analytical expression for the Radioelectric field, numerically calculated, graphed the theoretical result for the GaAs:Si/GaAs:Be doped superlattices. The Radioelectric field depends on the frequency and amplitude of the external EMW, the parameters of doped superlattices and especially the quantum number $m$ characterizing the effect of phonon confinement. The remarkable contribution of different confined potential of the doped superlattices and the confined optical phonon leads to unusual properties of the Radioelectric field comparing with that in the normal bulk semiconductor as well as other two-dimensional systems in cases of both unconfined phonon and confined phonon. The Radioelectric field has multiple resonant peaks as the condition $\varepsilon_{n'} - \varepsilon_n \pm \hbar\left(\omega_0^2 - v^2\frac{\pi^2 m^2}{L^2}\right)^{1/2} \pm \hbar\Omega = 0$ is satisfied. The Radioelectric field also increases as the increasing of the quantum number $m$ and amplitude $F$ of the laser radiation. We conclude that the mechanism of this behaviour of the Radioelectric field is not only the confined electron but also the effect of phonon confinement.

## Acknowledgments

This work was completed with financial support from the National Foundation for Science and Technology Development of Vietnam (Nafosted 103.01-2015.22), Program of Basic Research in Natural Science (TN.15.07) and Vietnam International Education Development (Project 911).

## References

[1] Friedman L 1985 *Phys. Rev. B* **32** 955-961.
[2] Bau N Q and Hieu N V 2013 *Superlattices and Microstructure* **63** 121-130.
[3] Wang X F and Lei X L 1994 *Phys. Rev. B* **49** 4780-89.
[4] Bau N Q, Hung D M and Hung L T 2010 *Progress In Electromagnetics Research Letters* **15** 175-185.
[5] Bau N Q and Hoi B D 2012 *Journal of the Korean Physical Society* **60** 59-64.
[6] Shmelev G M, Tsurkan G I, and Epshtein E M 1982 *Phys. Stat. sol. (b)* **109** K53-58.
[7] Kryuchkov S V, Kukhar' E I and Sivashova E S 2008 *Physics of the Solid State* **50** 1150-56.
[8] Malevich V L and Epshtein E M 1976 *Izvestiya vuz. Fizika* **19** 230-237.
[9] Bhat J S, Mulimani B G, and Kubakaddi S S 1992 *J. Appl. Phys.* **72** 4966-68.
[10] Rudin S and Reinecke T L 1990 *Phys. Rev. B* **41** 7713-17.
[11] Long D T and Bau N Q 2015 *J. Phys.: Conf. Ser.* **627** 012019 1-5.

# Ginzburg-Landau functional for three order parameter problem

**Nguyen Van Hinh**[1,2] **and Nguyen Tri Lan**[2]
[1]Faculty of Fundamental Science, Ha Noi University of Industry, Minh Khai, Bac Tu Liem, Hanoi, Vietnam
[2]Institute of Physics Vietnamese Academy of Science and Technology, 10 Dao Tan, Ba Dinh, Hanoi, Vietnam

E-mail: nguyenvanhinh78@gmail.com

**Abstract.** A model is presented utilizing a generic Hamiltonian with equal pairings in channels based on quantum field theory and functional integral formalism, to show the correlation among the order parameters which are described in multi-component Ginzburg-Landau functional. In the vicinity of the phase transition, the further perturbative expansions of the functional around the mean-field theory in the auxiliary fields are carried out with the aim of leading to a possible solution for the coexistence of many phases. The work is motivated by the recent theoretical researches and experimental evidences of the coexistence of superconductivity and ferromagnetism in U and Ce compounds.

## 1. Introduction

Conventional superconductivity was discovered in 1911 but till 1957 it was transparently understood thanks to appearance of BCS theory. Alternatively, this theory also caused a thought that superconductivity was incompatible with ferromagnetism for a long time. The physical reason for this incompatibility is the same as in the case of magnetic impurities in a superconductor[1]. Namely, in the framework of the standard BCS theory at low enough temperatures superconductivity is formed under condensation of singlet Cooper pairs. A singlet Cooper pair is a bound state of two electrons having opposite momenta and spins and which is formed indirectly through electron-phonon-electron interaction (an electron emits a phonon, an other electron absorbs this phonon). While, in ferromagnetic ground state all electron spins align parallel to produce a net spontaneous magnetization. Thus, when a standard ferromagnetic phase arises within a conventional superconductor, the spontaneous magnetization $M$ will break down singlet Cooper pairs i.e., destroy conventional superconductivity.

In the background, the discovery of $UGe_2$ in 2000, and after that, $UIr, URhGe$, and $UCoGe$ with coexistence of ferromagnetism (FM) and superconductivity (SC) came as a big surprise and it appeared to require novel concepts to interpret. According to previous theories, ferromagnetism is induced by spin moments of localized 4f electrons, whereas superconductivity comes from Cooper pairs that formed by conduction electrons. Such discovery together with a number of reliable experimental data about coherence length and superconducting gap [2, 3, 4, 5], is in favor of the conclusion that 5f- electrons from U-atoms responsible for both FM and SC. Cooper pairs in these metallic compounds be long to spin-triplet and magnetic-fluctuation induced pairing is a possible mechanism. In recent years, beside experimental investigations which examine the dependence of the phase transition on the applied pressure and magnetic field, there are also theoretical researches that concentrate on finding out phase transition mechanism, natures of phases and the dependence of temperature of phase transition and spontaneous magnetization moment on the parameters of materials. Different mechanisms have been proposed such as: coupled charge density waves and spin density waves [7], magnon exchange [8], electron interaction mediated by ferromagnetically aligned localized moments [9,10], screened phonon interactions[11], d-electron exchange [12], M-trigger [13, 14]....However, there have been no reliable answer for these important issues thus, we will confine our consideration to a phenomenological level.

The above consideration motivates transforming the fermionic field theory to an effective one based on the coupling fields which are expressed in terms of order parameter fields for different channels. The main purpose of this paper is formulation of a multi-component G-L functional which can describe the coexistence of many phases. In our research, through HS transformation, a microscopic Hamiltonian will be split into possible channels, then we can get a functional which only depends on order parameters. Thereby, we will arrive at a generic performance of Ginzburg-Landau (G-L) functional for three order parameter system through calculations based on Green function. Based on the specific problems of ferromagnetic superconductivity or antiferromagnetic superconductivity, we will draw the formal performance for G-L functional and create a graph showing the dependence of order parameters.

## 2. The model Hamiltonian of the system

### 2.1. The Hamiltonian

Our starting point is an interacting electron gas model [15]. In the terms of second quantization, the Hamiltonian of the system can be written as:

$$H = H_0 + H_1, \tag{1}$$

where,

$$H_0 = \sum_{\sigma}\sum_{\mathbf{k}} \epsilon_\sigma(\mathbf{k})\psi^\dagger_\sigma(\mathbf{k})\psi_\sigma(\mathbf{k}), \tag{2}$$

$H_0$ is unperturbed Hamiltonian describing the system of free electrons; $\psi_\sigma(\mathbf{k})$ and $\psi^\dagger_\sigma(\mathbf{k})$ are the annihilation and creation operators of the electrons with the spin projection $\sigma=\uparrow,\downarrow$ respectively; $\epsilon_\sigma(\mathbf{k})$ is the dispersion of the free electrons; and

$$H_1 = \sum_{\sigma_1,\sigma_2,\sigma'_1,\sigma'_2}\sum_{\mathbf{k},\mathbf{k}',\mathbf{q}} V_{\sigma_1\sigma_2\sigma'_1\sigma'_2}(\mathbf{k},\mathbf{k}',\mathbf{q})\psi^\dagger_{\sigma_1}(\mathbf{k}-\frac{\mathbf{q}}{2})\psi^\dagger_{\sigma_2}(\mathbf{k}'+\frac{\mathbf{q}}{2})\psi_{\sigma'_2}(\mathbf{k}'-\frac{\mathbf{q}}{2})\psi_{\sigma'_1}(\mathbf{k}+\frac{\mathbf{q}}{2}); \tag{3}$$

$H_1$ is a generic effective two-body interaction term of interacting electrons written in the normal order, which may cover the contributions of other interactions in the system.

### 2.2. Grand partition function

In a previous short paper[15], the authors Nguyen Tri Lan and Nguyen Toan Thang have shown that there are, at least, three inequivalent choices of pairing up the fermionic operators to construct the fermionic bilinear term of the generic two-body interaction. Those are pairings in the direct channel $(\psi^\dagger\psi)_{\sigma\sigma'}(\mathbf{k},-\mathbf{q},\tau)$, in the exchange channel $(\psi^\dagger\psi)_{\sigma_2\sigma_1'}(\mathbf{k}',\mathbf{k},\mathbf{q},\tau)$, and in the cooper channel $(\psi\psi)_{\sigma_2'\sigma_1'}(\mathbf{k}',\mathbf{k},\mathbf{q})$ and $(\psi\psi)^\dagger_{\sigma_2\sigma_1}(\mathbf{k}',\mathbf{k},-\mathbf{q})$. Nevertheless, the "right" choice of the decoupling field should be only motivated by physical reasoning, i.e one has to proceed to derive an effective theory based on the coupling field later. For this problem, the generic effective two-body interaction term $H_1\left[\psi^\dagger,\psi\right]$ can be broken down to a summation of all possible fermionic bilinear terms including all of three channels with arbitrary parameters $\{\gamma_i\}$, where $i\in\{\text{d,ex,C}\}$ as

$$H_1\left[\psi^\dagger,\psi\right]=\gamma_d^2H_1^{\rm d}\left[\psi^\dagger,\psi\right]-\gamma_{ex}^2H_1^{\rm ex}\left[\psi^\dagger,\psi\right]+\gamma_C^2H_1^{\rm C}\left[\psi^\dagger,\psi\right]. \tag{4}$$

By introducing HS decoupling on the two-body interaction term,the authors are able to express the grand partition function as a functional integral of the auxiliary fields $\{\varphi_i\}$ as follow

$$\mathcal{Z}=\frac{1}{\mathcal{W}}\int\left[\mathcal{D}\varphi\right]\left[\mathcal{D}\varphi^*\right]\exp\left\{-S_0\left[\{\varphi_i\}\right]\right\}\exp\left\{\ln\det\left(\frac{1}{2}\left[\mathcal{G}_0\right]^{-1}\right)\right\}. \tag{5}$$

The logarithmic contributions in Eq.(5) can be expanded as if it is function (a consequence trace operator), i.e.,

$$\mathrm{Tr}\ln\left(\left[\mathcal{G}_0\right]^{-1}\right)\ =\ln\left[G_0^e\right]^{-1}+\ln\left[G_0^h\right]^{-1}+\sum_{N\geq1}\mathrm{Tr}\left[\mathcal{G}\right]^N, \tag{6}$$

where the trace operator is understood as spin, momentum and frequency diagonal operator whose matrix elements give the free Green's function of the free electrons.

Performing matrix multiplications, then totalizing elements on the main diagonal (see Appendix A), in the energy-momentum space, we can write:

$$\begin{aligned}\mathrm{Tr}\left[\mathcal{G}\right]^N\ &=(-1)^{N-1}\tfrac{1}{N}\textstyle\sum_{\sigma_i=\uparrow,\downarrow}\sum_{\alpha_i=e,h}\sum_{p_i,\omega_{n_i}}G^{\alpha_1}_{\sigma_1\sigma_1}(\mathbf{p_1},\omega_{n_1})\Delta^{\alpha_1\alpha_2}_{\sigma_1\sigma_2}(\mathbf{p_1},\omega_{n_1};\mathbf{p_2},\omega_{n_2})\times\\ &G^{\alpha_2}_{\sigma_2\sigma_2}(\mathbf{p_2},\omega_{n_2})\Delta^{\alpha_2\alpha_3}_{\sigma_2\sigma_3}(\mathbf{p_2},\omega_{n_2};\mathbf{p_3},\omega_{n_3})\ldots G^{\alpha_N}_{\sigma_N\sigma_N}(\mathbf{p_N},\omega_{n_N})\Delta^{\alpha_N\alpha_1}_{\sigma_N\sigma_1}(\mathbf{p_N},\omega_{n_N};\mathbf{p_1},\omega_{n_1})\end{aligned} \tag{7}$$

where

$$\Delta^{\alpha_i\alpha_j}_{\sigma_i\sigma_j}\ =\delta_{\alpha_i\alpha_j}\Phi^{\alpha_i\alpha_j}_{\sigma_i\sigma_j}+\left(1-\delta_{\alpha_i\alpha_j}\right)\mathcal{F}^{\alpha_i\alpha_j}_{\sigma_i\sigma_j}. \tag{8}$$

As it was discussed in [15], the independence of physical quantities on HS coupling's parameters $\{\gamma_i\}$ can be expressed in mathematical language by all order zero-derivatives of the grand thermodynamic potential, i.e., for any set of HS coupling's parameters $\{\gamma_i\}$ following conditions must be satisfied

$$\begin{aligned}\frac{\mathrm{d}}{\mathrm{d}\gamma_i}\Omega\ &=-\beta\frac{\mathrm{d}}{\mathrm{d}\gamma_i}\ln\mathcal{Z}\ =-\beta\left\langle\frac{\mathrm{d}}{\mathrm{d}\gamma_i}\ln\det\left(\frac{1}{2}\left[\mathcal{G}_0\right]^{-1}\right)\right\rangle\\ &\equiv0.\end{aligned} \tag{9}$$

To avoid the complexity and the complication in further calculations, these parameters can be presented, without the loss of generality, by two new independent parameters $\phi$ and $\theta$ as follows

Journal of Physics: Conference Series **726** (2016) 012016 doi:10.1088/1742-6596/726/1/012016

$$\left\{ \begin{array}{ll} \gamma_{\rm d} & = \cosh\theta\sin\phi, \\ \gamma_{\rm ex} & = \sinh\theta, \\ \gamma_{\rm C} & = \cosh\theta\cos\phi. \end{array} \right. . \tag{10}$$

In the terms of parameters $\phi$ and $\theta$, the equations (9) are rewritten as

$$\left\langle \frac{\rm d}{{\rm d}\phi} \ln\det\left( [\mathcal{G}_0]^{-1} \right) \right\rangle = 0, \tag{11}$$

and

$$\left\langle \frac{\rm d}{{\rm d}\theta} \ln\det\left( [\mathcal{G}_0]^{-1} \right) \right\rangle = 0. \tag{12}$$

In general, it is impossible to evaluate the self-consistent system of equations(11) and (12) without some further approximation. A good and convenient idea of such approximation is some mean-field theory obtained by seeking for solution to the set of saddle point equations of the effective action in Eq. (5). We have

$$\frac{\delta}{\delta\varphi_i} \left( S_0\left[\{\varphi_i\}\right] - \ln\det\left( \frac{1}{2} [\mathcal{G}]^{-1} \right) \right) = 0. \tag{13}$$

Eliminating parameters $\theta, \phi$ from Eq. (11) , Eq. (12) and Eq. (13) (see Appendix B ), we obtain the system of three equations

$$\sum_{\substack{\sigma_2 \\ \sigma_2'}} \sum_{\substack{p_2.p_2' \\ \omega_{n2},\omega_{n2}'}} (\varphi_1)^*_{\sigma_2\sigma_2'}(p_2,\omega_{n2};p_2',\omega_{n2}') (V_1)^{-1}_{\sigma_2\sigma_2'\sigma_l\sigma_k}(p_2',\omega_{n2}';p_l,\omega_{n_l}) =$$

$$= \sum_{N\geq 1} (-1)^{N-1} \frac{1}{N} \Bigg\{ \sum_{\substack{\sigma_3 \\ \alpha_3}} \sum_{\substack{p_3 \\ \omega_{n_3}}} \cdots \sum_{\substack{\sigma_N \\ \alpha_N}} \sum_{\substack{p_N \\ \omega_{n_N}}} G^{\alpha_l}_{\sigma_l\sigma_l}(\mathbf{p_l},\omega_{n_l}) \underbrace{\delta_{\alpha_l\alpha_k}(-1)^{\alpha_l} i\cosh\theta\sin\phi}_{\sigma_l=\sigma_1,\sigma_k=\sigma_2}$$

$$G^{\alpha_k}_{\sigma_k\sigma_k}(\mathbf{p_k},\omega_{n_k})\Delta^{\alpha_k\alpha_3}_{\sigma_k\sigma_3}(\mathbf{p_k},\omega_{n_k};\mathbf{p_3},\omega_{n_3})\ldots G^{\alpha_N}_{\sigma_N\sigma_N}(\mathbf{p_N},\omega_{n_N})\Delta^{\alpha_N\alpha_1}_{\sigma_N\sigma_1}(\mathbf{p_N},\omega_{n_N};\mathbf{p_1},\omega_{n_1}) + \cdots$$

$$+ \sum_{\substack{\sigma_2 \\ \alpha_2}} \sum_{\substack{p_2 \\ \omega_{n2}}} \cdots \sum_{\substack{\sigma_{N-1} \\ \alpha_{N-1}}} \sum_{\substack{p_{N-1} \\ \omega_{n_{N-1}}}} G^{\alpha_k}_{\sigma_k\sigma_k}(\mathbf{p_k},\omega_{n_k})\Delta^{\alpha_k\alpha_2}_{\sigma_k\sigma_2}(p_k,\omega_{n_k};\mathbf{p_2},\omega_{n_2})$$

$$G^{\alpha_2}_{\sigma_2\sigma_2}(\mathbf{p_2},\omega_{n_2})\Delta^{\alpha_2\alpha_3}_{\sigma_2\sigma_3}(\mathbf{p_2},\omega_{n_2};\mathbf{p_3},\omega_{n_3})\ldots G^{\alpha_l}_{\sigma_l\sigma_l}(\mathbf{p_l},\omega_{n_l}) \underbrace{\delta_{\alpha_l\alpha_k}(-1)^{\alpha_l} i\cosh\theta\sin\phi}_{\sigma_l=\sigma_N,\sigma_k=\sigma_1} \Bigg\}, \tag{14}$$

$$\sum_{\substack{\sigma_2 \\ \sigma_1'}} \sum_{\substack{p_2.p_1' \\ \omega_{n2},\omega_{n1}'}} (\varphi_2)^*_{\sigma_2\sigma_1'}(p_2,\omega_{n2};p_1',\omega_{n1}') (V_2)^{-1}_{\sigma_2\sigma_1'\sigma_l\sigma_k}(p_1',\omega_{n1}';p_l,\omega_{n_l}) =$$

$$= \sum_{N\geq 1} (-1)^{N-1} \frac{1}{N} \Bigg\{ \sum_{\substack{\sigma_3 \\ \alpha_3}} \sum_{\substack{p_3 \\ \omega_{n_3}}} \cdots \sum_{\substack{\sigma_N \\ \alpha_N}} \sum_{\substack{p_N \\ \omega_{n_N}}} G^{\alpha_l}_{\sigma_l\sigma_l}(\mathbf{p_l}, \omega_{n_l}) \underbrace{\delta_{\alpha_l\alpha_k}(-1)^{\alpha_l} \sinh\theta}_{\sigma_l=\sigma_1,\sigma_k=\sigma_2}$$

$$G^{\alpha_k}_{\sigma_k\sigma_k}(\mathbf{p_k}, \omega_{n_k}) \Delta^{\alpha_k\alpha_3}_{\sigma_k\sigma_3}(\mathbf{p_k}, \omega_{n_k}; \mathbf{p_3}, \omega_{n_3}) \ldots G^{\alpha_N}_{\sigma_N\sigma_N}(\mathbf{p_N}, \omega_{n_N}) \Delta^{\alpha_N\alpha_1}_{\sigma_N\sigma_1}(\mathbf{p_N}, \omega_{n_N}; \mathbf{p_1}, \omega_{n_1}) + \cdots$$

$$+ \sum_{\substack{\sigma_2 \\ \alpha_2}} \sum_{\substack{p_2 \\ \omega_{n_2}}} \cdots \sum_{\substack{\sigma_{N-1} \\ \alpha_{N-1}}} \sum_{\substack{p_{N-1} \\ \omega_{n_{N-1}}}} G^{\alpha_k}_{\sigma_k\sigma_k}(\mathbf{p_k}, \omega_{n_k}) \Delta^{\alpha_k\alpha_2}_{\sigma_k\sigma_2}(\mathbf{p}_k, \omega_{n_k}; \mathbf{p_2}, \omega_{n_2})$$

$$G^{\alpha_2}_{\sigma_2\sigma_2}(\mathbf{p_2}, \omega_{n_2}) \Delta^{\alpha_2\alpha_3}_{\sigma_2\sigma_3}(\mathbf{p_2}, \omega_{n_2}; \mathbf{p_3}, \omega_{n_3}) \ldots G^{\alpha_l}_{\sigma_l\sigma_l}(\mathbf{p_l}, \omega_{n_l}) \underbrace{\delta_{\alpha_l\alpha_k}(-1)^{\alpha_l} \sinh\theta}_{\sigma_l=\sigma_N,\sigma_k=\sigma_1} \Bigg\}, \tag{15}$$

$$\sum_{\substack{\sigma_1' \\ \sigma_2'}} \sum_{\substack{p_1' \cdot p_2' \\ \omega_{n_1'}, \omega_{n_2}'}} (\varphi_3)^*_{\sigma_2\sigma_2'}(p_1', \omega_{n_1'}; p_2', \omega_{n_2}') (V_3)^{-1}_{\sigma_1'\sigma_2'\sigma_l\sigma_k}(p_2', \omega_{n_2}'; p_l, \omega_{n_l}) =$$

$$= \sum_{N\geq 1} (-1)^{N-1} \frac{1}{N} \Bigg\{ \sum_{\substack{\sigma_3 \\ \alpha_3}} \sum_{\substack{p_3 \\ \omega_{n_3}}} \cdots \sum_{\substack{\sigma_N \\ \alpha_N}} \sum_{\substack{p_N \\ \omega_{n_N}}} G^{\alpha_l}_{\sigma_l\sigma_l}(\mathbf{p_l}, \omega_{n_l}) \underbrace{(1 - \delta_{\alpha_l\alpha_k})(-2i\cosh\theta\cos\phi)}_{\sigma_l=\sigma_1,\sigma_k=\sigma_2}$$

$$G^{\alpha_k}_{\sigma_k\sigma_k}(\mathbf{p_k}, \omega_{n_k}) \Delta^{\alpha_k\alpha_3}_{\sigma_k\sigma_3}(\mathbf{p_k}, \omega_{n_k}; \mathbf{p_3}, \omega_{n_3}) \ldots G^{\alpha_N}_{\sigma_N\sigma_N}(\mathbf{p_N}, \omega_{n_N}) \Delta^{\alpha_N\alpha_1}_{\sigma_N\sigma_1}(\mathbf{p_N}, \omega_{n_N}; \mathbf{p_1}, \omega_{n_1}) + \cdots$$

$$+ \sum_{\substack{\sigma_2 \\ \alpha_2}} \sum_{\substack{p_2 \\ \omega_{n_2}}} \cdots \sum_{\substack{\sigma_{N-1} \\ \alpha_{N-1}}} \sum_{\substack{p_{N-1} \\ \omega_{n_{N-1}}}} G^{\alpha_k}_{\sigma_k\sigma_k}(\mathbf{p_k}, \omega_{n_k}) \Delta^{\alpha_k\alpha_2}_{\sigma_k\sigma_2}(\mathbf{p}_k, \omega_{n_k}; \mathbf{p_2}, \omega_{n_2})$$

$$G^{\alpha_2}_{\sigma_2\sigma_2}(\mathbf{p_2}, \omega_{n_2}) \Delta^{\alpha_2\alpha_3}_{\sigma_2\sigma_3}(\mathbf{p_2}, \omega_{n_2}; \mathbf{p_3}, \omega_{n_3}) \ldots G^{\alpha_l}_{\sigma_l\sigma_l}(\mathbf{p_l}, \omega_{n_l}) \underbrace{(1 - \delta_{\alpha_l\alpha_k})(-2i\cosh\theta\cos\phi)}_{\sigma_l=\sigma_N,\sigma_k=\sigma_1} \Bigg\}. \tag{16}$$

which describe many-body relations between the physical quantities of the system where many kinds of fluctuations of corresponding order parameters are considered. Further analysis of this self-consistent system of equations shows the concrete constraints of diagram cancellation between diagram families generated by different HS couplings, which will be presented in a forthcoming paper.

## 3. Three - component Ginzburg-Landau functional

Expanding effective action in the Hubbard-Stratonovich auxiliary fields $\{\varphi_i\}$ to quartic one which only includes terms allowed by symmetry of system and retains minimum numbers of the simplest terms to get the results with multiple meanings, we will obtain G-L free energy functional with the participation of several order parameters describing relationship of density wave, spin wave and superconductivity phases.

$$\begin{aligned} f_{GL}\left[\{\varphi_i\}\right] &= a_1 |\varphi_1|^2 + a_2 |\varphi_2|^2 + a_3 |\varphi_3|^2 + a_4 (\varphi_1^* \varphi_2) + a_5 (\varphi_2^* \varphi_1) \\ &+ b_1 (\varphi_1^*) |\varphi_3|^2 + b_2 (\varphi_1) |\varphi_3|^2 + b_3 (\varphi_2^*) |\varphi_3|^2 + b_4 (\varphi_2) |\varphi_3|^2 \\ &+ c_1 |\varphi_1|^4 + c_2 |\varphi_2|^4 + c_3 |\varphi_3|^4 + c_4 |\varphi_1|^2 |\varphi_2|^2 + c_5 |\varphi_2|^2 |\varphi_3|^2 + c_6 |\varphi_3|^2 |\varphi_1|^2 \end{aligned}$$

$$
\begin{aligned}
&+\left[c_7 |\varphi_1|^2 + c_8 |\varphi_2|^2 + c_9 |\varphi_3|^2\right] (\varphi_1^* \varphi_2) \\
&+\left[c_{10} |\varphi_1|^2 + c_{11} |\varphi_2|^2 + c_{12} |\varphi_3|^2\right] (\varphi_2^* \varphi_1) \\
&+c_{13} (\varphi_1^* \varphi_2) (\varphi_1^* \varphi_2) + c_{14} (\varphi_2^* \varphi_1) (\varphi_2^* \varphi_1)
\end{aligned} \tag{17}
$$

Here, all three quadratic terms $|\varphi_1|^2, |\varphi_2|^2, |\varphi_3|^2$ must be equivalent, and all three quartic terms $|\varphi_1|^4, |\varphi_2|^4, |\varphi_3|^4$ too, they describe each special channel respectively. The other terms describe interaction between channels that result in the coexistance of equilibrium phases.

Thanks to decoupling of two-body interaction terms using H-S transformations, we have above functional in general manner. In fact, auxiliary fields $\{\varphi_i\}$ are not any specific phases because above spin indexes, $\sigma, \sigma'$ are common for all possible phases. In addition, above inversed potential $V^{-1}$ which is included in decoupling method in almost physical systems does not always exist. Depending on how spin indexes are split, we will have singlet superconducting order or triplet superconducting order, ferromagnetic order or antiferromagnetic order. If the order is singlet, the system can fully exist an antiferromagnetic phase plus singlet superconducting phase as in $CehIn_5$ and $CeIrIn_5$. If the order is triplet, the system can fully exist an ferromagnetic phase plus triplet superconducting phase as in $UGe_2$ without depending on whether electronic is localized or not.

The remaining part of this paper will apply the model (17) to study the conditions causing Meissner phase in the presence of ferromagnetic order. In addition, we will establish a phase diagram corresponding to the ferromagnetic superconductor model when no external magnetic field exists in $UGe_2$.

## 4. Unconventional Superconductivity in $UGe_2$

### *4.1. Experimental studies*

The experimental researches about $UGe_2$ [3, 16, 17, 18] show that at ambient pressure $UGe_2$ is an itinerant ferromagnet having Curie temperature $T_{curie} = 52K$, and the spontaneous moment $\mu_s = 1.4\mu B/U -$ atom. An easy axis is the a-axis in orthorhombic crystal [7]. When the pressure increases the system passes through two quantum phase transition, one is from ferromagnetic phase to FS phase at $P = 1GPa$, and the other from ferromagnetic phase to paramagnetic phase at higher pressure $P_c = 1.6$GPa. The superconducting phase exists entirely within the ferromagnetic domain at low temperature and pressure interval between $1.0 \div 1.6$ GPa with a maximum $T_{sc} = 0.8K$ near 1.2 GPa. In ferromagnetic domain, there are two distinct ferromagnetic phases usually denoted by FM2 (highly polarized phase) and FM1 (weakly polarized phase). The line $T_x$ separate two phases ends at $T_x = T_{sc}$ and the pressure $P_x = 1.2$ GPa. The order of transition from FM1 to paramagnetic changes from second to first at the tricritical point $T_{cr}$ on the $T(P)$ diagram, and at critical pressure $P_c$. The more $P$ increases, the more both $T_{FM}$ and $T_{FS}$ drop down to disappear almost simultaneously around $P \sim 1.7$ GPa.

### *4.2. Theoretical studies*

With the experimental evidences about $UGe_2$, as presented above, people believe that the ferromagnetism and superconductivity are caused by itinerant 5f-electrons in the same band. The Cooper pairs formed from the electrons to be of spin-triplet type, and magnetic-fluctuations are responsible for inducing pairing. Driving from the general multi-components G-L functional (17), we formulate specific functional which is capable of describing the coexistence of ferromagnetism and superconductivity in $UGe_2$. We know that, $UGe_2$ is ferromagnet having orthorhombic structure with the magnetic easy axis is the a-axis. The order parameters in orthorhombic ferromagnet have general form as the same one allowed by symmetry given by Machida and Ohmi [19], and others [20, 13, 14]. According to these authors the Ginzburg-Landau (GL) free energy density part which describes the interaction between the ferromagnetic order parameter

Journal of Physics: Conference Series **726** (2016) 012016 doi:10.1088/1742-6596/726/1/012016

$\mathbf{M}$ and the superconducting order parameter $\psi$ has to be coupled linearly and quadratically, namely, $i\gamma\mathbf{M}.(\psi \times \psi^*)$ and $\mathbf{M}^2.\psi.\psi^*$ (corresponding to $(\varphi_2^*)|\varphi_3|^2 + (\varphi_2)|\varphi_3|^2$ and $|\varphi_2|^2|\varphi_3|^2$ in the model (17)) respectively. If we choose a coordinate system with $x//b, y//c, z//a$ where a-axis direction is easy to be magnetized, then $\mathbf{M} = (0, 0, M)$ and $\psi = (\psi_1, \psi_2, 0)$, implying that the Cooper pair spin orientation points to the $\mathbf{M}$ direction. Therefor, we considerably decreased complexity of analysis of the free energy in the below part.

Ignoring effects of anisotropy from both Cooper pairs and crystals, simultaneously restricting in consideration of the uniform order parameters, i.e., neglecting the $x$-dependence of $\psi$ and $M$, in the traditional approach, G-L energy functional of the triplet ferromagnetic superconductor is split into a sum of three terms:

$$f_{GL}(\psi, \mathbf{M}) = f_S(\psi) + f_F(\mathbf{M}) + f_I(\psi, \mathbf{M}), \tag{18}$$

where

$$f_S(\psi) = a_s|\psi|^2 + \frac{b_s}{2}|\psi|^4, \tag{19}$$

$$f_F(\mathbf{M}) = a_f|\mathbf{M}|^2 + \frac{b_f}{2}|\mathbf{M}|^4, \tag{20}$$

and

$$f_I(\psi, \mathbf{M}) = i\gamma_0\mathbf{M}.(\psi \times \psi^*) + \delta_0|\psi|^2|\mathbf{M}|^2. \tag{21}$$

In Eq. (18): The first term $f_S(\psi)$ given by Eq. (19), describes the superconductivity for $M = H = 0$ ($H$ is external magnetic field). $\psi = (\psi_1, \psi_2, 0)$ is the complex vector order parameter with three-components which are complex numbers. The complex numbers $\psi_1, \psi_2$ will be often expressed in terms of moduli and phase angles ($\psi_1 = \varphi_1 e^{i\theta_1}, \psi_2 = \varphi_2 e^{i\theta_2}$), where $\varphi = (\varphi_1^2 + \varphi_2^2)^{1/2}$ is magnitude of the complex vector order parameter $\psi = (\psi_1, \psi_2, 0)$; The second term $f_F(M)$ given by Eq. (20) is a part describing the free energy of a standard isotropic ferromagnet, where $\mathbf{M}$ is the ferromagnetic order parameter; and finally the term $f_I(\psi, \mathbf{M})$ given by Eq. (21) describes the interaction between the ferromagnetic order parameter $\mathbf{M}$ and the superconducting order parameter $\psi$. Where $\delta_0 > 0$ and the parameter $\gamma_0$ for ferromagnetic superconductors may take both positive and negative values.

In Eq. (19), choose $b_s > 0$ and $a_s = \alpha_s(T - T_s)$, where $\alpha_s$ is a positive material parameter and $T_s$ is the critical temperature of a standard second order phase transition in case there is no magnet field. In Eq. (20) $b_f > 0$, $T_f$ is critical temperature of the ferromagnet and $a_f = \alpha_f(T - T_f)$ depend on the material parameter $\alpha_f > 0$ and the temperature. In general, the values of the material parameters $(T_s, T_f, a_s, b_s, a_f, b_f, \gamma_0, \delta_0)$ depends on selection of the concrete substance and thermodynamic parameters as temperature $T$ and pressure $P$.

We assume that only $T_f$ significantly depends on the pressure $P$. Expanding $T_f(P)$ in a Taylor series about the critical pressure point $P_c$ at which it reach to zero. We keep only the first term, enough for get non-trivial result, yields $T_f(P) = T_f^{'}(P_c - P) = T_f(0)\,(1 - P/P_c)$[20]. From (20) we find

$$M = (\alpha_f/b_f)^{1/2}\,[T_f(P) - T]^{1/2}\,. \tag{22}$$

Setting $M_0 = [\alpha_f T_{f0}/b_f]^{1/2}$ which is the value of the magnetization $M$ corresponding to a pure magnetic subsystem $\psi = 0$ at $T = P = 0$ and $T_{f0} = T_f(0)$. Dividing both sides of Eq.(18) by $(b_f M_0^4)$, then calculating the right hand side, we obtain

$$f = \; r\phi^2 + \tfrac{1}{2}\phi^4 + tm^2 + \tfrac{1}{2}m^4 + 2\gamma m\phi_1\phi_2 sin\theta + \delta\phi^2 m^2, \tag{23}$$

where $f = f_{GL}/\left(b_f M_0^4\right)$, $m = M/M_0$ and $\phi_j = |\varphi_j| / \left[(b_f/b_s)^{1/4} M_0\right]$, $\phi = (\phi_1^2 + \phi_2^2)^{1/2}$ and $\theta = \theta_2 - \theta_1$ is the phase angle between the complex $\psi_1$ and $\psi_2$.

The parameters $t$ and $r$ in Eq. (23) are given by

$$r = \kappa\left(\tilde{T} - \tilde{T}_s\right);\quad t = \tilde{T} - \tilde{T}_f(P) = \tilde{T} - 1 + \tilde{P}, \tag{24}$$

where $\kappa = \alpha_s b_f^{1/2}/\alpha_f b_s^{1/2}$ , $\gamma = \gamma_0/\left[\alpha_f T_{f0}/b_s\right]^{1/2}$ , $\delta = \delta_0/\left(b_s b_f\right)^{1/2}$ whereas $\tilde{T} = T/T_{f0}, \tilde{T}_f(P) = T_f(P)/T_{f0}, \tilde{T}_s = T_s/T_{f0}$ and $\tilde{P} = P/P_c$ are the reduced temperatures and pressure.

Imposing the conditions of equilibrium and stable phases (the free energy has to have minima, i.e, find out conditions in which multi-function has minima) on the system which has the free energy given by Eq.(23), for $r > 0$, i.e., $T > T_s$ there are three stable phases:

- the normal phase (N): it exists for $\phi = m = 0$ and the stability conditions: $t \geq 0, r \geq 0$;
- the pure ferromagnetic phase (FM): it exists for $\phi = 0, m = (-t)^{1/2} > 0$, where $t < 0$ and is stable provided $r \geq 0$ and $r \geq \delta t + \gamma\,(-t)^{1/2}$;
- the phase of coexistence of ferromagnetic order and superconductivity (FS): given by $\sin\theta = -1$ ( for $\sin\theta = 1$, we has other phase domain thermodynamically equivalent, so we are not consideration here),$\phi_1 = \phi_2 = \phi/\sqrt{2}$, where the conductivity parameter satisfy the equation

$$\phi^2 = \kappa(\tilde{T}_s - \tilde{T}) + \gamma m - \delta m^2 \;\; \geq 0, \tag{25}$$

and the magnetization $m$ is root of the equation

$$2\left(1 - \delta^2\right)m^3 + 3\gamma\delta m^2 + 2(t - \delta r - \frac{\gamma^2}{2})m + \gamma r = 0. \tag{26}$$

The stability conditions for the FS phase domain given by Eqs. (25) and (26) are

$$\kappa(\tilde{T}_s - \tilde{T}) + 2\gamma m - \delta m^2 \;\; \geq 0,$$

$$\gamma m \quad \geq 0,$$

and

$$\left(3-\delta^2\right)m^2+\gamma\delta m+\tilde{T}+t-\delta r \quad \geq 0.$$

From Eq.(25) and existence condition of FM phase, it is derived that the second order phase transition line $T_{FS}(P)$ separating the FM and FS phases is given by the solution of the equation

$$\tilde{T}_{FS}(\tilde{P}) \quad = \tilde{T}_s+\tfrac{\delta}{\kappa}t(\tilde{T}_{FS},\tilde{P})+\tfrac{\gamma}{\kappa}\left[-t(\tilde{T}_{FS},\tilde{P})\right]^{1/2} \tag{27}$$

The coefficients of this equation are find out with the help of experimental data. $T_{FS}(P)$ curve has a maximum at point $(T_m, P_m)$ which we obtain by both derivative sides of Eq.(27) with respect to $\tilde{P}$ . We have

$$\frac{\partial\tilde{T}_{FS}}{\partial\tilde{P}} \quad = \frac{\frac{\partial\tilde{T}_s}{\partial\tilde{P}}+\frac{\delta}{\kappa}-\frac{\gamma}{2\kappa}\left[-t(\tilde{T}_{FS},\tilde{P})\right]^{-1/2}}{1-\frac{\delta}{\kappa}+\frac{\gamma}{\kappa}\left[-t(\tilde{T}_{FS},\tilde{P})\right]^{-1/2}} \tag{28}$$

Setting $\frac{\partial\tilde{T}_{FS}}{\partial\tilde{P}} \;=\; 0$, and note that $\frac{\partial\tilde{T}_s}{\partial\tilde{P}} \;=\; 0$ since $\tilde{T}_s$ does not depend on $\tilde{P}$ we derive $t(\tilde{T}_m,\tilde{P}_m) \;=\; -\frac{\gamma^2}{4\delta^2}$. Substituting this expression back in Eq.(27), we obtain coordinate of maximum point of $T_{FS}(P)$ curve as

$$\begin{cases} \tilde{T}_m = & \tilde{T}_s+\frac{\gamma^2}{4\kappa\delta} \\ \tilde{P}_m = & 1-\tilde{T}_m-\frac{\gamma^2}{4\delta^2}. \end{cases} . \tag{29}$$

On the other hand, the $T_{FS}(P)$ curve cut $\tilde{P}$-axis at two points having coordinate:

$$\begin{cases} (\tilde{P}=1,\tilde{T}_{FS}=0) \\ (\tilde{P}=1-\frac{\gamma^2}{\delta^2},\tilde{T}_{FS}=0). \end{cases} . \tag{30}$$

In order to outline a $T-P$ diagram of $UGe_2$ we need information about the values of $P_c, T_{f0}, T_s, \gamma/\kappa, \delta/\kappa$. From experimental phase diagrams [17,18,19,20], we have: $P_c = 1.6$GPa, $T_{f0} = 52$K, $T_m \approx 0.8$K and $P_m \approx 1.2$GPa. The experimental evidences show that it has not pure superconducting phase coexisting with ferromagnetism in $UGe_2$. This indicates that critical temperature $T_s$ of the pure superconductivity state is too small, and so considered as $T_s = 0$. Using these given parameters, we plot the $T-P$ diagram of $UGe_2$ shown in Fig.C1.

## 5. Conclusions and Discussions

Through calculations based on quantum field theory and functional integral formalism, the multi-component G-L functional is established. Based on the specific problem of ferromagnetic superconductivity, we transform the correlation among the order parameters described in multi-component Ginzburg-Landau functional into the correlation among order parameters $T$ and $P$ (physical quantities can be obtained directly from the experimental phase diagrams). $T-P$ diagram of $UGe_2$ outlined on the basis of theoretical calculations and shown in Fig.C1 has a agreement with the main experimental findings, although $P_m$ corresponding to the maximum (found at $\sim 1.45$ GPa ) is about 0.25 GPa higher than experimental data [3]. If the experimental plots are accurate, this difference may result from not including contribution of anisotropy of the spin-triplet Cooper pair and crystal or from any effect which is outside the scope of our current model. The relative importance of this effect on the phase diagram needs to be investigated. However, in this context, the introduction of specific problem of ferromagnetic superconductivity only aims to illustrate that the multi-component G-L functional (17) can describe the coexistence of many phases in many-body system. Also, we obtain the system of three equations (14), (15) and (16) which describes many-body relations between the physical quantities of the system where many kinds of fluctuations of corresponding order parameters are considered. This self-consistent system of equations provides a general framework, which may serve as starting points for any further investigations considering the coexistence of many phases in many-body system and will be presented in a forthcoming paper.

## Appendix A. Trace of matrix whose matrix elements are free Green's function of the free electron

$$\mathrm{Tr}\ln\left([\mathcal{G}_0]^{-1}\right) = \mathrm{Tr}\ln\begin{pmatrix} [G_0^e]^{-1} - \Phi & \mathcal{F} \\ \mathcal{F}^* & \left[G_0^h\right]^{-1} + \Phi \end{pmatrix}$$

$$= \ln[G_0^e]^{-1} + \ln\left[G_0^h\right]^{-1} + \mathrm{Tr}\sum_{n\geq 1}(-1)^{n-1}\frac{1}{n}\left[\begin{pmatrix} G_0^e & 0 \\ 0 & G_0^h \end{pmatrix}\begin{pmatrix} -\Phi & \mathcal{F} \\ \mathcal{F}^* & \Phi \end{pmatrix}\right]^n. \quad \text{(A.1)}$$

Setting

$$[\mathcal{G}]^n = (-1)^{n-1}\frac{1}{n}\left[\begin{pmatrix} G^e & 0 \\ 0 & G^h \end{pmatrix}\begin{pmatrix} \Phi^{ee} & \mathcal{F}^{eh} \\ \mathcal{F}^{he} & \Phi^{hh} \end{pmatrix}\right]^n,$$

where

$$\Phi^{ee} = -\Phi, \quad \mathcal{F}^{eh} = \mathcal{F},$$
$$\Phi^{hh} = \Phi, \quad \mathcal{F}^{he} = \mathcal{F}^*,$$
$$G^e = G_0^e \quad G^h = G_0^h.$$

First-order expansion

$$[\mathcal{G}]^1 = \begin{pmatrix} G^e & 0 \\ 0 & G^h \end{pmatrix}\begin{pmatrix} \Phi^{ee} & \mathcal{F}^{eh} \\ \mathcal{F}^{he} & \Phi^{hh} \end{pmatrix}$$
$$= \begin{pmatrix} G^e\Phi^{ee} & G^e\mathcal{F}^{eh} \\ G^h\mathcal{F}^{he} & G^h\Phi^{hh} \end{pmatrix},$$

with

$$G^e\Phi^{ee} = \begin{pmatrix} [G^e]_{\uparrow\uparrow} & 0 \\ 0 & [G^e]_{\downarrow\downarrow} \end{pmatrix}\begin{pmatrix} \Phi^{ee}_{\uparrow\uparrow} & \Phi^{ee}_{\uparrow\downarrow} \\ \Phi^{ee}_{\downarrow\uparrow} & \Phi^{ee}_{\downarrow\downarrow} \end{pmatrix}$$
$$= \begin{pmatrix} [G^e]_{\uparrow\uparrow}\Phi^{ee}_{\uparrow\uparrow} & [G^e]_{\uparrow\uparrow}\Phi^{ee}_{\uparrow\downarrow} \\ [G^e]_{\downarrow\downarrow}\Phi^{ee}_{\downarrow\uparrow} & [G^e]_{\downarrow\downarrow}\Phi^{ee}_{\downarrow\downarrow} \end{pmatrix};$$

$$G^h\Phi^{hh} = \begin{pmatrix} \left[G^h\right]_{\uparrow\uparrow} & 0 \\ 0 & \left[G^h\right]_{\downarrow\downarrow} \end{pmatrix} \begin{pmatrix} \Phi^{hh}_{\uparrow\uparrow} & \Phi^{hh}_{\uparrow\downarrow} \\ \Phi^{hh}_{\downarrow\uparrow} & \Phi^{hh}_{\downarrow\downarrow} \end{pmatrix} = \begin{pmatrix} \left[G^h\right]_{\uparrow\uparrow}\Phi^{hh}_{\uparrow\uparrow} & \left[G^h\right]_{\uparrow\uparrow}\Phi^{hh}_{\uparrow\downarrow} \\ \left[G^h\right]_{\downarrow\downarrow}\Phi^{hh}_{\downarrow\uparrow} & \left[G^h\right]_{\downarrow\downarrow}\Phi^{hh}_{\downarrow\downarrow} \end{pmatrix}.$$

Tracing matrix we obtain

$$\mathrm{Tr}\,[\mathcal{G}]^1 = [G^e]_{\uparrow\uparrow}\Phi^{ee}_{\uparrow\uparrow} + [G^e]_{\downarrow\downarrow}\Phi^{ee}_{\downarrow\downarrow} + \left[G^h\right]_{\uparrow\uparrow}\Phi^{hh}_{\uparrow\uparrow} + \left[G^h\right]_{\downarrow\downarrow}\Phi^{hh}_{\downarrow\downarrow}$$
$$= \sum_{\sigma_1} G^e_{\sigma_1\sigma_1}\Phi^{ee}_{\sigma_1\sigma_1} + \sum_{\sigma_1} G^h_{\sigma_1\sigma_1}\Phi^{hh}_{\sigma_1\sigma_1} = \sum_{\alpha_1}\sum_{\sigma_1} G^{\alpha_1}_{\sigma_1\sigma_1}\Phi^{\alpha_1\alpha_1}_{\sigma_1\sigma_1} = \sum_{\alpha_1}\sum_{\sigma_1} G^{\alpha_1}_{\sigma_1\sigma_1}\triangle^{\alpha_1\alpha_1}_{\sigma_1\sigma_1}. \quad \text{(A.2)}$$

In the same way, we also obtain:
quadratic expansion

$$\mathrm{Tr}\,[\mathcal{G}]^2 = -\frac{1}{2}\{ \sum_{\sigma_1\sigma_2} (G^e_{\sigma_1\sigma_1}\Phi^{ee}_{\sigma_1\sigma_2}G^e_{\sigma_2\sigma_2}\Phi^{ee}_{\sigma_2\sigma_1}) + \sum_{\sigma_1\sigma_2} \left(G^h_{\sigma_1\sigma_1}\Phi^{hh}_{\sigma_1\sigma_2}G^h_{\sigma_2\sigma_2}\Phi^{hh}_{\sigma_2\sigma_1}\right)$$
$$+ \sum_{\sigma_1\sigma_2} \left(G^e_{\sigma_1\sigma_1}\mathcal{F}^{eh}_{\sigma_1\sigma_2}G^h_{\sigma_2\sigma_2}\mathcal{F}^{he}_{\sigma_2\sigma_1}\right) + \sum_{\sigma_1\sigma_2} \left(G^h_{\sigma_1\sigma_1}\mathcal{F}^{he}_{\sigma_1\sigma_2}G^e_{\sigma_2\sigma_2}\mathcal{F}^{eh}_{\sigma_2\sigma_1}\right) \}$$
$$= -\frac{1}{2}\{ \sum_{\alpha_1}\sum_{\sigma_1\sigma_2} (G^{\alpha_1}_{\sigma_1\sigma_1}\Phi^{\alpha_1\alpha_1}_{\sigma_1\sigma_2}G^{\alpha_1}_{\sigma_2\sigma_2}\Phi^{\alpha_1\alpha_1}_{\sigma_2\sigma_1}) + \sum_{\substack{\alpha_1,\alpha_2\\ \alpha_1\neq\alpha_2}} \sum_{\sigma_1\sigma_2} (G^{\alpha_1}_{\sigma_1\sigma_1}\mathcal{F}^{\alpha_1\alpha_2}_{\sigma_1\sigma_2}G^{\alpha_2}_{\sigma_2\sigma_2}\mathcal{F}^{\alpha_2\alpha_1}_{\sigma_2\sigma_1}) \}, \quad \text{(A.3)}$$

cubic expansion

$$\mathrm{Tr}\,[\mathcal{G}]^3 = \frac{1}{3}\Big\{\sum_{\sigma_i} G^e_{\sigma_1\sigma_1}\Phi^{ee}_{\sigma_1\sigma_2}G^e_{\sigma_2\sigma_2}\Phi^{ee}_{\sigma_2\sigma_3}G^e_{\sigma_3\sigma_3}\Phi^{ee}_{\sigma_3\sigma_1} + \sum_{\sigma_i} G^h_{\sigma_1\sigma_1}\Phi^{hh}_{\sigma_1\sigma_2}G^h_{\sigma_2\sigma_2}\Phi^{hh}_{\sigma_2\sigma_3}G^h_{\sigma_3\sigma_3}\Phi^{hh}_{\sigma_3\sigma_1}$$
$$+3\sum_{\sigma_i} G^e_{\sigma_1\sigma_1}\Phi^{ee}_{\sigma_1\sigma_2}G^e_{\sigma_2\sigma_2}\mathcal{F}^{eh}_{\sigma_2\sigma_3}G^h_{\sigma_3\sigma_3}\mathcal{F}^{he}_{\sigma_3\sigma_1} + 3\sum_{\sigma_i} G^h_{\sigma_1\sigma_1}\Phi^{hh}_{\sigma_1\sigma_2}G^h_{\sigma_2\sigma_2}\mathcal{F}^{he}_{\sigma_2\sigma_3}G^e_{\sigma_3\sigma_3}\mathcal{F}^{eh}_{\sigma_3\sigma_1}\Big\}$$
$$= \frac{1}{3}\sum_{\sigma_i=\uparrow,\downarrow}\sum_{\alpha_i=e,h} G^{\alpha_1}_{\sigma_1\sigma_1}\Delta^{\alpha_1\alpha_2}_{\sigma_1\sigma_2}G^{\alpha_2}_{\sigma_2\sigma_2}\Delta^{\alpha_2\alpha_3}_{\sigma_2\sigma_3}G^{\alpha_3}_{\sigma_3\sigma_3}\Delta^{\alpha_3\alpha_1}_{\sigma_3\sigma_1}, \quad \text{(A.4)}$$

quartic expansion

$$\mathrm{Tr}\,[\mathcal{G}]^4 = -\frac{1}{4}\Big\{(G^e\Phi^{ee})^4 + 4\,(G^e\Phi^{ee})^2\left(G^e\mathcal{F}^{eh}G^h\mathcal{F}^{he}\right) + 4\,(G^e\Phi^{ee})\left(G^h\Phi^{hh}\right)\left(G^e\mathcal{F}^{eh}G^h\mathcal{F}^{he}\right)$$
$$+\left(G^h\Phi^{hh}\right)^4 + 4\left(G^h\Phi^{hh}\right)^2\left(G^h\mathcal{F}^{he}G^e\mathcal{F}^{eh}\right) + 2\left(G^h\mathcal{F}^{he}G^e\mathcal{F}^{eh}\right)^2\Big\}$$
$$= -\frac{1}{4}\Big\{\sum_{\sigma_i}\sum_{\alpha=e,h} G^\alpha_{\sigma_1\sigma_1}\Phi^{\alpha\alpha}_{\sigma_1\sigma_2}G^\alpha_{\sigma_2\sigma_2}\Phi^{\alpha\alpha}_{\sigma_2\sigma_3}G^\alpha_{\sigma_3\sigma_3}\Phi^{\alpha\alpha}_{\sigma_3\sigma_4}G^\alpha_{\sigma_4\sigma_4}\Phi^{\alpha\alpha}_{\sigma_4\sigma_1}$$
$$+4\sum_{\sigma_i}\sum_{\substack{\alpha,\beta=e,h\\ \alpha\neq\beta}} G^\alpha_{\sigma_1\sigma_1}\Phi^{\alpha\alpha}_{\sigma_1\sigma_2}G^\alpha_{\sigma_2\sigma_2}\Phi^{\alpha\alpha}_{\sigma_2\sigma_3}G^\alpha_{\sigma_3\sigma_3}\mathcal{F}^{\alpha\beta}_{\sigma_3\sigma_4}G^\beta_{\sigma_4\sigma_4}\mathcal{F}^{\beta\alpha}_{\sigma_4\sigma_1}$$

$$+2\sum_{\sigma_i}\sum_{\substack{\alpha,\beta=e,h\\ \alpha\neq\beta}} G^{\alpha}_{\sigma_1\sigma_1}\Phi^{\alpha\alpha}_{\sigma_1\sigma_2}G^{\alpha}_{\sigma_2\sigma_2}\mathcal{F}^{\alpha\beta}_{\sigma_2\sigma_3}G^{\beta}_{\sigma_3\sigma_3}\Phi^{\beta\beta}_{\sigma_3\sigma_4}G^{\beta}_{\sigma_4\sigma_4}\mathcal{F}^{\beta\alpha}_{\sigma_4\sigma_1}$$

$$+\sum_{\sigma_i}\sum_{\substack{\alpha,\beta=e,h\\ \alpha\neq\beta}} G^{\alpha}_{\sigma_1\sigma_1}\mathcal{F}^{\alpha\beta}_{\sigma_1\sigma_2}G^{\beta}_{\sigma_2\sigma_2}\mathcal{F}^{\beta\alpha}_{\sigma_2\sigma_3}G^{\alpha}_{\sigma_3\sigma_3}\mathcal{F}^{\alpha\beta}_{\sigma_3\sigma_4}G^{\beta}_{\sigma_4\sigma_4}\mathcal{F}^{\beta\alpha}_{\sigma_4\sigma_1}\Big\}$$

$$=-\frac{1}{4}\sum_{\sigma_i=\uparrow,\downarrow}\sum_{\alpha_i=e,h} G^{\alpha_1}_{\sigma_1\sigma_1}\Delta^{\alpha_1\alpha_2}_{\sigma_1\sigma_2}G^{\alpha_2}_{\sigma_2\sigma_2}\Delta^{\alpha_2\alpha_3}_{\sigma_2\sigma_3}G^{\alpha_3}_{\sigma_3\sigma_3}\Delta^{\alpha_3\alpha_4}_{\sigma_3\sigma_4}G^{\alpha_4}_{\sigma_4\sigma_4}\Delta^{\alpha_4\alpha_1}_{\sigma_4\sigma_1} \tag{A.5}$$

By using inductive method, we get

$$\begin{aligned}
\mathrm{Tr}\,[\mathcal{G}]^N &= (-1)^{N-1}\frac{1}{N}\sum_{\sigma_i=\uparrow,\downarrow}\sum_{\alpha_i=e,h}\sum_{p_i,\omega_{n_i}} G^{\alpha_1}_{\sigma_1\sigma_1}(\mathbf{p_1},\omega_{n_1})\Delta^{\alpha_1\alpha_2}_{\sigma_1\sigma_2}(\mathbf{p_1},\omega_{n_1};\mathbf{p_2},\omega_{n_2})\\
&\quad G^{\alpha_2}_{\sigma_2\sigma_2}(\mathbf{p_2},\omega_{n_2})\Delta^{\alpha_2\alpha_3}_{\sigma_2\sigma_3}(\mathbf{p_2},\omega_{n_2};\mathbf{p_3},\omega_{n_3})\dots G^{\alpha_N}_{\sigma_N\sigma_N}(\mathbf{p_N},\omega_{n_N})\Delta^{\alpha_N\alpha_1}_{\sigma_N\sigma_1}(\mathbf{p_N},\omega_{n_N};\mathbf{p_1},\omega_{n_1})\\
&= (-1)^{N-1}\frac{1}{N}\sum_{\sigma_i=\uparrow,\downarrow}\sum_{\alpha_i=e,h}\sum_{\substack{\omega_n,\omega_{\nu_i}\\ \mathbf{k},\mathbf{q}}} G^{\alpha_1}_{\sigma_1\sigma_1}(\mathbf{k},\omega_n)\Delta^{\alpha_1\alpha_2}_{\sigma_1\sigma_2}(\mathbf{q_1},\omega_{\nu_1})G^{\alpha_2}_{\sigma_2\sigma_2}(\mathbf{k}-\mathbf{q_1},\omega_n-\omega_{\nu_1})\\
&\quad\times\Delta^{\alpha_2\alpha_3}_{\sigma_2\sigma_3}(\mathbf{q_2},\omega_{\nu_2})\dots G^{\alpha_N}_{\sigma_N\sigma_N}(\mathbf{k}-\mathbf{q_1}-\dots-\mathbf{q_{N-1}},\omega_n-\omega_{\nu_1}-\dots-\omega_{\nu_{N-1}})\\
&\quad\Delta^{\alpha_N\alpha_1}_{\sigma_N\sigma_1}(-\mathbf{q_1}-\dots-\mathbf{q_{N-1}},-\omega_{\nu_1}-\dots-\omega_{\nu_{N-1}}).
\end{aligned} \tag{A.6}$$

## Appendix B. Eliminating parameters $\theta,\phi$

field equation (13) for $\varphi_1$ can be written

$$\sum_{\substack{\sigma_2\\ \sigma_2'}}\sum_{\substack{p_2.p_2'\\ \omega_{n_2},\omega_{n_2}'}} (\varphi_1)^*_{\sigma_2\sigma_2'}(p_2,\omega_{n_2};p_2',\omega_{n_2}')\,(V_1)^{-1}_{\sigma_2\sigma_2'\sigma_l\sigma_k}(p_2',\omega_{n_2}';p_l,\omega_{n_l}) =$$

$$\begin{aligned}
&=\sum_{N\geq 1}(-1)^{N-1}\frac{1}{N}\Bigg\{\sum_{\substack{\sigma_3\\ \alpha_3}}\sum_{\substack{p_3\\ \omega_{n_3}}}\cdots\sum_{\substack{\sigma_N\\ \alpha_N}}\sum_{\substack{p_N\\ \omega_{n_N}}} \underbrace{G^{\alpha_l}_{\sigma_l\sigma_l}(\mathbf{p_l},\omega_{n_l})\frac{\partial}{\partial(\varphi_1)_{\sigma_l\sigma_k}}\Delta^{\alpha_l\alpha_k}_{\sigma_l\sigma_k}(\mathbf{p_l},\omega_{n_l};\mathbf{p_k},\omega_{n_k})}_{\sigma_l=\sigma_1,\sigma_k=\sigma_2}\\
&\quad\times G^{\alpha_k}_{\sigma_k\sigma_k}(\mathbf{p_k},\omega_{n_k})\Delta^{\alpha_k\alpha_3}_{\sigma_k\sigma_3}(\mathbf{p_k},\omega_{n_k};\mathbf{p_3},\omega_{n_3})\dots G^{\alpha_N}_{\sigma_N\sigma_N}(\mathbf{p_N},\omega_{n_N})\Delta^{\alpha_N\alpha_1}_{\sigma_N\sigma_1}(\mathbf{p_N},\omega_{n_N};\mathbf{p_1},\omega_{n_1})\\
&\quad+\cdots\\
&\quad+\sum_{\substack{\sigma_2\\ \alpha_2}}\sum_{\substack{p_2\\ \omega_{n_2}}}\cdots\sum_{\substack{\sigma_{N-1}\\ \alpha_{N-1}}}\sum_{\substack{p_{N-1}\\ \omega_{n_{N-1}}}} G^{\alpha_k}_{\sigma_k\sigma_k}(\mathbf{p_k},\omega_{n_k})\Delta^{\alpha_k\alpha_2}_{\sigma_k\sigma_2}(\mathbf{p}_k,\omega_{n_k};\mathbf{p_2},\omega_{n_2})G^{\alpha_2}_{\sigma_2\sigma_2}(\mathbf{p_2},\omega_{n_2})\\
&\quad\times\Delta^{\alpha_2\alpha_3}_{\sigma_2\sigma_3}(\mathbf{p_2},\omega_{n_2};\mathbf{p_3},\omega_{n_3})\dots\underbrace{G^{\alpha_l}_{\sigma_l\sigma_l}(\mathbf{p_l},\omega_{n_l})\frac{\partial}{\partial(\varphi_1)_{\sigma_l\sigma_k}}\Delta^{\alpha_l\alpha_k}_{\sigma_l\sigma_k}(\mathbf{p_l},\omega_{n_l};\mathbf{p_k},\omega_{n_k})}_{\sigma_l=\sigma_N,\sigma_k=\sigma_1}\Bigg\}
\end{aligned} \tag{B.1}$$

Multiplying both sides by $\tanh\theta(\Phi_1)_{\sigma_l\sigma_k}(\mathbf{p_l},\omega_{n_l};\mathbf{p_k},\omega_{n_k})$, we have:

$$\sum_{\substack{\sigma_2\\ \sigma_2'}}\sum_{\substack{p_2.p_2'\\ \omega_{n_2},\omega_{n_2}'}}\left(\begin{array}{c}(\varphi_1)^*_{\sigma_2\sigma_2'}(p_2,\omega_{n_2};p_2',\omega_{n_2}')\,(V_1)^{-1}_{\sigma_2\sigma_2'\sigma_l\sigma_k}(p_2',\omega_{n_2}';p_l,\omega_{n_l})\\ \times(\Phi_1)_{\sigma_l\sigma_k}(\mathbf{p_l},\omega_{n_l};\mathbf{p_k},\omega_{n_k})\end{array}\right)\tanh\theta=$$

$$=\sum_{N\geq1}(-1)^{N-1}\frac{1}{N}\Bigg\{\sum_{\substack{\sigma_3\\ \alpha_3}}\sum_{\substack{p_3\\ \omega_{n_3}}}\cdots\sum_{\substack{\sigma_N\\ \alpha_N}}\sum_{\substack{p_N\\ \omega_{n_N}}}G^{\alpha_l}_{\sigma_l\sigma_l}(\mathbf{p_l},\omega_{n_l})$$

$$\times\left[\underbrace{\tanh\theta(\Phi_1)_{\sigma_l\sigma_k}(\mathbf{p_l},\omega_{n_l};\mathbf{p_k},\omega_{n_k})\frac{\partial}{\partial(\varphi_1)_{\sigma_l\sigma_k}}\Delta^{\alpha_l\alpha_k}_{\sigma_l\sigma_k}(\mathbf{p_l},\omega_{n_l};\mathbf{p_k},\omega_{n_k})}_{\sigma_l=\sigma_1,\sigma_k=\sigma_2}\right]\times$$

$$\times G^{\alpha_k}_{\sigma_k\sigma_k}(\mathbf{p_k},\omega_{n_k})\Delta^{\alpha_k\alpha_3}_{\sigma_k\sigma_3}(\mathbf{p_k},\omega_{n_k};\mathbf{p_3},\omega_{n_3})\ldots G^{\alpha_N}_{\sigma_N\sigma_N}(\mathbf{p_N},\omega_{n_N})\Delta^{\alpha_N\alpha_1}_{\sigma_N\sigma_1}(\mathbf{p_N},\omega_{n_N};\mathbf{p_1},\omega_{n_1})$$

$$+\cdots$$

$$+\sum_{\substack{\sigma_2\\ \alpha_2}}\sum_{\substack{p_2\\ \omega_{n_2}}}\cdots\sum_{\substack{\sigma_{N-1}\\ \alpha_{N-1}}}\sum_{\substack{p_{N-1}\\ \omega_{n_{N-1}}}}G^{\alpha_k}_{\sigma_k\sigma_k}(\mathbf{p_k},\omega_{n_k})\Delta^{\alpha_k\alpha_2}_{\sigma_k\sigma_2}(\mathbf{p}_k,\omega_{n_k};\mathbf{p_2},\omega_{n_2})G^{\alpha_2}_{\sigma_2\sigma_2}(\mathbf{p_2},\omega_{n_2})$$

$$\times\Delta^{\alpha_2\alpha_3}_{\sigma_2\sigma_3}(\mathbf{p_2},\omega_{n_2};\mathbf{p_3},\omega_{n_3})\ldots G^{\alpha_l}_{\sigma_l\sigma_l}(\mathbf{p_l},\omega_{n_l})$$

$$\times\left[\underbrace{\tanh\theta(\Phi_1)_{\sigma_l\sigma_k}(\mathbf{p_l},\omega_{n_l};\mathbf{p_k},\omega_{n_k})\frac{\partial}{\partial(\varphi_1)_{\sigma_l\sigma_k}}\Delta^{\alpha_l\alpha_k}_{\sigma_l\sigma_k}(\mathbf{p_l},\omega_{n_l};\mathbf{p_k},\omega_{n_k})}_{\sigma_l=\sigma_N,\sigma_k=\sigma_1}\right]\Bigg\}.\quad\text{(B.2)}$$

Similarly, multiplying both sides of the field equation (13) for $\varphi_2$ by $\coth\theta(\Phi_2)_{\sigma_l\sigma_k}(\mathbf{p_l},\omega_{n_l};\mathbf{p_k},\omega_{n_k})$, and for $\varphi_3$ by $\tanh\theta(\varphi_3)^{\alpha_l}_{\sigma_l\sigma_k}(\mathbf{p_l},\omega_{n_l};\mathbf{p_k},\omega_{n_k})$ we have:

$$\sum_{\substack{\sigma_2\\ \sigma_1'}}\sum_{\substack{p_2.p_1'\\ \omega_{n_2},\omega_{n_1}'}}\left(\begin{array}{c}(\varphi_2)^*_{\sigma_2\sigma_1'}(p_2,\omega_{n_2};p_1',\omega_{n_1}')\,(V_2)^{-1}_{\sigma_2\sigma_1'\sigma_l\sigma_k}(p_1',\omega_{n_1}';p_l,\omega_{n_l})\\ (\Phi_2)_{\sigma_l\sigma_k}(\mathbf{p_l},\omega_{n_l};\mathbf{p_k},\omega_{n_k})\end{array}\right)\coth\theta=$$

$$=\sum_{N\geq1}(-1)^{N-1}\frac{1}{N}\Bigg\{\sum_{\substack{\sigma_3\\ \alpha_3}}\sum_{\substack{p_3\\ \omega_{n_3}}}\cdots\sum_{\substack{\sigma_N\\ \alpha_N}}\sum_{\substack{p_N\\ \omega_{n_N}}}G^{\alpha_l}_{\sigma_l\sigma_l}(\mathbf{p_l},\omega_{n_l})$$

$$\times\left[\underbrace{\coth\theta(\Phi_2)_{\sigma_l\sigma_k}(\mathbf{p_l},\omega_{n_l};\mathbf{p_k},\omega_{n_k})\frac{\partial}{\partial(\varphi_2)_{\sigma_l\sigma_k}}\Delta^{\alpha_l\alpha_k}_{\sigma_l\sigma_k}(\mathbf{p_l},\omega_{n_l};\mathbf{p_k},\omega_{n_k})}_{\sigma_l=\sigma_1,\sigma_k=\sigma_2}\right]$$

$$\times G^{\alpha_k}_{\sigma_k\sigma_k}(\mathbf{p_k},\omega_{n_k})\Delta^{\alpha_k\alpha_3}_{\sigma_k\sigma_3}(\mathbf{p_k},\omega_{n_k};\mathbf{p_3},\omega_{n_3})\ldots G^{\alpha_N}_{\sigma_N\sigma_N}(\mathbf{p_N},\omega_{n_N})\Delta^{\alpha_N\alpha_1}_{\sigma_N\sigma_1}(\mathbf{p_N},\omega_{n_N};\mathbf{p_1},\omega_{n_1})$$

$$+\cdots$$

$$+\sum_{\substack{\sigma_2\\ \alpha_2}}\sum_{\substack{p_2\\ \omega_{n_2}}}\cdots\sum_{\substack{\sigma_{N-1}\\ \alpha_{N-1}}}\sum_{\substack{p_{N-1}\\ \omega_{n_{N-1}}}}G^{\alpha_k}_{\sigma_k\sigma_k}(\mathbf{p_k},\omega_{n_k})\Delta^{\alpha_k\alpha_2}_{\sigma_k\sigma_2}(\mathbf{p}_k,\omega_{n_k};\mathbf{p_2},\omega_{n_2})G^{\alpha_2}_{\sigma_2\sigma_2}(\mathbf{p_2},\omega_{n_2})$$

$$\times\Delta^{\alpha_2\alpha_3}_{\sigma_2\sigma_3}(\mathbf{p_2},\omega_{n_2};\mathbf{p_3},\omega_{n_3})\dots G^{\alpha_l}_{\sigma_l\sigma_l}(\mathbf{p_l},\omega_{n_l})$$

$$\times\left[\underbrace{\coth\theta(\Phi_2)_{\sigma_l\sigma_k}(\mathbf{p_l},\omega_{n_l};\mathbf{p_k},\omega_{n_k})\frac{\partial}{\partial(\varphi_2)_{\sigma_l\sigma_k}}\Delta^{\alpha_l\alpha_k}_{\sigma_l\sigma_k}(\mathbf{p_l},\omega_{n_l};\mathbf{p_k},\omega_{n_k})}_{\sigma_l=\sigma_N,\sigma_k=\sigma_1}\right]\Bigg\},\quad \text{(B.3)}$$

and

$$\sum_{\substack{\sigma'_1\\ \sigma'_2}}\sum_{\substack{p'_1.p'_2\\ \omega_{n'_1},\omega'_{n_2}}}\begin{pmatrix}(\varphi_3)^*_{\sigma_2\sigma'_2}(p'_1,\omega_{n'_1};p'_2,\omega'_{n_2})(V_3)^{-1}_{\sigma'_1\sigma'_2\sigma_l\sigma_k}(p'_2,\omega'_{n_2};p_l,\omega_{n_l})\\ \times(\varphi_3)^{\alpha_l}_{\sigma_l\sigma_k}(\mathbf{p_l},\omega_{n_l};\mathbf{p_k},\omega_{n_k})\end{pmatrix}\tanh\theta=$$

$$=\sum_{N\geq1}(-1)^{N-1}\frac{1}{N}\Bigg\{\sum_{\substack{\sigma_3\\ \alpha_3}}\sum_{\substack{p_3\\ \omega_{n_3}}}\cdots\sum_{\substack{\sigma_N\\ \alpha_N}}\sum_{\substack{p_N\\ \omega_{n_N}}}G^{\alpha_l}_{\sigma_l\sigma_l}(\mathbf{p_l},\omega_{n_l})$$

$$\times\left[\underbrace{\tanh\theta(\varphi_3)^{\alpha_l}_{\sigma_l\sigma_k}(\mathbf{p_l},\omega_{n_l};\mathbf{p_k},\omega_{n_k})\frac{\partial}{\partial(\varphi_3)_{\sigma_l\sigma_k}}\Delta^{\alpha_l\alpha_k}_{\sigma_l\sigma_k}(\mathbf{p_l},\omega_{n_l};\mathbf{p_k},\omega_{n_k})}_{\sigma_l=\sigma_1,\sigma_k=\sigma_2}\right]\times$$

$$\times G^{\alpha_k}_{\sigma_k\sigma_k}(\mathbf{p_k},\omega_{n_k})\Delta^{\alpha_k\alpha_3}_{\sigma_k\sigma_3}(\mathbf{p_k},\omega_{n_k};\mathbf{p_3},\omega_{n_3})\dots G^{\alpha_N}_{\sigma_N\sigma_N}(\mathbf{p_N},\omega_{n_N})\Delta^{\alpha_N\alpha_1}_{\sigma_N\sigma_1}(\mathbf{p_N},\omega_{n_N};\mathbf{p_1},\omega_{n_1})$$
$$+\cdots$$

$$+\sum_{\substack{\sigma_2\\ \alpha_2}}\sum_{\substack{p_2\\ \omega_{n_2}}}\cdots\sum_{\substack{\sigma_{N-1}\\ \alpha_{N-1}}}\sum_{\substack{p_{N-1}\\ \omega_{n_{N-1}}}}G^{\alpha_k}_{\sigma_k\sigma_k}(\mathbf{p_k},\omega_{n_k})\Delta^{\alpha_k\alpha_2}_{\sigma_k\sigma_2}(\mathbf{p}_k,\omega_{n_k};\mathbf{p_2},\omega_{n_2})G^{\alpha_2}_{\sigma_2\sigma_2}(\mathbf{p_2},\omega_{n_2})$$

$$\times\Delta^{\alpha_2\alpha_3}_{\sigma_2\sigma_3}(\mathbf{p_2},\omega_{n_2};\mathbf{p_3},\omega_{n_3})\dots G^{\alpha_l}_{\sigma_l\sigma_l}(\mathbf{p_l},\omega_{n_l})$$

$$\times\left[\underbrace{\tanh\theta(\varphi_3)^{\alpha_l}_{\sigma_l\sigma_k}(\mathbf{p_l},\omega_{n_l};\mathbf{p_k},\omega_{n_k})\frac{\partial}{\partial(\varphi_3)_{\sigma_l\sigma_k}}\Delta^{\alpha_l\alpha_k}_{\sigma_l\sigma_k}(\mathbf{p_l},\omega_{n_l};\mathbf{p_k},\omega_{n_k})}_{\sigma_l=\sigma_N,\sigma_k=\sigma_1}\right]\Bigg\}.\quad \text{(B.4)}$$

Adding respective sides together for three equations, we obtain:

$$\sum_{\substack{\sigma_2\\ \sigma'_2}}\sum_{\substack{p_2.p'_2\\ \omega_{n_2},\omega'_{n_2}}}\begin{pmatrix}(\varphi_1)^*_{\sigma_2\sigma'_2}(p_2,\omega_{n_2};p'_2,\omega'_{n_2})(V_1)^{-1}_{\sigma_2\sigma'_2\sigma_l\sigma_k}(p'_2,\omega'_{n_2};p_l,\omega_{n_l})\\ \times(\Phi_1)_{\sigma_l\sigma_k}(\mathbf{p_l},\omega_{n_l};\mathbf{p_k},\omega_{n_k})\end{pmatrix}\tanh\theta+$$

$$\sum_{\substack{\sigma_2\\ \sigma'_1}}\sum_{\substack{p_2.p'_1\\ \omega_{n_2},\omega'_{n_1}}}\begin{pmatrix}(\varphi_2)^*_{\sigma_2\sigma'_1}(p_2,\omega_{n_2};p'_1,\omega'_{n_1})(V_2)^{-1}_{\sigma_2\sigma'_1\sigma_l\sigma_k}(p'_1,\omega'_{n_1};p_l,\omega_{n_l})\\ \times(\Phi_2)_{\sigma_l\sigma_k}(\mathbf{p_l},\omega_{n_l};\mathbf{p_k},\omega_{n_k})\end{pmatrix}\coth\theta+$$

$$\sum_{\substack{\sigma'_1\\ \sigma'_2}}\sum_{\substack{p'_1.p'_2\\ \omega_{n'_1},\omega'_{n_2}}}\begin{pmatrix}(\varphi_3)^*_{\sigma_2\sigma'_2}(p'_1,\omega_{n'_1};p'_2,\omega'_{n_2})(V_3)^{-1}_{\sigma'_1\sigma'_2\sigma_l\sigma_k}(p'_2,\omega'_{n_2};p_l,\omega_{n_l})\\ \times(\varphi_3)^{\alpha_l}_{\sigma_l\sigma_k}(\mathbf{p_l},\omega_{n_l};\mathbf{p_k},\omega_{n_k})\end{pmatrix}\tanh\theta=$$

$$= \sum_{N\geq 1} (-1)^{N-1} \frac{1}{N} \Bigg\{ \sum_{\substack{\sigma_3 \\ \alpha_3}} \sum_{\substack{p_3 \\ \omega_{n_3}}} \cdots \sum_{\substack{\sigma_N \\ \alpha_N}} \sum_{\substack{p_N \\ \omega_{n_N}}} G^{\alpha_l}_{\sigma_l\sigma_l}(\mathbf{p_l}, \omega_{n_l})$$

$$\times \left[ \underbrace{\tanh\theta(\Phi_1)_{\sigma_l\sigma_k} \frac{\partial \triangle^{\alpha_l\alpha_k}_{\sigma_l\sigma_k}}{\partial(\varphi_1)_{\sigma_l\sigma_k}} + \coth\theta(\Phi_2)_{\sigma_l\sigma_k} \frac{\partial \triangle^{\alpha_l\alpha_k}_{\sigma_l\sigma_k}}{\partial(\varphi_2)_{\sigma_l\sigma_k}} + \tanh\theta(\varphi_3)_{\sigma_l\sigma_k} \frac{\partial \triangle^{\alpha_l\alpha_k}_{\sigma_l\sigma_k}}{\partial(\varphi_3)_{\sigma_l\sigma_k}}}_{\frac{\partial}{\partial\theta}\triangle^{\alpha_l\alpha_k}_{\sigma_l\sigma_k}} \right]$$

$$\times G^{\alpha_k}_{\sigma_k\sigma_k}(\mathbf{p_k}, \omega_{n_k}) \Delta^{\alpha_k\alpha_3}_{\sigma_k\sigma_3}(\mathbf{p_k}, \omega_{n_k}; \mathbf{p_3}, \omega_{n_3}) \ldots G^{\alpha_N}_{\sigma_N\sigma_N}(\mathbf{p_N}, \omega_{n_N}) \Delta^{\alpha_N\alpha_1}_{\sigma_N\sigma_1}(\mathbf{p_N}, \omega_{n_N}; \mathbf{p_1}, \omega_{n_1})$$

$$+ \cdots$$

$$+ \sum_{\substack{\sigma_2 \\ \alpha_2}} \sum_{\substack{p_2 \\ \omega_{n_2}}} \cdots \sum_{\substack{\sigma_{N-1} \\ \alpha_{N-1}}} \sum_{\substack{p_{N-1} \\ \omega_{n_{N-1}}}} G^{\alpha_k}_{\sigma_k\sigma_k}(\mathbf{p_k}, \omega_{n_k}) \Delta^{\alpha_k\alpha_2}_{\sigma_k\sigma_2}(p_k, \omega_{n_k}; \mathbf{p_2}, \omega_{n_2})$$

$$\times G^{\alpha_2}_{\sigma_2\sigma_2}(\mathbf{p_2}, \omega_{n_2}) \Delta^{\alpha_2\alpha_3}_{\sigma_2\sigma_3}(\mathbf{p_2}, \omega_{n_2}; \mathbf{p_3}, \omega_{n_3}) \ldots\ldots G^{\alpha_l}_{\sigma_l\sigma_l}(\mathbf{p_l}, \omega_{n_l})$$

$$\times \left[ \underbrace{\tanh\theta(\Phi_1)_{\sigma_l\sigma_k} \frac{\partial \triangle^{\alpha_l\alpha_k}_{\sigma_l\sigma_k}}{\partial(\varphi_1)_{\sigma_l\sigma_k}} + \coth\theta(\Phi_2)_{\sigma_l\sigma_k} \frac{\partial \triangle^{\alpha_l\alpha_k}_{\sigma_l\sigma_k}}{\partial(\varphi_2)_{\sigma_l\sigma_k}} + \tanh\theta(\varphi_3)_{\sigma_l\sigma_k} \frac{\partial \triangle^{\alpha_l\alpha_k}_{\sigma_l\sigma_k}}{\partial(\varphi_3)_{\sigma_l\sigma_k}}}_{\frac{\partial}{\partial\theta}\triangle^{\alpha_l\alpha_k}_{\sigma_l\sigma_k}} \right] \Bigg\}$$

$$= \sum_{N\geq 1} (-1)^{N-1} \frac{1}{N} \Bigg\{ \sum_{\substack{\sigma_3 \\ \alpha_3}} \sum_{\substack{p_3 \\ \omega_{n_3}}} \cdots \sum_{\substack{\sigma_N \\ \alpha_N}} \sum_{\substack{p_N \\ \omega_{n_N}}} G^{\alpha_l}_{\sigma_l\sigma_l}(\mathbf{p_l}, \omega_{n_l}) \left[ \underbrace{\frac{\partial}{\partial\theta} \triangle^{\alpha_l\alpha_k}_{\sigma_l\sigma_k}(\mathbf{p_l}, \omega_{n_l}; \mathbf{p_k}, \omega_{n_k})}_{\sigma_l=\sigma_1, \sigma_k=\sigma_2} \right]$$

$$\times G^{\alpha_k}_{\sigma_k\sigma_k}(\mathbf{p_k}, \omega_{n_k}) \Delta^{\alpha_k\alpha_3}_{\sigma_k\sigma_3}(\mathbf{p_k}, \omega_{n_k}; \mathbf{p_3}, \omega_{n_3}) \ldots G^{\alpha_N}_{\sigma_N\sigma_N}(\mathbf{p_N}, \omega_{n_N}) \Delta^{\alpha_N\alpha_1}_{\sigma_N\sigma_1}(\mathbf{p_N}, \omega_{n_N}; \mathbf{p_1}, \omega_{n_1})$$

$$+ \cdots$$

$$+ \sum_{\substack{\sigma_2 \\ \alpha_2}} \sum_{\substack{p_2 \\ \omega_{n_2}}} \cdots \sum_{\substack{\sigma_{N-1} \\ \alpha_{N-1}}} \sum_{\substack{p_{N-1} \\ \omega_{n_{N-1}}}} G^{\alpha_k}_{\sigma_k\sigma_k}(\mathbf{p_k}, \omega_{n_k}) \Delta^{\alpha_k\alpha_2}_{\sigma_k\sigma_2}(p_k, \omega_{n_k}; \mathbf{p_2}, \omega_{n_2})$$

$$\times G^{\alpha_2}_{\sigma_2\sigma_2}(\mathbf{p_2}, \omega_{n_2}) \Delta^{\alpha_2\alpha_3}_{\sigma_2\sigma_3}(\mathbf{p_2}, \omega_{n_2}; \mathbf{p_3}, \omega_{n_3}) \ldots G^{\alpha_l}_{\sigma_l\sigma_l}(\mathbf{p_l}, \omega_{n_l}) \left[ \underbrace{\frac{\partial}{\partial\theta} \triangle^{\alpha_l\alpha_k}_{\sigma_l\sigma_k}(\mathbf{p_l}, \omega_{n_l}; \mathbf{p_k}, \omega_{n_k})}_{\sigma_l=\sigma_N, \sigma_k=\sigma_1} \right] \Bigg\}. \quad \text{(B.5)}$$

Taking the sum of both sides of equation (B.5) with respect to $\alpha_l, \alpha_k; \sigma_l, \sigma_k; p_l, p_k; \omega_{n_l}, \omega_{n_k}$ ( short-hand notation $a \equiv (\alpha_l, \alpha_k; \sigma_l, \sigma_k; p_l, p_k; \omega_{n_l}, \omega_{n_k})$), we obtain:

$$\sum_a \sum_{\substack{\sigma_2 \\ \sigma'_2}} \sum_{\substack{p_2 . p'_2 \\ \omega_{n_2}, \omega'_{n_2}}} \left( \begin{array}{c} (\varphi_1)^*_{\sigma_2\sigma'_2}(p_2, \omega_{n_2}; p'_2, \omega'_{n_2}) (V_1)^{-1}_{\sigma_2\sigma'_2\sigma_l\sigma_k}(p'_2, \omega'_{n_2}; p_l, \omega_{n_l}) \\ \times (\Phi_1)_{\sigma_l\sigma_k}(\mathbf{p_l}, \omega_{n_l}; \mathbf{p_k}, \omega_{n_k}) \end{array} \right) \tanh\theta +$$

$$\sum_a \sum_{\substack{\sigma_2 \\ \sigma'_1}} \sum_{\substack{p_2 . p'_1 \\ \omega_{n_2}, \omega'_{n_1}}} \left( \begin{array}{c} (\varphi_2)^*_{\sigma_2\sigma'_1}(p_2, \omega_{n_2}; p'_1, \omega'_{n_1}) (V_2)^{-1}_{\sigma_2\sigma'_1\sigma_l\sigma_k}(p'_1, \omega'_{n_1}; p_l, \omega_{n_l}) \\ \times (\Phi_2)_{\sigma_l\sigma_k}(\mathbf{p_l}, \omega_{n_l}; \mathbf{p_k}, \omega_{n_k}) \end{array} \right) \coth\theta +$$

$$\sum_a \sum_{\substack{\sigma_1' \\ \sigma_2'}} \sum_{\substack{p_1' \cdot p_2' \\ \omega_{n_1'}, \omega_{n_2}'}} \left( \begin{array}{c} (\varphi_3)^*_{\sigma_2\sigma_2'}(p_1', \omega_{n_1'}; p_2', \omega_{n2}') (V_3)^{-1}_{\sigma_1'\sigma_2'\sigma_l\sigma_k}(p_2', \omega_{n2}'; p_l, \omega_{n_l}) \\ \times (\varphi_3)^{\alpha_l}_{\sigma_l\sigma_k}(\mathbf{p_l}, \omega_{n_l}; \mathbf{p_k}, \omega_{n_k}) \end{array} \right) \tanh\theta =$$

$$= \sum_{N\geq 1} (-1)^{N-1} \frac{1}{N} \Bigg\{$$

$$\sum_{\substack{\sigma_1 \\ \alpha_1}} \sum_{\substack{p_1 \\ \omega_{n_1}}} \cdots \sum_{\substack{\sigma_N \\ \alpha_N}} \sum_{\substack{p_N \\ \omega_{n_N}}} \left( \begin{array}{c} G^{\alpha_1}_{\sigma_1\sigma_1}(\mathbf{p_1}, \omega_{n_1}) \frac{\partial}{\partial\theta} \Delta^{\alpha_1\alpha_2}_{\sigma_1\sigma_2}(\mathbf{p_1}, \omega_{n_1}; \mathbf{p_2}, \omega_{n_2}) \\ \ldots \times G^{\alpha_N}_{\sigma_N\sigma_N}(\mathbf{p_N}, \omega_{n_N}) \Delta^{\alpha_N\alpha_1}_{\sigma_N\sigma_1}(\mathbf{p_N}, \omega_{n_N}; \mathbf{p_1}, \omega_{n_1}) \end{array} \right)$$

$$+ \cdots$$

$$+ \sum_{\substack{\sigma_1 \\ \alpha_1}} \sum_{\substack{p_1 \\ \omega_{n_1}}} \cdots \sum_{\substack{\sigma_N \\ \alpha_N}} \sum_{\substack{p_N \\ \omega_{n_N}}} \left( \begin{array}{c} G^{\alpha_1}_{\sigma_1\sigma_1}(\mathbf{p_1}, \omega_{n_1}) \Delta^{\alpha_1\alpha_2}_{\sigma_1\sigma_2}(\mathbf{p_1}, \omega_{n_1}; \mathbf{p_2}, \omega_{n_2}) \\ \ldots \times G^{\alpha_N}_{\sigma_N\sigma_N}(\mathbf{p_N}, \omega_{n_N}) \frac{\partial}{\partial\theta} \Delta^{\alpha_N\alpha_1}_{\sigma_N\sigma_1}(\mathbf{p_N}, \omega_{n_N}; \mathbf{p_1}, \omega_{n_1}) \end{array} \right) \Bigg\}$$

$$= \frac{\partial}{\partial\theta} \ln\det\left(\frac{1}{2}[\mathcal{G}_0]^{-1}\right). \quad \text{(B.6)}$$

Since $\left\langle \frac{\partial}{\partial\theta} \ln\det\left(\frac{1}{2}[\mathcal{G}_0]^{-1}\right)\right\rangle = 0$ and note that $\left\langle (\varphi_i)^* \left(V_i^{-1}\right) (\varphi_i)^* \right\rangle = 0$, we derive:

$$\left\langle \varphi_1^* V_1^{-1} \varphi_1 \right\rangle \tanh\theta + \left\langle \varphi_2^* V_2^{-1} \varphi_2 \right\rangle \coth\theta + \left\langle \varphi_3^* V_2^{-1} \varphi_3 \right\rangle \tanh\theta \ = 0. \quad \text{(B.7)}$$

Form above equation, we obtain:

$$\sinh\theta \ = \left[ -\frac{\left\langle \varphi_2^* V_2^{-1} \varphi_2 \right\rangle}{\left\langle \varphi_1^* V_1^{-1} \varphi_1 \right\rangle + \left\langle \varphi_2^* V_2^{-1} \varphi_2 \right\rangle + \left\langle \varphi_3^* V_3^{-1} \varphi_3 \right\rangle} \right]^{1/2}, \quad \text{(B.8)}$$

$$\coth\theta \ = \left[ \frac{\left\langle \varphi_1^* V_1^{-1} \varphi_1 \right\rangle + \left\langle \varphi_3^* V_3^{-1} \varphi_3 \right\rangle}{\left\langle \varphi_1^* V_1^{-1} \varphi_1 \right\rangle + \left\langle \varphi_2^* V_2^{-1} \varphi_2 \right\rangle + \left\langle \varphi_3^* V_3^{-1} \varphi_3 \right\rangle} \right]^{1/2}. \quad \text{(B.9)}$$

Similar to calculations with parameter $\theta$, for parameter $\phi$ we also have:

$$\sin\phi \ = \left[ \frac{\left\langle \varphi_1^* V_1^{-1} \varphi_1 \right\rangle}{\left\langle \varphi_1^* V_1^{-1} \varphi_1 \right\rangle + \left\langle \varphi_3^* V_3^{-1} \varphi_3 \right\rangle} \right]^{1/2}, \quad \text{(B.10)}$$

$$\cos\phi \ = \left[ \frac{\left\langle \varphi_3^* V_3^{-1} \varphi_3 \right\rangle}{\left\langle \varphi_1^* V_1^{-1} \varphi_1 \right\rangle + \left\langle \varphi_3^* V_3^{-1} \varphi_3 \right\rangle} \right]^{1/2}. \quad \text{(B.11)}$$

for convenience the following short-hand notations have been used

$$\left\langle \varphi_1^* V_1^{-1} \varphi_1 \right\rangle = \left\langle \sum_a \sum_{\substack{\sigma_2, p_2, \omega_{n2} \\ \sigma_2', p_2', \omega_{n2}'}} \left( \begin{array}{c} (\varphi_1)^*_{\sigma_2\sigma_2'}(p_2, \omega_{n2}; p_2', \omega_{n2}') (V_1)^{-1}_{\sigma_2\sigma_2'\sigma_l\sigma_k}(p_2', \omega_{n2}'; p_l, \omega_{n_l}) \\ \times (\varphi_1)_{\sigma_l\sigma_k}(\mathbf{p_l}, \omega_{n_l}; \mathbf{p_k}, \omega_{n_k}) \end{array} \right) \right\rangle,$$

Journal of Physics: Conference Series **726** (2016) 012016 doi:10.1088/1742-6596/726/1/012016

$$\left\langle \varphi_2^* V_2^{-1} \varphi_2 \right\rangle = \left\langle \sum_a \sum_{\substack{\sigma_2, p_2, \omega_{n_2}. \\ \sigma_1', p_1', \omega_{n_1}'}} \begin{pmatrix} (\varphi_2)^*_{\sigma_2 \sigma_1'} (p_2, \omega_{n_2}; p_1', \omega_{n_1}') (V_2)^{-1}_{\sigma_2 \sigma_1' \sigma_l \sigma_k} (p_1', \omega_{n_1}'; p_l, \omega_{n_l}) \\ (\varphi_2)_{\sigma_l \sigma_k} (\mathbf{p_l}, \omega_{n_l}; \mathbf{p_k}, \omega_{n_k}) \end{pmatrix} \right\rangle,$$

$$\left\langle \varphi_3^* V_2^{-1} \varphi_3 \right\rangle = \left\langle \sum_a \sum_{\substack{\sigma_1', p_1', \omega_{n_1'}. \\ \sigma_2', p_2', \omega_{n_2}'}} \begin{pmatrix} (\varphi_3)^*_{\sigma_2 \sigma_2'} (p_1', \omega_{n_1'}; p_2', \omega_{n_2}') (V_3)^{-1}_{\sigma_1' \sigma_2' \sigma_l \sigma_k} (p_2', \omega_{n_2}'; p_l, \omega_{n_l}) \\ \times (\varphi_3)_{\sigma_l \sigma_k} (\mathbf{p_l}, \omega_{n_l}; \mathbf{p_k}, \omega_{n_k}) \end{pmatrix} \right\rangle.$$

## Appendix C. Figures and figure captions

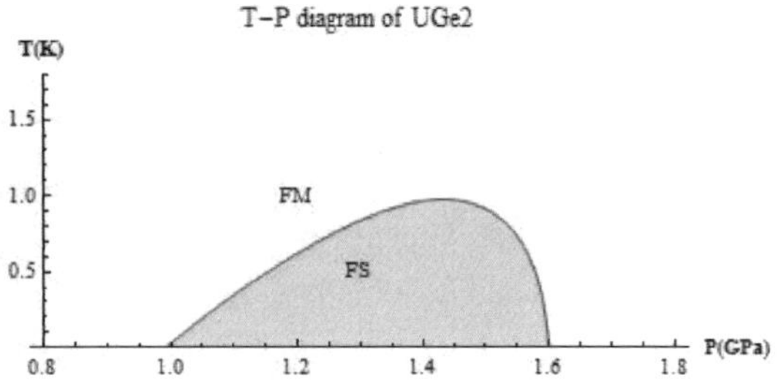


**Figure C1.** An illustration of T P phase diagram of UGe2 calculated for $T_s = 0, T_{f0} = 52$K, $P_c = 1.6$GPa,$\gamma/\kappa = 0.10496, \delta/\kappa = 0.17902$. The FS phase domain is shaded. The solid line shows the second order FM-FS phase transition.

## References


[1] Abrikosov A A, Gorkov L P and Eksp Zh 1960 *Teor. Fiz.* **39** 1781 [1961 *Sov. Phys. JETP* **12** 1243 ]

[2] Saxena S S, Agarwal P, Ahilan K, Grosche F M, Haselwimmer R K W, Steiner M J, Pugh E, Walker I R, Julian S R, Monthoux P, Lonzarich G G, Huxley A, Sheikin I, Braithwaite D and Flouquet J 2000 *Nature* **406** 587

[3] Huxley A, Sheikin I, Ressouche E, Kernavanois N, Braithwaite D, Calemczuk R and Flouquet J 2001 *Phys. Rev* B **63** 144519

[4] Pfleiderer C, Uhlatz M, Hayden S M, Vollmer R, Lohneysen H V, Berhoeft N R and Lonzarich G G 2001 *Nature* **412** 58

[5] Aoki D, Huxley A, Ressouche E, Braithwaite D, Flouquet J, Brison J P, Lhotel E, and Paulsen C 2001 *Nature* **413** 613

[6] Kamihara Y, Watanabe T, Hirano M and Hosono H, 2008 *J. Am. Chem. Soc.* **130** 3296

[7] Watanabe S and Miyake K *cond-mat* 0110492

[8] Karchev N 2003 *Phys.Rev.* B **67** 54416

[9] Abrikosov A A 2001 *J. Phys. Condens. Matter.* **13** L943

[10] Mineev, Chambel T et al 2004 *Phys. Rev.* B 144521

[11] Shimahara H and Kohmoto M 2002 *Europhys. Lett.* **57** 247

[12] Spalek J and Wrobel P *cond-mat* 0202043

[13] Shopova D V and Uzunov D I 2005 *Phys. Rev.* **72** 024531 [ 2003 *Phys. Lett* A **313** 139]

[14] Shopova D V and Uzunov D I 2003 *J. Phys. Studies* **4** 426 [ 2003 *Compt. Rend Acad. Bulg. Sci.* **56** 35; 2005*Bulg. J. of Phys.* **32** 81; Shopova D V, Tsvetkov T E and Uzunov D I 2005 *Cond. Matter Phys.* **8** 181]

[15] Lan Nguyen Tri and Thang Nguyen Toan 2009 *Comm. Phys.* **19** 53-64

[16] Saxena S S, Agarwal P, Ahilan K, Grosche F M, Haselwimmer R K W, Steiner M J, Pugh E, Walker I R, Julian S R, Monthoux P, Lonzarich G G, Huxley A, Sheikin I, Braithwaite D and Flouquet J 2000*Nature* **406** 587

[17] Tateiwa N, Kobayashi T C, Hanazono K, Amaya A, Haga Y, Settai R and Onuki Y 2001 *J. Phys. Condensed Matter* **13** L17

[18] Harada A, Kawasaki S, Mukuda H, Kitaoka Y, Haga Y, Yamamoto E, Onuki Y, Itoh K M, Haller E E and Harima H 2007 *Phys. Rev.* B **75** 140502

[19] Machida K and Ohmi T 2001 *Phys. Rev. Lett.* **86** 850

[20] Samokhin K V and Walker M B 2002 *Phys. Rev.* B **66** 174501 [ Samokhin K V and Walker M B 2002 *Phys. Rev.* B **66** 024512; Walker M B and Samokhin K V 2002 *Phys. Rev. Lett.* **88** 207001] Phys. Rev. B 79 (2009) 064501

[21] Kitaoka Y, Kotegawa H, Harada A, Kawasaki S, Kawasaki Y, Haga Y, Yamamoto E, Onuki Y, Itoh K M, Haller E E, and Harima H 2005 *J. Phys. Condensed Matter* **17** S975

[22] Shopova D V and Uzunov D I 2009 *Phys. Rev.* **79** B 064501

# Cooper pair of superconductivity in the coordinate representation and $q$-deformed harmonic oscillator

**Man Van Ngu[1], Ngo Gia Vinh[2], Nguyen Tri Lan[3], Luu Thi Kim Thanh[4] and Nguyen Ai Viet[3]**

[1] Hung Yen Industrial Colleage, Hung Yen Province, Vietnam
[2] Bac Ninh Department of Education and Training, Bac Ninh Province, Vietnam
[3] Institute of Physics, Vietnam Academy of Science and Technology, 10 Dao Tan street, Ba Dinh, Hanoi, Vietnam
[4] Hanoi Pedagogical University No.2, Vinh Phuc Province, Vietnam

E-mail: nguqlkt@yahoo.com

**Abstract.** In this work we study the similarity between the wave functions of $q$-deformed harmonic oscillator and wave functions of Cooper pair. The wave functions of Cooper pairs in coordinate-space have an "onion-like" layered structure with exponent decay (Boltzmann) envelope modulation. The ground state wave function of $q$-deform harmonic oscillator has the form of oscillate functions with Gaussian decay envelope modulation. The corresponding between Boltzmann and Gaussian forms of envelope functions and their quantum similarity are discussed.

## 1. Introduction

In the last decades, the mathematical methods bases on quantum group theory and its $q$–deformed algebras have found their applications in many fields of interest from physics to chemistry, and even to other sciences of complexity [1, 2, 3, 4, 5, 6, 7, 8, 9, 10, 11, 12]. The main idea of the $q$–deformed algebras applicability is of its possibility to transform the perturbative methods approach to the non–perturbative ones, which would simplify mathematical representations and open new physical insights of the systems under consideration. In general panorama of quantum group theory and its $q$–deformed algebras, it would remind us the pioneering research of W. Heisenberg who has provided the famous commutation relations in quantum mechanics and developed $q$–deformed formalism. In $q$–deformed algebras, the key parameter is $q$ that can take any value in the range from 0 to 1 in which the formalism falls back to non–deformed when parameter $q$ approaches to one.

In the our previous work [13] we have investigated the relations between first, second quantization representations of quantum mechanics and $q$–deform algebra. In the case of harmonic oscillation, the axiom of first quantization (the commutation relation between coordinate and momentum operators) and the axiom of second quantization (the commutation relation between creation and annihilation operators) are equivalent, so that harmonic oscillator would be considered as a good object for investigating the link of two these representations in quantum mechanics. We have also shown that in the case of $q$–deformed harmonic oscillation, the breakdown of the axiom of second quantization leads to the breakdown of the axiom of first quantization, and visa versa. Using the coordinate representation we study fine structures of the vacuum state wavefunction depending on the deformation parameter $q$. A comparison with fine structures of Cooper pair in the coordinate representation [11] is also performed. We have shown that the real part $\mathrm{Re}\Psi_0$ of ground state wavefunction $\Psi_0$ of $q$-deformed harmonic oscillator is very similar the wave function $\Psi_C$ of Cooper pair in superconductivity.

In present work, we continue to develop the main idea of investigating the relation between wavefunctions of $q$–deformed harmonic oscillator and of Cooper pairs in space representations, and to answer the question which deformation parameter $q$ will be equivalent to Cooper pairs? We have explored the change of the envelope functions and studied the physical meaning of that given transition.

To simplify the mathematical expression, we will use the system of natural units with $\hbar = c = 1$.

## 2. $q$–deformed harmonic oscillator

In this section, we recall some basic of $q$–deformed harmonic oscillator [13] that would provide the consistency of ongoing representation. We start with the $q$-deformed harmonic oscillator which is represented via Hamiltonian as

$$H = \frac{\omega}{2}\left(aa^\dagger + a^\dagger a\right), \tag{1}$$

where the creation and annihilation operators in the Hamiltonian (1) satisfy the commutation relations

$$\begin{aligned} \left[a, a^\dagger\right]_q &= aa^\dagger - qa^\dagger a \\ &= 1, \end{aligned} \tag{2}$$

in which $q$– deformation parameter takes values in the range $[0,1]$.

It is easy to obtain the energy spectrum of $q$–deformed harmonic oscillator taking the form

$$E_n = \frac{\omega}{2}\left([n]_q + [n+1]_q\right),$$

where

$$[n]_q = \frac{1-q^n}{1-q},$$

are the $q$-integers that differ from natural numbers. The properties of $q$–numbers are out of the scope of this work.

The values of ratio $\frac{[n]_q}{n}$ depend on deformation parameter $q$ are presented in the figure 1.

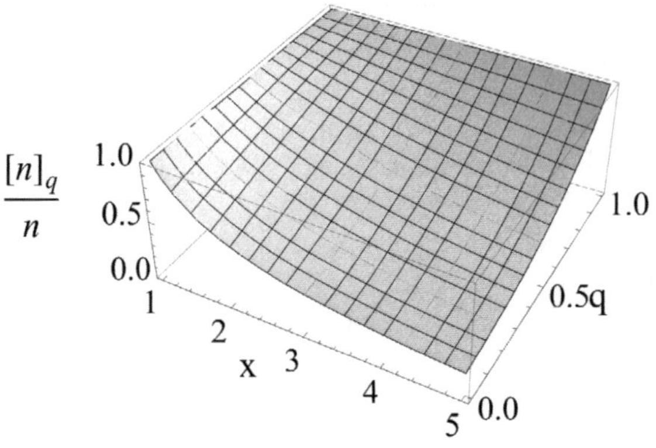


**Figure 1.** The values of ratio $\frac{[n]_q}{n}$ depend on deformation parameter $q$.

This ratio tends to zero as

$$\lim_{q\to 0} \frac{[n]_q}{n} = 0. \tag{3}$$

The maximum value of $[n]_{qmax} \to \infty$ when $q \to 1$ and we return to case of standard harmonic oscillator. By introducing new deformation parameter $\alpha$

$$\alpha = \sqrt{-\frac{1}{2} \ln q}, \tag{4}$$

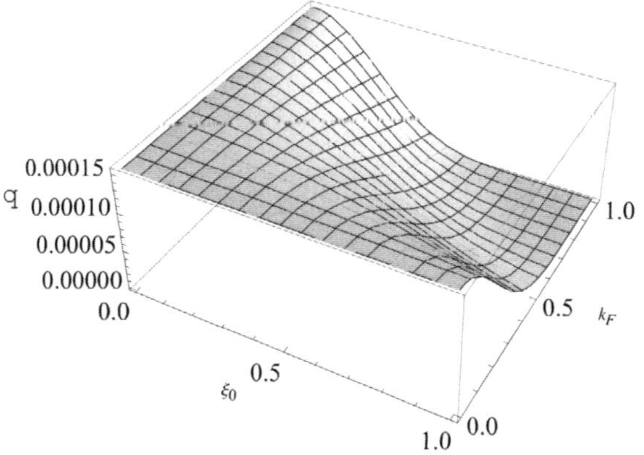


**Figure 2.** Drawing of deformed parameter $q$ as the function of new one $\alpha = \frac{3}{2}a$, where $a$ is defined in the second line of equation (7), $q = \exp(-2\alpha^2)$.

taking values in the interval $[0, \infty]$, the ground state wave function of $q$-deformed harmonic oscillator is found and given in the form

$$\Psi_D(x) = C_D \exp\left(-\frac{x^2}{2} + i\frac{3}{2}\alpha x\right), \tag{5}$$

where $C_D$ is the normalized constant. It is obvious that in the case of $q \to 1$, i.e. $\alpha \to 0$ standard harmonic oscillator is recovered.

The density operator of $q$-deformed harmonic oscillator is defined by

$$\rho(x) = \Psi_0^+(x)\,\Psi_0(x)\,,$$

so that the expectation value of ground state is

$$\langle 0\,|\rho|\,0\rangle = \langle 0\,|\rho_0|\,0\rangle = \frac{1}{2}.$$

Values of density vacuum fluctuations for both harmonic oscillator and $q$-deformed harmonic oscillator are the same. Real and imaginary parts of ground state wave function are presented in figure 3.

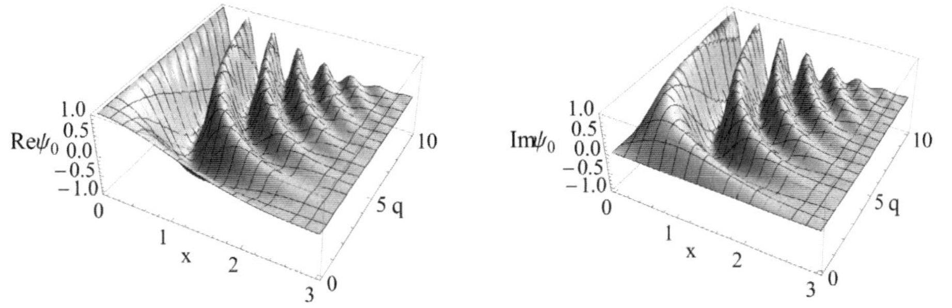


**Figure 3.** Real part Re and imaginary part Im of ground state wave function.

As expected, when $q = 1$, the wave function falls back to harmonic oscillator case with real part of Gaussian form, and imaginary part of zero.

## 3. Comparison with Cooper pairs in superconductivity

From condensed matter physics we have learn that the real part of wave function plays important role. In this section we try to explore the physical meaning of real part of ground state wave function of $q$-deformed harmonic oscillator. For comparison we take the well known case of Cooper pairs in superconductivity. In a long history of the BCS theory, the Cooper pair usually analyzed in the momentum space. The first investigation of Cooper pair in coordinate space was done in the work [11], where it was shown that this leads to a spherically symmetrical quasi–atomic wave function with an identical "onion–like" layered structure for each of the electrons constituting the Cooper pair.

The internal structure of Cooper pair wave function (also called the singlet pairing function or the Gorkov's wave function) is given by

$$\Psi_C(r) = C_C \cos(k_F r)\, K(r/\pi\xi_0)\,, \tag{6}$$

where $C_C$ is the normalized constant, $k_F$ is the Fermi wave vector at the surface of the Fermi sea and $K_0$ is the zero–order modified Bessel function with an asymptotic form that is similar to an exponential for large $x$.

Denote dimensionless variable $x$ and parameter $a$

$$\begin{aligned} x &= \tfrac{r}{\pi\xi_0}, \\ a &= \pi k_F \xi_0, \end{aligned} \tag{7}$$

the Cooper pair wave function can be rewritten as

Journal of Physics: Conference Series **726** (2016) 012017 doi:10.1088/1742-6596/726/1/012017

$$\Psi_C(r) = C_C \cos(ax) K_0(x) . \tag{8}$$

The Cooper wave function in coordinate representation corresponds to the standing waves with a spatial modulation $\alpha$. Typically values of these parameters are $1/k_F \sim 0.1nm$ and $\xi_0 \sim 100\,nm$. These waves rapidly oscillate around $k_F$, modulate slowly varying envelope with a characteristic scale off $\xi_0$. Taking the value $k_F\xi_0 = 20/\pi$, or $a = \pi k_F \xi_0 = 20$, and corresponding $C_C \simeq$, the wave function of Cooper pair is plotted in the figure 4.

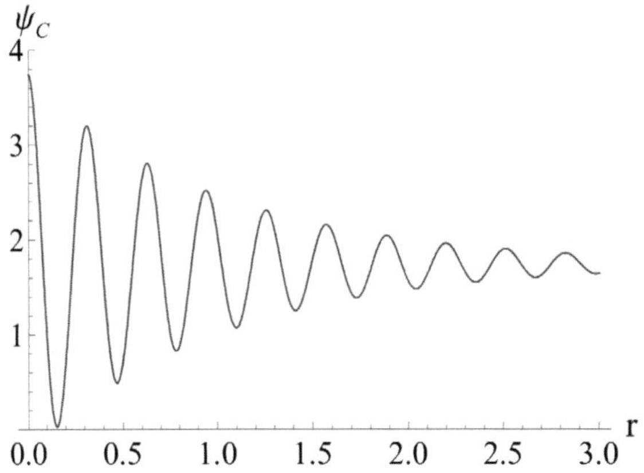


**Figure 4.** The wave function of Cooper pair in superconductivity.

Graphically, we could realize that the real part of ground state wave function of $q$–deformed harmonic oscillator (see figure 3) is very similar with the wave function of Cooper pair.

The real part Re of ground state wave function of $q$-deformed harmonic oscillator with $\alpha = 3a/2 = 30$ and exponential envelope function $\lambda = 10/3$ is presented in figure 5.

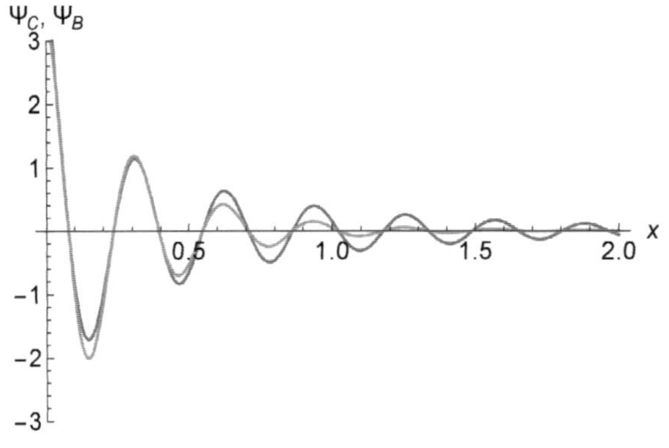


**Figure 5.** The real part Re of ground state wave function of $q$-deformed harmonic oscillator with $\alpha = 3a/2 = 30$ and exponential envelope function $\lambda = 10/3$.

The similarity between real parts of wave functions of $q$-deformed harmonic oscillator and of Cooper pair seems to be not simply accidental, that it supports a serious question of why the forms of the wave functions are so similar? We are going to find the reasons of this graphical similarity in below.

## 4. Transition between Gaussian and Boltzmann forms of envelope functions

As an attempt to extract physical meanings of $q$–deformation algebra, we, in this section, study the similarity of envelop functions in different cases by mean of graphical comparison. First we realize that the zero–order modified Bessel function $K_0$ is very look like the exponential function with $\lambda \simeq 10/3$ [13]. In following figures 6, 7 and 8, the Cooper pair's $F_C(x)$, Bortzmann–like $F_B(x)$ and Gaussian–like $F_G(x)$ envelope functions are pairwise presented.

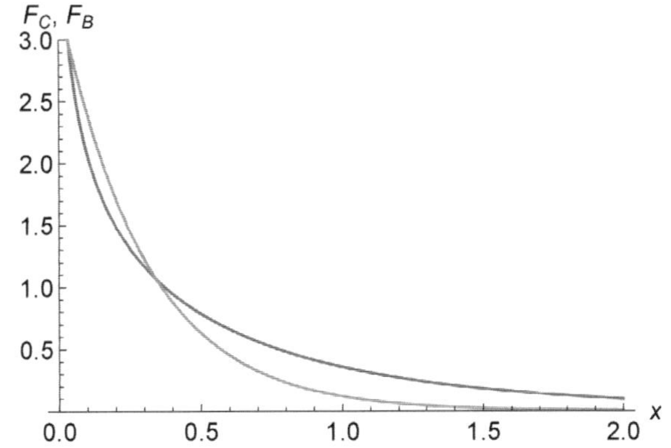


**Figure 6.** The Cooper pair's $F_C(x)$ and Bortzmann–like $F_B(x)$ envelope functions.

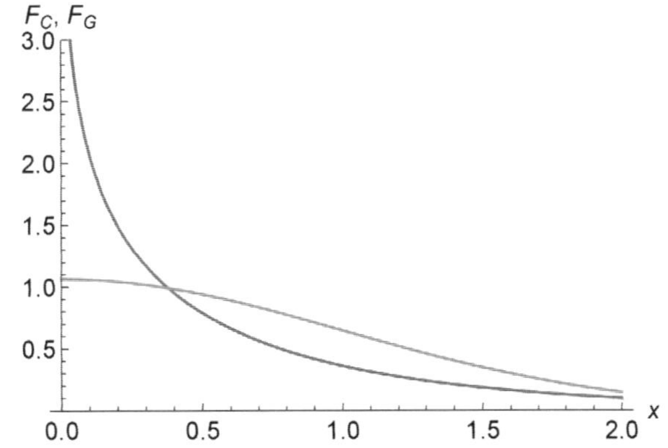


**Figure 7.** The Cooper pair's $F_C(x)$ and Gaussian–like $F_G(x)$ envelope functions.

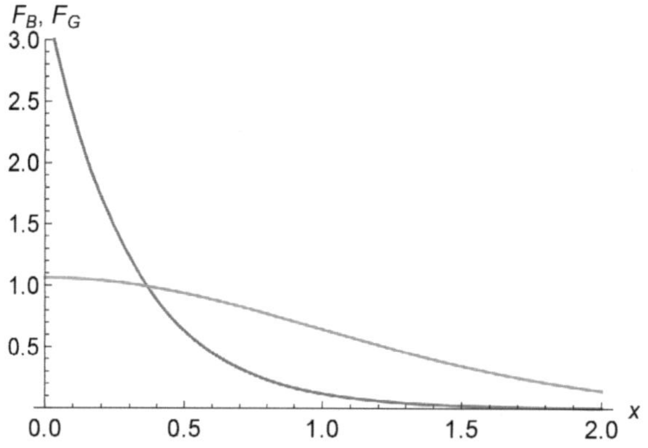


**Figure 8.** The Gaussian–like $F_G(x)$ and Bortzmann–like $F_B(x)$ envelope functions.

The phenomena of Boltzmann–Gaussian transition of the envelope functions are quite common in study of complexity (for example in economy see [14, 15]). That might be the links to the symmetry of the system where the Boltzmann form of the envelope function is corresponding to the translation symmetry of the system and Gaussian one corresponds to other symmetry. It is well–known in the superconducting state, the Cooper pair is formed and can move in system as free quasi–particle without resistance, i.e. some translation symmetry for Cooper pair is occurred. This interesting topic would be the subject of our next research.

## 5. Discussion

In current work, we have presented the relation between wave functions of $q$–deformed harmonic oscillator and Cooper pairs in space representations and the interchangeability of their set–parameters, $\{q, \alpha\} \longleftrightarrow \{k_F, \xi_0\}$. We have also found that a strong deformation where $q \simeq 0$ is corresponding to Copper pairs in superconductivity.

Different types of the envelope functions such as Cooper–, Boltzmann–, Gaussian–, Mean Gaussian–likes have been investigated and compared. We have explored the transition from Boltzmann form to Gaussian form of the envelope functions of normal and superconducting states, and pointed out some possible physical meanings and underlying physics of that transition.

We consider necessity to rewrite the Hamiltonian of deformed harmonic oscillator to describe the phenomenon of phase transitions. The connections between the $q$-deformed harmonic oscillators, coherent states, and the physical meanings of the envelope functions will be the topic of our future consideration.

## Acknowledgments

This research is funded by Vietnam National Foundation for Science and Technology Development (NAFOSTED) under grant number 103.01-2015.42.

## References

[1] Angelova M, Dobrev V and Frank A 2004 *EPJ* D **31** 27
[2] Mizrahi S S, Camargo Lima J P and Dodonov V V 2004 *J. Phys.* A **37** 3707
[3] Eremin V V and Meldianov A A 2008 *arXiv:0810.1967v1*
[4] Bagheri Harouni M, Roknizadeh R and Naderi M H 2009 *J. Phys.* A **42** 045403
[5] Abe S, Beck C and Cohen E G D 2007 *Phys. Rev.* E **76** 031102
[6] Abe S 2009 *Phys. Rev.* E **79** 041116
[7] Lavagno A and Swamy P P 2010 *Phys. A* **389** 993
[8] Algin A and Senay M 2012 *Phys. Rev.* E **85** 041123
[9] Andrade F M and Silva E O 2013 *Phys. Lett.* B **719** 467

[10] Tkachuck V M 2013 *Phys. Rev.* A **86** 062112
[11] Kadin A M 2007 *JSNM* **20** 285
[12] Vinh N G, Ngu M V, Lan N T, Thanh L T K and Viet N A 2015 *J. Phys.: Conf. Series* **627** 012022
[13] Ngu M V, Vinh N G, Lan N T, Thanh L T K and Viet N A 2015 *J. Phys.: Conf. Series* **627** 012021
[14] Anh C T, Lan N T and Viet N A 2014 *J. Phys.: Conf. Series* **537** 012005
[15] Anh C T, Lan N T and Viet N A 2015 *J. Phys.: Conf. Series* **627** 012025

Journal of Physics: Conference Series **726** (2016) 012018 doi:10.1088/1742-6596/726/1/012018

# The possible conection between $q$-defomed hamornic oscillator formation and anharmonicity

**Ngo Gia Vinh**[1], **Man Van Ngu**[2], **Nguyen Tri Lan**[3], **Luu Thi Kim Thanh**[4] **and Nguyen Ai Viet**[3]

[1] Bac Ninh Department of Education and Traning, Bac Ninh, Vietnam
[2] Hung Yen Industrial Colleage, Hung Yen, Vietnam
[3] Institute of Physics, VAST, 10 Dao Tan, Ba Dinh, Hanoi, Vietnam
[4] Hanoi Pedagogical University No.2, Nguyen Van Linh Street Vinh Phuc, Vietnam

E-mail: `ngogiavinh@bacninh.edu.vn`

**Abstract.** In our previous article, the connection between $q$–deformed harmonic oscillation and Morse–like asymmetric potential is investigated. In present work, a possibility of the connection between $q$-deformed harmonic oscillator and anharmonic symmetric potential is in detail considered. For simplicity, we take the inverse square cosine–hyperbolic form of potential, i.e Pöschl–Teller potential. The relation between the deformation parameter $q$ and the set of parameters of anharmonic symmetric potential was found. The correspondence of two types of connections between $q$-deformed harmonic oscillator with asymmetric and symmetric potentials are discussed.

## 1. Introduction

In the last twenty years, quantum group and deformed Heisenberg algebras with $q$-deformed harmonic oscillator have been a subject of intensive investigation. This approach has found some useful applications in various branches of physics and chemistry [1, 2, 3, 4, 5, 6, 7]. The method of $q$–deformed quantum mechanics was developed on the base of Heisenberg commutation relation (the Heisenberg algebra) within some typical generalizations. The main parameter of this method is the deformation parameter $q$ which is usually considered to variety in the range $0 < q < 1$, and the models have been constructed that the behaviors of studying objects reduce to theirs conventional counterparts as $q \to 1$.

The Morse potential finds an important role in describing the interaction among atoms in diatomic and even in polyatomic molecules [8, 9, 10, 11, 12, 14] of atomic and molecular physics. Despite its quite simple form, the Morse potential describes very well the vibrations of diatomic molecules. This is because that four–particle complex system (two heavy atomic nuclei with positive charge and two light electrons with negative charge) can be reduced to relative motion of two atomic nuclei in an effective potential which is average Coulomb interaction of nuclei and electron clouds. The Morse–like potential models just work with a simple one-dimensional three-parameter effective potential, and find many applications in condensed matter, bio–physics, nano–science and quantum optics.

The Morse potential in algebraic approach can be written in terms of the generators of $SU(2)$. The quantum relation between $q$-deformed harmonic oscillator and the Morse potential was considered in [10], where then the anharmonic vibrations in the Morse potential have been described as the levels of $q$-deformed harmonic oscillator. The extended $SU(2)$ model ($q$-Morse potential) has been also developed to compare with phenomenological Dunham expansion and experimental data for numbers of diatomic molecules [10]. In this work, by considering deformed algebra as mathematical object and atomic effective potential as physical model, we use this relation in inverse way to investigate properties of $q$-deformed harmonic oscillator on the base of the Morse potential.

In one hand, the potential of harmonic oscillation is parabolic with infinity equal–step levels. In other hand, we show that the potential of $q$–deformed harmonic oscillator can be described as Morse-like anharmonic potential with finite unequal–step levels. The relation between the deformation parameter $q$ and the set of parameters of Morse–like anharmonic potential was found. We have also investigated the partition function and some thermodynamic properties of $q$-deformed harmonic oscillator.

In the our previous work [14] we have shown that mathematical deformation properties can be represented and understood in the language of physical object, which can be described by an anharmonic potential.

The asymmetric representation of deformed harmonic oscillators was investigated in [14] with the Morse potential. And as a further step, in this work we study the symmetric representation of deformed harmonic oscillators with corresponding potential.

## 2. Harmonic oscillator and $q$–deformed harmonic oscillator

In $q$-deformed harmonic oscillator, creation $a^\dagger$ and annihilation $a$ operators of $q$–deformed harmonic oscillator satisfy the commutation relation

$$[a, a^\dagger]_q = aa^\dagger - qa^\dagger a = 1, \tag{1}$$

where $q$ is deformation parameter taking values in [0, 1].

In the second quantization representation, the Hamiltonian operator of $q$-deformed harmonic oscillator is written as

$$H = \frac{\omega}{2}\left(aa^\dagger + a^\dagger a\right). \tag{2}$$

As the results of simple algebraic manipulations, energy spectrum of $q$-deformed harmonic oscillator is obtained as follows

$$E_n = \frac{\omega}{2}\left([n]_q + [n+1]_q\right), \tag{3}$$

where $[n]_q = \frac{1-q^n}{1-q}$ is the $q$-integer which differs from natural numbers. For very small derivation from unity $\varepsilon = 1 - q$, the energy spectrum becomes quadratic if the higher order contribution $C = O\left((\varepsilon^2)\right)$ is neglected

$$E_n = \hbar\omega\left(n + \frac{1}{2} - \frac{n^2}{2}\varepsilon + C\right). \tag{4}$$

In result (4), the energy levels are represented by a system of parallel lines are not equidistant. The extent depends on the deformation parameter $q$.

## 3. Physical model for $q$–deformed harmonic oscillator

In the case of $q = 1 (\varepsilon = 0)$, the energy levels return to non–deformed expression, i.e.

$$E_n = \hbar\omega\left(n + \frac{1}{2}\right), \tag{5}$$

Journal of Physics: Conference Series **726** (2016) 012018 doi:10.1088/1742-6596/726/1/012018

we obtained the energy levels of the harmonic oscillator, the gaps between the energy levels are constant (Figure 1).

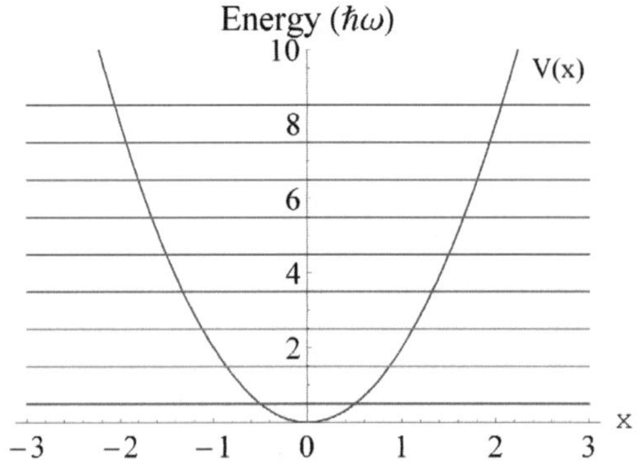


**Figure 1.** Energy spectrum has infinity equal-step levels form.

In the range of parameter values $0 < q < 1$, the energy levels are represented by a system of parallel lines are not equidistant.

In all previous work, we have studied the deformation parameter $q$ through Morse potential, which is the asymmetric anharmonic potential. In this ongoing one, the deformation parameter $q$ with symmetric anharmonic potential form (Figure 2) will be under consideration.

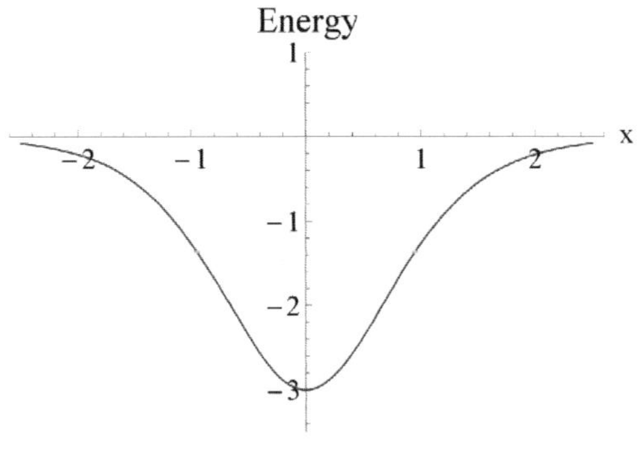


**Figure 2.** The potential is symmetric anharmonic.

To easily compare potential energy shapes, we observe their appearance on the same coordinate system (Figure 3). The symmetric potential is anharmonic and energy spectrum has finite unequal-step levels, for comparison also plotted harmonic spectrum of corresponding parabolic potential.

## 4. Symmetric representations

The symmetric potential has an important role in describing the interaction of atoms in diatomic, and even polyatomic molecules. In the case, the symmetric potential is anharmonic, energy spectrum has finite unequal–step levels as presented in the Figure 4, where energy spectrum of symmetric anharmonic and corresponding parabolic potentials are compared.

The symmetric anharmonic potential chosen in study is Pöschl – Teller potential

$$V_s(x) = -\frac{U_0}{\cosh^2(\alpha x)}, \tag{6}$$

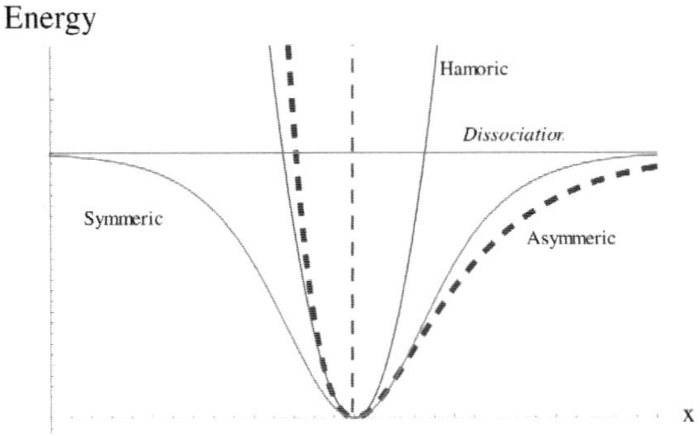


**Figure 3.** The harmonic potential, anharmonic potential and symmetric potential.

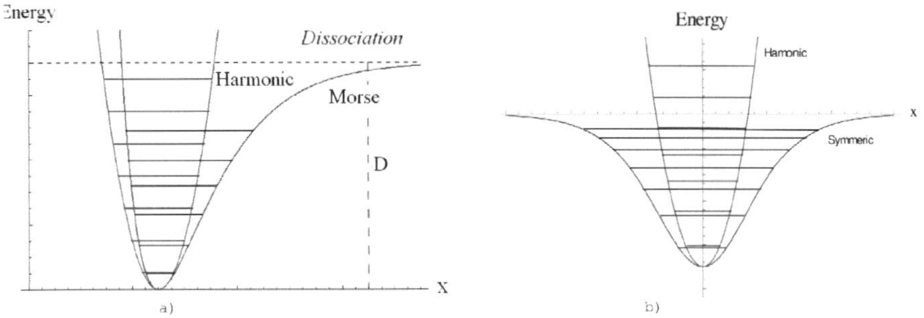


**Figure 4.** Energy Levels: a) Morse potential and Harmonic potential. b) Symmetric and Harmonic potential.

which is an effective potential with a set of two parameters $U_0, \alpha$ taken from experimental data.

The Schrödinger equation of symmetric anharmonic potential (6) has been exactly and analytically solved [13]. The eigenvalues of this Hamiltonian read

$$E_s = -\frac{\alpha^2\hbar^2}{8m}\left[-(1+2n)+\sqrt{1+\frac{8U_0}{\alpha^2\hbar^2}}\right]^2. \tag{7}$$

By introducing new quantity

$$\omega_s = \frac{\alpha^2\hbar}{2m}\sqrt{1+\frac{8U_0}{\alpha^2\hbar^2}}, \tag{8}$$

the energy level of system under consideration can be expressed as

$$E_s = \hbar\omega_s\left[n+\frac{1}{2}-\frac{\varepsilon}{2}\left(n+\frac{1}{2}\right)^2+C_s\right], \tag{9}$$

where $\omega_s$ is the frequency of symmetric oscillator.

Following the tasks presented in our last work [14], the deformation parameter $q_s$ of the given potential is found as

$$q_s = 1-\frac{2}{\sqrt{1+\frac{8U_0}{\alpha^2\hbar^2}}}. \tag{10}$$

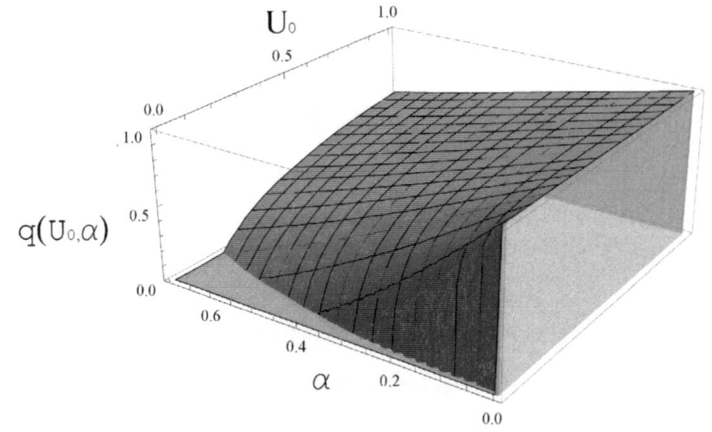


**Figure 5.** $q$-deformed symmetric representation depending parameters $U_0, \alpha$.

It is obvious that this system has a finite number of energy levels with maximum number $n_{S.max}$

$$\begin{aligned} n_{S.max} &= \left[\frac{1}{(1-q_S)}\right] \\ &= \left[\frac{1}{2}\sqrt{1+\frac{8U_0}{\alpha^2\hbar^2}}\right], \end{aligned} \tag{11}$$

where the notation $[f]$ is the integer part of the number $f$.

Using the above relationship, we propose the deformation model (SPD model), based on the symmetric potential, for investigating the properties of $q$–deformed harmonic oscillator. The main idea is that the role of parabolic potential for harmonic oscillator would be replaced by the symmetric representation potential for $q$-deformed harmonic oscillation.

In above proposed model, every given value of deformation parameter $q_S$ in the interval from zero to unity, $q_S \in [0,1]$, can be described by a symmetric potential with the largest number $n_{S.max}$ determined by the expression (11). Here, we note the well–defined one–to–one correspondence between $q_S$ and $n_{S.max}$. The values of largest number depending on deformation parameter $q_S$ and on parameters of material system are plotted in figure 6 and in figure 7.

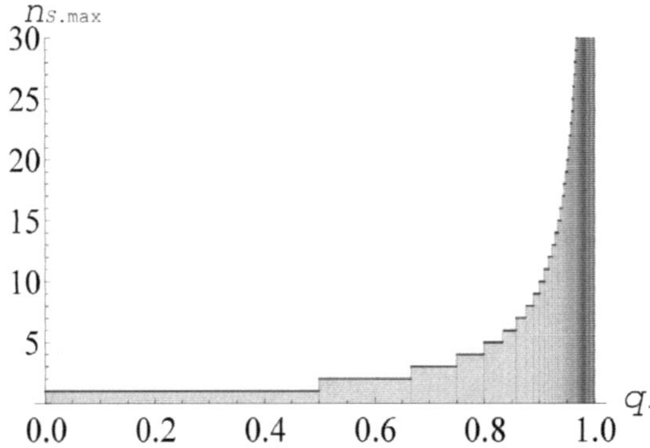


**Figure 6.** The values of largest number $n_{S.max}$ depending on deformation parameter $q_S$.

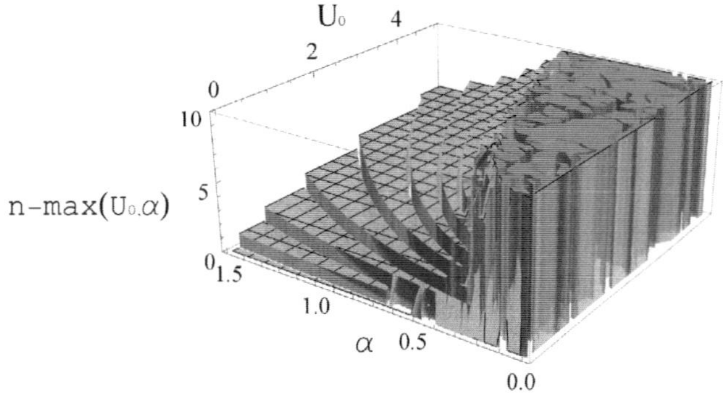


**Figure 7.** The values of largest number $n_{S.max}$ depending on given input parameters $U_0, \alpha$.

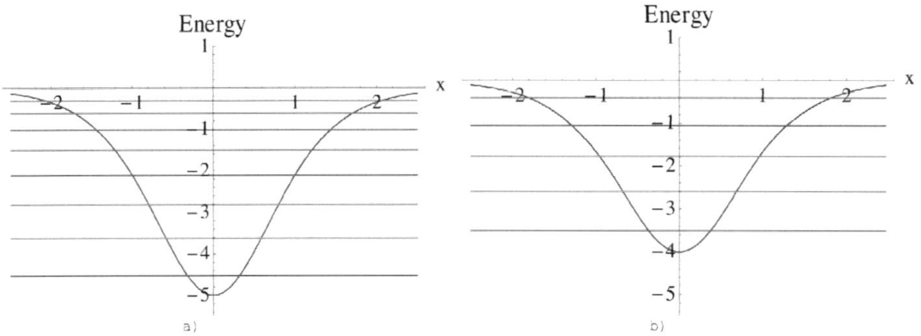


**Figure 8.** Dependence of number of energy levels on $U_0$ while $\alpha$ is fixed ($\alpha = 1$), a) $U_0 = 5$, b) $U_0 = 4$.

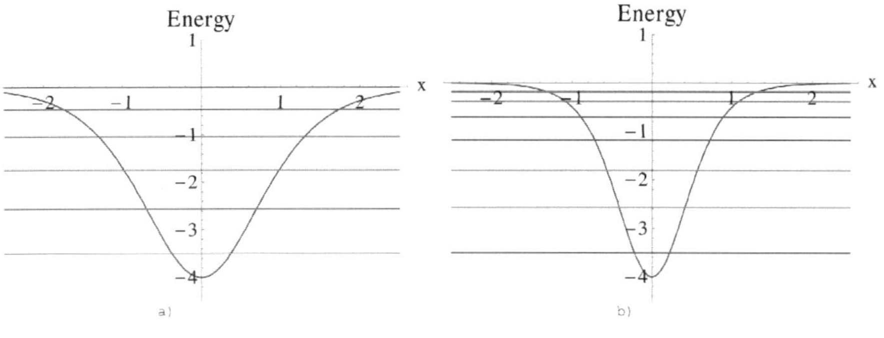


**Figure 9.** Dependence of number of energy levels on $\alpha$ while $U_0$ is fixed ($U_0 = 4$), a) $\alpha = 1.2$, b) $\alpha = 1.8$.

It is shown that the total number of energy levels rapidly grows up when $q$ tends to unity. In the limit of weak deformation $q \to 1$, $n_{\max} \to \infty$ . In contrast, when $q$ tends to 0.5 only one level can be found, i.e. in the strong deformation limit $q \to 1/2$, $n_{\max} \to 1$. The actual working range of deformation parameter $q$ is not to be in $q \in [0,1]$, but in more narrow half range $0.5 < q < 1$, where the physical system has more than one energy level.

The last but not least, the main parameters of harmonic, $q$–deformed harmonic, symmetric and asymmetric potentials are compared in Table 1.

| | Harmonic | $q$-deformed Harmonic | Asymmetric | Symmetric |
|---|---|---|---|---|
| Potential | $-\frac{1}{2}kx^2$ | | $D\left(1-e^{-k(x-x_0)}\right)$ | $-\frac{U_0}{\cosh^2(\alpha x)}$ |
| $\varepsilon$ | 0 | $1-q$ | $\frac{\hbar k}{\sqrt{2mD}}$ | $\frac{2}{\sqrt{1+\frac{8mU_0}{\alpha^2\hbar^2}}}$ |
| Frequency | $\sqrt{k/m}$ | $\hbar\omega$ | $\hbar\sqrt{\frac{2D}{mk^2}}$ | $\frac{\alpha^2\hbar^2}{2m}\sqrt{1+\frac{8mU_0}{\alpha^2\hbar^2}}$ |
| $q$ | 1 | $1-\varepsilon$ | $1-\frac{\hbar k}{\sqrt{2mD}}$ | $1-\frac{2\alpha\hbar}{\sqrt{\alpha^2\hbar^2+8mU_0}}$ |
| $n_{max}$ | $\infty$ | $\left[\frac{1}{(1-q)}\right]$ | $\left[\sqrt{\frac{8Dk^2}{\hbar^2}}-1\right]$ | $\left[\frac{1}{2}\sqrt{1+\frac{8mU_0}{\alpha^2\hbar^2}}\right]$ |

**Table 1.** Comparison of asymmetric and symmetric representations.

## 5. Discussion

Comparing the energy spectrum of $q$-deformed harmonic oscillator (4) and energy spectrum of symmetric anharmonic potential (7) and neglecting the higher order contribution $O\left(\left(\varepsilon^2\right)\right) = O\left(\left((1-q)^2\right)\right)$, we realize the relations

$$\omega_s \leftrightarrow \omega, \tag{12}$$

$$\frac{2}{N} \leftrightarrow \varepsilon. \tag{13}$$

This is the relation between $q$-deformed harmonic oscillator and symmetric anharmonic vibrations, providing the analogy of $q$–deformation and symmetric anharmonicity. The change of energy spectrum from linear to quadratic form is reasoned by deformations of commutation relations in the mathematical algebraic approach, while in case of the symmetric potential by appearance of periodic distribution of atoms or molecules.

The main results of this work are the consequences of the expression of energy levels (7), on which we proposed a new symmetric potential deformation SPD model for investigating properties of $q$-deformed harmonic oscillator via the symmetric potential (6). Instead of parabolic potential for harmonic oscillation, the symmetric anharmonic potential is used to study the $q$-deformed harmonic oscillation, and as it is expected, the mathematical deformation properties now would be described and understood in the language of physical anharmonic behaviors.

In our proposed SPD model, $q$-deformed harmonic oscillator can be described by the corresponding symmetric Pöschl–Teller potential. With every given value of deformation parameter in the interval $q_S \in [0,1]$, we would determine the largest number $n_{S.max}$ and the deformed energy spectrum $E_q(n)$ of $q$–deformed harmonic oscillation.

In one hand, the asymmetric representation of deformed harmonic oscillators by using a symmetric Morse potential [14] can be applied to describe interaction between two atoms molecules, where the space among them is clearly non–symmetric. In other hand, the possible applications of symmetric representation for deformed harmonic oscillators are atomic chain, symmetric atoms, where the space is translationally symmetric. This models could be expanded and generalized to investigate many other physical problems such as composite bosons with deformation which is the topic of our ongoing research.

## Acknowledgments
This research is funded by Vietnam National Foundation for Science and Technology Development (NAFOSTED)under grant number 103.01-2015.42.

## References
[1] Eremin V V and Meldianov A A 2008 *arXiv:0810.1967v1*
[2] Mirzrahi S S, Camargo I P and Dodonov V V 2004 *J. Phys.* A **37** 3707
[3] Abe S 2009 *Phys. Rev.* E **79** 041116
[4] Lavagno A and Swamy P P 2010 *Phys.* A **389** 993
[5] Algin A and Senay M 2012 *Phys. Rev.* E **85** 041123
[6] Andrade F M and Silva E O 2013 *Phys. Lett.* B **719** 467
[7] Tkachuck V M 2013 *Phys. Rev.* A **86** 062112
[8] Angelova M and Franck A 2005 *Phys. of automic nuclei.* **68** 1689-1697
[9] Dong S H, Lemus R and Frank A 2002 *J. Quant. Phys.* **86** 433
[10] Angelova M, Dobrev V and Franck A 2004 *Eur. Phys. J.* D **31** 27
[11] Kaplan I G 2003 *Handbook of Molecular Physics and Quantum Chemistry* (Wiley)
[12] Lima E F and Hornos J E M 2005 *J. Phys.* B **38** 815
[13] Landau L D and Lifshitz E M 1991 vol 3 *of Course of Theoretical Phys.*
[14] Vinh N G, Ngu M V, Lan N T, Viet N A and Thanh L T K 2015 *J. Phys.: Conf. Series* **627** 012022

# Energy spectrum inverse problem of $q$-deformed harmonic oscillator and WBK approximation

**Nguyen Anh Sang[1], Do Thi Thu Thuy[1], Nguyen Thi Ha Loan[1], Nguyen Tri Lan[2] and Nguyen Ai Viet[2]**

[1]Hanoi Pedagogical University 2, Nguyen Van Linh Street, Phuc Yen Town, Vinh Phuc, Vietnam.

[2]Institute of Physics,Vietnam Academy of Science and Technology, 10 Dao Tan, Ha Noi, Viet Nam

E-mail: anhsangk13@gmail.com

**Abstract.** Using the connection between q-deformed harmonic oscillator and Morse–like anharmonic potential we investigate the energy spectrum inverse problem. Consider some energy levels of energy spectrum of $q$–deformed harmonic oscillator are known, we construct the corresponding Morse-like potential then find out the deform parameter $q$. The application possibility of using the WKB approximation in the energy spectrum inverse problem was discussed for the cases of parabolic potential (harmonic oscillator), Morse–like potential ($q$–deformed harmonic oscillator). so we consider our deformed–three–levels simple model, where the set–parameters of Morse potential and the corresponding set–parameters of level deformations are easily and explicitly defined. For practical problems, we propose the deformed-three-levels simple model, where the set–parameters of Morse potential and the corresponding set-parameters of level deformations are easily and explicitly defined.

## 1. Introduction

Recently, quantum group and deformed Heisenberg algebras with $q$–deformed harmonic oscillator have been a subject of intensive investigation. This approach is found some applications in various branches of physics and chemistry [1–6]. The method of $q$–deformed quantum mechanics was developed on the base of Heisenberg commutation relation (the Heisenberg algebra). The main parameter of this method is the deformation parameter $q \in [0, 1]$.

The Morse potential has an important role in describing the interaction of atoms in diatomic and even polyatomic molecules [7–10] in atomic and molecular physics. Despite its quite simple form, the Morse potential describes very well the vibrations of diatomic molecules. This is because that four–particle complex system (two heavy atomic nuclei with positive charge and two light electrons with negative charge) can be reduced to relative motion between two atomic nuclei in an effective potential which is average Coulomb interaction of nuclei and electron clouds. Morse–like potential models just work with a simple one–dimensional three–parameter effective potential, found many applications in condensed matter, biophysics, nano–science and quantum optics.

The Morse potential in algebraic approach can be written in terms of the generators of SU(2). The quantum relation between $q$–deformed harmonic oscillator and Morse potential was considered in [8], where then the anharmonic vibrations in Morse potential have been described as the levels of $q$–deformed harmonic oscillator. The extended SU(2) model ($q$-Morse potential) also develop to compare with phenomenological Dunham's expansion and experimental data for numbers of diatomic molecules [8, 9]. In this work, considering deformed algebra is mathematical object and atomic effective potential is physical model, we use this relation in inverse way to investigate properties of $q$–deformed harmonic oscillator on the base of the Morse potential.

In the our previous works [11, 12] we have shown that the potential of harmonic oscillator is parabolic with infinity equal–step levels, and the potential of $q$-deformed harmonic oscillator might be described as Morse–like anharmonic potential with finite unequal–step levels. The relation between the deformation parameter $q$ and the set of parameters of Morse–like anharmonic potential was found.

In this work, using the connection between $q$–deformed harmonic oscillator and Morse–like anharmonic potential we investigate the inverse problem of the energy spectrum. Consider some energy levels of energy spectrum of $q$–deformed harmonic oscillator are known, we construct the corresponding Morse–like potential then find out the deform parameter $q$ and associate with it the new parameter $\delta$ characterized the level deformation. The possibility of using the WKB approximation in the energy spectrum inverse problem was discussed for the cases of parabolic potential (harmonic oscillator) and Morse–like potential ($q$–deformed harmonic oscillator).

## 2. Semiclassical Wentzel–Kramers–Brillouin WKB method

### *2.1. Bohr–Sommerfeld quantization conditions*

It is well–known that the Bohr–Sommerfeld quantization conditions are a procedure for selecting out certain discrete set of states of a classical integrable motion as allowed states. These are like the allowed orbits of the Bohr model or standing waves of de Broglie of the atom; the system can only be in one of these states, but not in any states in between.

A particle in a one–dimensional potential can be described by the Schrödinger equation:

$$\left[-\frac{\hbar^2}{2\mu}\frac{\partial^2}{\partial_x^2}+V(x)\right]\psi(x)=E\psi(x), \tag{1}$$

where $\mu$ is the particle mass, $\hbar$ is the Planck constant.

The semiclassical Wentzel–Kramers–Brillouin WKB method is based on the ansatz

$$\psi(x)=A(x)\exp\left[iS(x)/\hbar\right]. \tag{2}$$

In the limit as $\hbar\to 0$, the action $S(x)$ in the exponential satisfies the Hamilton–Jacobi equation for the action function

$$E=\frac{1}{2\mu}\left(\frac{dS}{dx}\right)^2+V(x), \tag{3}$$

we have

$$\begin{aligned}S'(x) &= \tfrac{dS}{dx}\\ &= \sqrt{2\mu\left[E-V(x)\right]}.\end{aligned} \tag{4}$$

Journal of Physics: Conference Series **726** (2016) 012019 doi:10.1088/1742-6596/726/1/012019

It is then shown that this usually leads to the Bohr–Sommerfeld quantum conditions on periodic orbits

$$\oint dx\,\sqrt{2\mu\,[E-V(x)]} = 2\pi\hbar\left(n+\frac{1}{2}\right), \tag{5}$$

with $n = 0,\, 1, 2\,, 3\,, \ldots$
For one–dimensional problems, the cyclic integral can be replaced by $2\int_{x_1}^{x_2}$, where $x_1$ and $x_2$ are the classical turning points of the motion and $E = V(x)$.

### 2.2. Application of WKB method to the case of harmonic potential

As an example consider the linear harmonic oscillator with

$$V_0\,(x) = \frac{1}{2}Kx^2, \tag{6}$$

where $K$ is the spring constant, and

$$x_{1,2} = \pm\sqrt{2E/K}. \tag{7}$$

The quantum condition reads

$$\int_{-\sqrt{2E/k}}^{+\sqrt{2E/k}} dx\sqrt{2\mu\,(E-Kx^2/2)} = 2\pi\hbar\left(n+\frac{1}{2}\right), \tag{8}$$

which can be solved to give

$$\begin{aligned} E_{0n} &= \hbar\left(n+\tfrac{1}{2}\right)\sqrt{\tfrac{K}{\mu}} \\ &- \hbar\omega_0\left(n+\tfrac{1}{2}\right), \end{aligned} \tag{9}$$

where $\omega_0 = \sqrt{K/\mu}$. This is one of a small number of cases in which the WKB method gives the exact quantum–mechanical energies.

### 2.3. Application WKB method to the case of Morse potential

The semiclassical (WKB) method applied to one–dimensional problems with bound states often reduces to the Sommerfeld–Wilson quantization conditions, the cyclic phase–space integrals

$$\oint dx\,\sqrt{2\mu\,[E-V_M(x)]} = 2\pi\hbar\left(n+\frac{1}{2}\right). \tag{10}$$

It turns out that this formula gives the exact bound–state energies for the Morse oscillator with

$$V_M(x) = D\left[e^{-2k(x-x_0)} - 2e^{-k(x-x_0)}\right], \tag{11}$$

where $x_0$ is the minimum position, $D$ is the depth and $k$ is the width of potential.
The requisite integral can be reduced to

$$2\int_{x_1}^{x_2} dx\,\sqrt{E-V_M(x)}, \tag{12}$$

in which $x_1$ and $x_2$ are the classical turning points

$$x_{1,2} = x_0 + \frac{1}{k}\ln\left[\frac{D}{E}\left(-1 \pm \sqrt{1+\frac{E}{D}}\right)\right]. \tag{13}$$

The integral can be done "by hand", using the transformation, using the transformation $r = \exp\left[k\left(x - x_0\right)\right]$ followed by a contour integration in the complex plane, but can evaluate the integral explicitly, needing only the additional fact that $\ln\left(-1\right) = i\pi$. The result reads

$$\frac{2\pi\sqrt{2\mu}}{k}\left(\sqrt{D}+\sqrt{-E}\right) = 2\pi\hbar\left(n+\frac{1}{2}\right), \tag{14}$$

which can be rewritten

$$\left(\sqrt{D}+\sqrt{-E}\right) = \frac{\hbar k}{\sqrt{2\mu}}\left(n+\frac{1}{2}\right). \tag{15}$$

The above equation can be solved for $E$ to give

$$-E = \left[-\sqrt{D} + \left(n+\frac{1}{2}\right)\frac{k}{\sqrt{2\mu}}\hbar\right]^2. \tag{16}$$

This result is coincided with exact solution (see text book of Landau–Liftshitz: QM).
Denote the highest bound state is $n_{max} = \left[\sqrt{2\mu D}/\left(\hbar k\right)\right]$ where [...] represents the floor, which for positive numbers is simply the integer part.
We have energy spectrum

$$E_n = D\left[\frac{2}{n_{max}}\left(n+\frac{1}{2}\right) - \frac{1}{n_{max}^2}\left(n+\frac{1}{2}\right)^2 - 1\right], \tag{17}$$

with $n = 0,\ 1, 2\ , 3\ , ...,\ n_{max}$, or

$$\begin{aligned} E_{Mn} &= E_n - D \\ &= \hbar\omega_M\left[\left(n+\tfrac{1}{2}\right) - \tfrac{1}{2n_{max}}\left(n+\tfrac{1}{2}\right)^2\right], \end{aligned} \tag{18}$$

where $\omega_M$ is the Morse frequency, which equals

$$\begin{aligned} \omega_M &= \tfrac{2D}{\hbar n_{max}} \\ &= k\sqrt{\tfrac{2D}{\mu}}, \end{aligned} \tag{19}$$

The values of $D$, $k$, and $x_0$ of diatomic molecules are taken from experiments.
Actually, the Morse experimental parameters are the dissociation parameter $D$, the fundamental vibrational frequency $\omega_M$, the equilibrium internuclear distance $x_0$. and the reduced mass $\mu = m_1 m_2/\left(m_1 + m_2\right)$. The exponential parameter is given by $k = \omega_M\sqrt{\mu/2D}$ in appropriate units $k \simeq 1$. The Schrödinger equation for the Morse oscillator is exactly solvable, giving the vibrational eigenvalues

$$\epsilon_M = \omega_M\left(n+\frac{1}{2}\right) - \frac{\omega_M^2}{4D}\left(n+\frac{1}{2}\right)^2, \tag{20}$$

Unlike the harmonic oscillator, the Morse potential has a finite number of bound vibrational levels with $n_{max} \simeq 2D/\omega_M$.
Compare with the Dunham anharmonic parameter relations in the spectroscopy $\chi_e\omega_M = A = k^2/(2D)$and $\omega_M = 2n_{max}A = k\sqrt{2D/\mu}$ we have

$$\chi_e = \frac{k^2}{2D\omega_M} = \frac{k^2}{4D^2}\hbar n_{max} = k\sqrt{\frac{\mu}{2D}} \tag{21}$$

Introduce the energy differences as

$$\Delta E_{Mn} = \hbar\omega_M \left(E_{Mn} - E_{M(n-1)}\right), \tag{22}$$

and for lowest energy levels

$$\Delta E_{M1} = \hbar\omega_M \left(1 - \frac{1}{n_{max}}\right) \Delta E_{M2} = \hbar\omega_M \left(1 - \frac{2}{n_{max}}\right) \Delta E_{M3} = \hbar\omega_M \left(1 - \frac{3}{n_{max}}\right) ... \tag{23}$$

According to our Morse Potential for Deformation model (MPD model) [12], we can define a new deformation parameter $q_M$, which is satisfied the equation

$$\frac{1}{2n_{max}} = 1 - q_M, \tag{24}$$

we have

$$\Delta E_{M1} = \hbar\omega_M q_M,\ \Delta E_{M2} = \hbar\omega_M (2q_M - 1),\ \Delta E_{M3} = \hbar\omega_M (3q_M - 2)\ ... \tag{25}$$

From the two first equations we obtain the deformation parameter $q_M$

$$q_M = \frac{1}{2 - \frac{\Delta E_{M2}}{\Delta E_{M1}}}, \tag{26}$$

and the Morse frequency

$$\omega_M = \frac{1}{\hbar}(2\Delta E_{M1} - \Delta E_{M2}). \tag{27}$$

## 3. Harmonic approximation for Morse oscillator

In this part we find the harmonic approximation for Morse oscillator.
From definition of Morse potential easy to see that

$$\frac{dV_M(x)}{dx} = V_M^{(1)}(x) = D\left[-2ke^{-2k(x-x_0)} - 2ke^{-k(x-x_0)}\right], \tag{28}$$

and

$$\frac{d^2V_M(x)}{dx^2} = V_M^{(2)}(x) = D\left[4k^2e^{-2k(x-x_0)} - 2k^2e^{-k(x-x_0)}\right], \tag{29}$$

so

$$V_M(x_0) = -D,\ V_M^{(1)}(x_0) = 0,\ V_M^2(x_0) = 2Dk^2. \tag{30}$$

Using the Taylor series expansion around $x = x_0$, we found harmonic approximation potential for Morse oscillator

$$V_H(x) \simeq V_M(x_0) + V_M^{(1)}(x_0)(x - x_0) + \frac{1}{2}V_M^{(2)}(x_0)(x - x_0)^2 = -D + Dk^2(x - x_0)^2, \quad (31)$$

with the spring constant $K'/2 = Dk^2$and corresponding harmonic frequency $\omega_H = \sqrt{K'/\mu} = k\sqrt{2D/\mu}$. We have the relation

$$\omega_i = k_i\sqrt{2D_i/\mu}, \quad (32)$$

where $i = M, H, 0$.

## 4. Inverse problem of deformed harmonic oscillator

Suppose that the 3 first energy levels $E_0, E_1, E_2$ are known, so we have two energy separations $\Delta_2$ and $\Delta_2$

$$\Delta_1 = E_1 - E_0,\ \Delta_2 = E_2 - E_1. \quad (33)$$

Define the deformation parameter $q$

$$q = \frac{1}{2 - \frac{\Delta_2}{\Delta_1}}, \quad (34)$$

the oscillation frequency $\omega_q$

$$\omega_q = \frac{1}{\hbar}(2\Delta_1 - \Delta_2), \quad (35)$$

The highest bound state $N_q$ is

$$N_q = [1/2(1 - q)] = \left[\frac{(2 - \Delta_2/\Delta_1)}{2(1 - \Delta_2/\Delta_1)}\right], \quad (36)$$

where [...] represents the integer part.

If $\Delta_2 < \Delta_1$, then $0 < q < 1$, $N_q = /2(1 - q)$, we have the case of deformed harmonic oscillator with finite unequal-step levels, which can be characterized by a Morse-like potential

$$V_{Mq}(x) = D_q\left[e^{-2k_q(x-x_0)} - 2e^{-k_q(x-x_0)}\right]. \quad (37)$$

with the energy deep $D_q$

$$D_q = \frac{\hbar\omega_q}{(1 - q)} = \frac{(2\Delta_1 - \Delta_2)^2}{(\Delta_1 - \Delta_2)}, \quad (38)$$

The exponential parameter $k_q$

$$k_q = \omega_q\sqrt{\frac{\mu}{2D_q}} = \sqrt{1 - q}\sqrt{\frac{\omega_q\mu}{2\hbar}} = \frac{1}{\hbar}\sqrt{\frac{\mu}{2}(\Delta_1 - \Delta_2)}. \quad (39)$$

The corresponding parabolic potential $V_{Hq}(x)$ with the oscillation frequency $\omega_q = \frac{1}{\hbar}(2\Delta_1 - \Delta_2) = k_q\sqrt{2D_q/\mu}$ is

$$V_{Hq}(x) = -D_q + \frac{K_q}{2}(x - x_0)^2 = -D_q + \frac{1}{2}\mu\omega_q^2(x - x_0)^2. \quad (40)$$

If the separations are equal $\Delta_2 = \Delta_1$, then $q = 1$, $N = \infty$, we have the case of harmonic oscillator with infinite equal-step levels, which can be characterized by a parabolic potential $V_{H1}(x)$ with the oscillation frequency $\omega_1 = k_1\sqrt{2D_1/\mu}$

$$V_{H1}(x) = -D + \frac{1}{2}\mu\omega_1^2(x - x_0)^2. \quad (41)$$

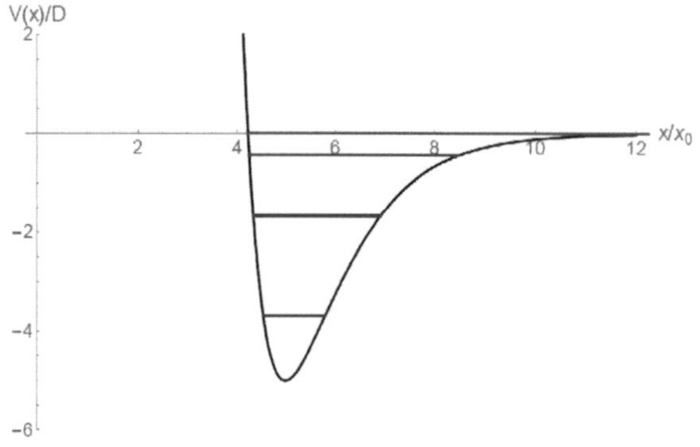


**Figure 1.** Morse potential

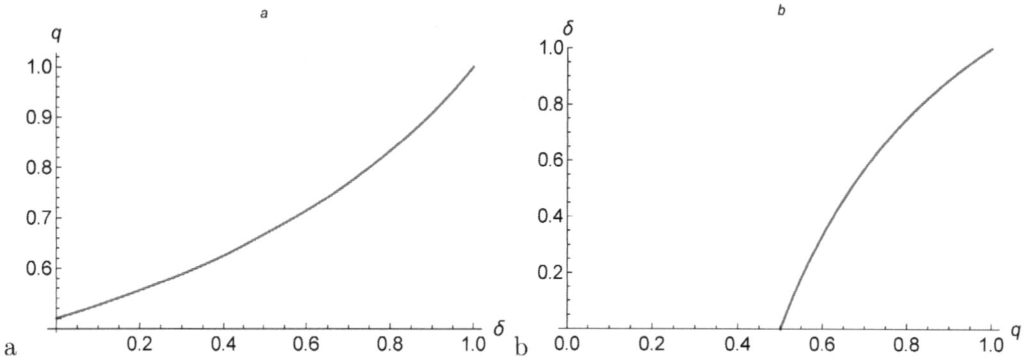


**Figure 2.** The relation between $q$ and $\delta$: a) $q$ on $\delta$, and b) $\delta$ on $q$.

## 5. Model example

Consider the first three energy levels $E_0$, $E_1$, $E_2$are known, we define two parameters $\Delta_1 = E_1 - E_0$, and $\Delta_2 = E_2 - E_1$. Introduce the ratio parameter $\delta$ as the main parameter of our model

$$\delta = \Delta E_2/\Delta E_1 = \Delta\varepsilon_2/\Delta\varepsilon_1. \tag{42}$$

For the case of Morse oscillator $0 < \delta < 1$ , example three levels Morse oscillator is presented in the figure 1. For the case of harmonic potential $\delta = 1$.

Taking also two values $r_0$ and $k_0 = (1/\hbar)\sqrt{\Delta_1\mu/2}$ are the equilibrium distance and imaginary impulse of anharmonic Morse oscillator (with q=1 is harmonic case), we will express all physical values via the energy difference $\{\Delta_1,\ \delta\}$or deformation$\{\Delta_1,\ q\}$set parameters.

We have the deformation parameter $q$

$$q = \frac{1}{2-\delta} \longleftrightarrow \delta = 2 - \frac{1}{q}, \tag{43}$$

the relation between q and $\delta$ are presented in the Fig. 2a and 2b

The value $n_{max}$ is

$$N = n_{max} = \left[\frac{(2-\delta)}{2\,(1-\delta)}\right] = \left[\frac{1}{2\,(1-q)}\right], \tag{44}$$

where [...]is the integer part, and are presented in the figure 3 as a functions on $\delta$ (upper steps), and on $q$ (lower steps).

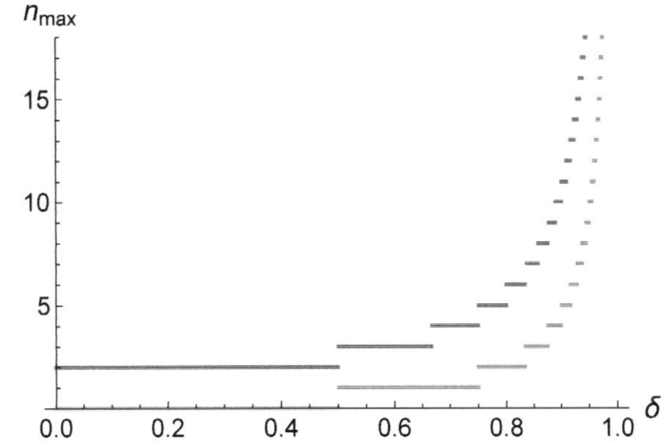


**Figure 3.** The relation value $n_{max}$ on $\delta$ (upper steps), and on $q$ (lower steps).

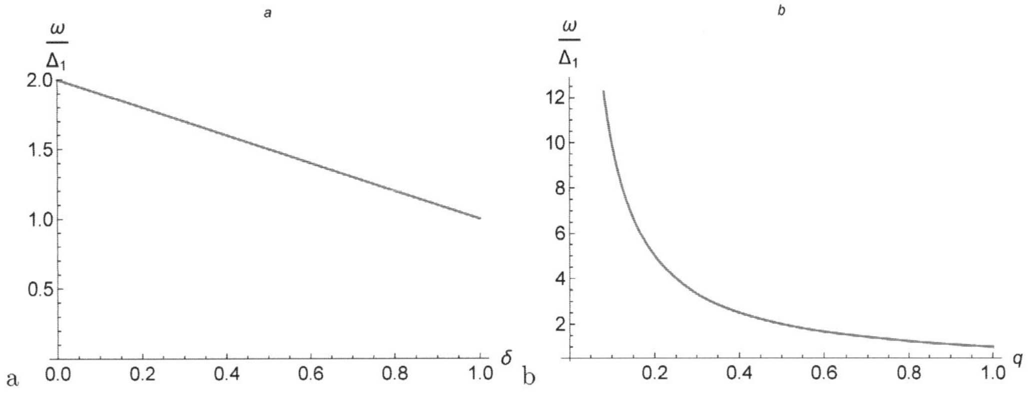


**Figure 4.** The frequency $\omega/\Delta_1$ : a) on $\delta$, and b) on $q$.

The frequency

$$\frac{\omega}{\Delta_1} = 2 - \delta = \frac{1}{q}, \tag{45}$$

and its values are presented in the figure 4 as a functions of (blue), and on q (yellow).
The energy deep $D_q$

$$\frac{D_q}{\Delta_1} = \frac{(2-\delta)^2}{(1-\delta)} = \frac{1}{q}\left[\frac{1}{(1-q)}\right], \tag{46}$$

are ploted in the figure 5.
The imaginary momentum-like width (effective mass) $k_q$

$$\frac{k_q}{k_0} = \sqrt{1-\delta} = \sqrt{\frac{1}{q} - 1}, \tag{47}$$

are presented in the figure 6.
We found the Morse-like potential

$$U_M(\chi) = \frac{V_M}{\hbar\omega_0} = \frac{(2-\delta)^2}{(1-\delta)}\left[e^{-2a(\chi-1)\sqrt{1-\delta}.} - 2e^{-a(\chi-1)\sqrt{1-\delta}}\right], \tag{48}$$

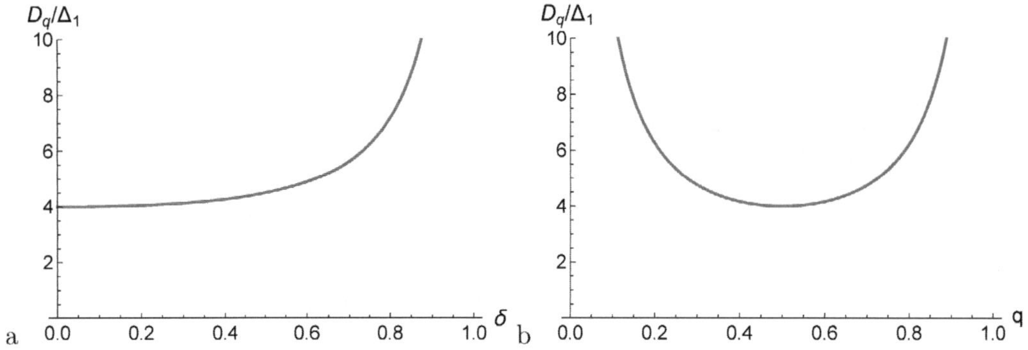


**Figure 5.** The energy deep of Morse potential $D_q/\Delta_1$: a) on $\delta$, and b) on $q$.

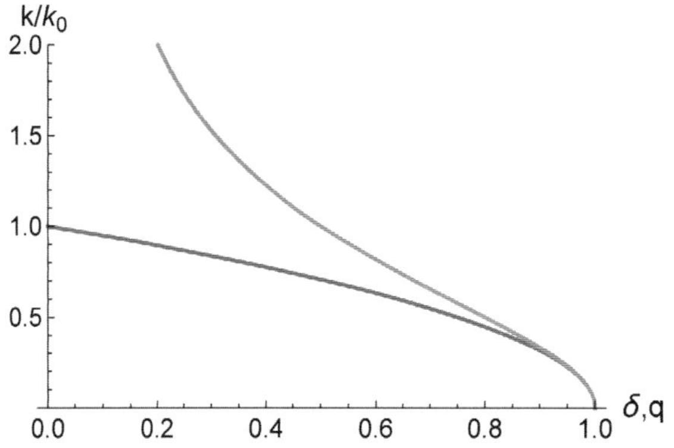


**Figure 6.** The parameter $k_q$ of Morse potential on $\delta$ (upper curve), and on $q$ (lower curve).

where $\chi = x/x_0$, $a = k_0 x_0$, and the corresponding harmonic potential

$$U_H(\chi) = \frac{V_H}{\hbar\omega_0} = -\frac{(2-\delta)^2}{(1-\delta)} + \frac{1}{2}(2-\delta)^2 \mu x_0^2 (\chi-1)^2 . \tag{49}$$

The Morse potential $U_M$ and the corresponding harmonic potential $U_H$ are presented in the figure 7.

The energy spectrum of Morse potential is

$$\varepsilon_n = \frac{E_n}{\Delta_1} = (2-\delta)\left[\left(n+\frac{1}{2}\right) - \frac{1}{2n_{max}}\left(n+\frac{1}{2}\right)^2\right], \; n = 0,\, 1, 2, 3, ..., n_{max}, \tag{50}$$

with first four levels

$$\begin{aligned} \varepsilon_0 &= (2-\delta)\left(\frac{1}{2} - \frac{1}{8n_{max}}\right), \\ \varepsilon_1 &= (2-\delta)\left(\frac{3}{2} - \frac{9}{8n_{max}}\right), \\ \varepsilon_2 &= (2-\delta)\left(\frac{5}{2} - \frac{25}{8n_{max}}\right), \\ \varepsilon_3 &= (2-\delta)\left(\frac{7}{2} - \frac{49}{8n_{max}}\right), \end{aligned} \tag{51}$$

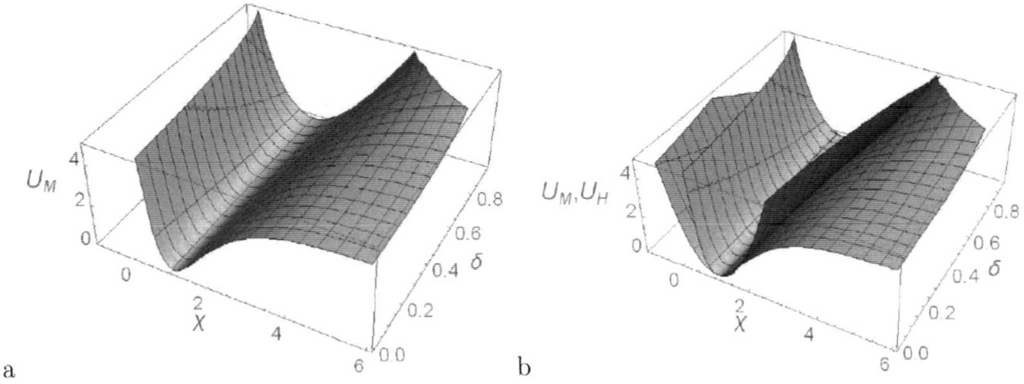


**Figure 7.** a) The Morse potential $U_M$ and b) with the corresponding harmonic potential $U_H$

The level difference $\Delta\varepsilon_n$ are

$$\Delta\varepsilon_n = \varepsilon_{n+1} - \varepsilon_n, \tag{52}$$

and for some lowest levels

$$\Delta\varepsilon_1 = (2-\delta)\left(1 - \frac{1}{n_{max}}\right), \ \Delta\varepsilon_2 = (2-\delta)\left(1 - \frac{2}{n_{max}}\right), \ \Delta\varepsilon_3 = (2-\delta)\left(1 - \frac{3}{n_{max}}\right) \ ... \tag{53}$$

In the harmonic limit $\delta \to 1$,$q \to 1$, $n_{max} \to \infty$, we have $\varepsilon_0 \to 1/2$,$\varepsilon_1 \to 3/2$ ,$\varepsilon_2 \to 5/2$ , ... and $\Delta\varepsilon_1 \to 1$,$\Delta\varepsilon_2 \to 1$,$\Delta\varepsilon_3 \to 1$ , ...

In the strong deformation limit $\delta \to 0$, $q \to 1/2$, $n_{max} \to 1$ (see the figure 3), we remain only two levels $\varepsilon_0 \to 2\,(1/2 - 1/16) = 0.875$, $\varepsilon_1 \to 2\,(3/2 - 9/16) = 1.875$, and $\Delta\varepsilon_1 \to 1$, $\Delta\varepsilon_2 \to 0$.

In general, for Morse like potential in WKB approximation the classical turning points are defined by the expression

$$\chi_{1,2}(n) = 1 + \frac{1}{k_0 x_0} Ln \left[\frac{D}{E(n)}\left(-1 \pm \sqrt{1 + \frac{E(n)}{D}}\right)\right]. \tag{54}$$

Given energy spectra $E(n)$, we define $\chi_{1,2}(n)$, so and Morse–like potential.

## 6. Discussion

In this work using the model proposed in [12] with the defined connection between q-deformed harmonic oscillator and Morse–like anharmonic potential, we investigate the energy spectrum inverse problem. Consider some lowest levels of energy spectrum of q-deformed harmonic oscillator are known, we construct the corresponding Morse-like potential with well defined deform parameter $q$ and also new level deformation parameter $\delta$. The possibility of using the WKB approximation in the energy spectrum inverse problem was discussed for both cases of parabolic potential (harmonic oscillator) and Morse-like potential (q-deformed harmonic oscillator).

Some most important relations between the set-parameters $\{\Delta_1, \delta, q\}$ of level deformations and set-parameters $\{D, k, x_0\}$ of Morse potential are derived explicitly. In the week-deform limit $\delta \to 1$,$q \to 1$; $n_{max} \to \infty$, we back to the harmonic case with many unique step levels. In the strong-deform limit $\delta \to 0$, $q \to 1/2$; $n_{max} \to 1$, we go to the two-levels problem, where only the ground ($n = 0$) and first ($n = 1$) levels can be existed and the other ($n \geq 2$) levels are collapsed.

In principal, with a given energy spectra $E(n)$ we can define the classical turning points $\chi_{1,2}(n)$ and then construct the effective potential by using the WKB approximation, but it is a inverse-like problem and usually difficult to solve. In practical, because for many problems of atom and molecular physics, quantum optics, condensed matter, ... only the ground and one or two first excited levels are considered, so we propose to use our deformed-three-levels simple model, where the set-parameters of Morse potential and the corresponding set-parameters of level deformations are easily and explicitly defined.

Our level deformation model might be useful in investigation the problem of entanglement entropy, when the environment or vacuum effects are taken in to account. This will be studied in the our next work.

## Acknowledgments

This research is funded by Vietnam National Foundation for Science and Technology Development (NAFOSTED) under grant number 103.01-2015.42.

## References

[1] Mirzrahi S S, Camargo J P and Dodonov V V 2004 *J. Phys. A* **37** 3707
[2] Abe S 2009 *Phys. Rev. E* **79** 041116
[3] Lavagno A and Swamy P P 2010 *Physica A* **389** 993
[4] Algin A and Senay M 2012 *Phys. Rev. E* **85** 041123
[5] Andrade F M and Silva E O 2013 *Phys. Lett. B* **719** 467
[6] Tkachuck V M 2013 *Phys. Rev. A* **86** 062112
[7] Dong S H, Lemus R and Frank A 2002 *Int. J. Quant. Phys.* **86** 433
[8] Angelova M, Dobrev V and Franck A 2004 *Eu. Phys. Jour. D* **31** 27; Angelova M 2002 arXiv:cond-mat/020124v1
[9] Kaplan I G 2003 Handbook of Molecular Physics and Quantum Chemistry (Wiley)
[10] Lima E F and Hornos J E M 2005 *J. Phys. B: At. Mol. Opt. Phys.* **38** 815
[11] Man Van Ngu, Ngo Gia Vinh, Nguyen Tri Lan, Luu Thi Kim Thanh, and Nguyen Ai Viet 2015 *Journal of Physics: Conference Series* **627** 012021
[12] Ngo Gia Vinh, Man Van Ngu, Do Thi Nga, Nguyen Tri Lan, Luu Thi Kim Thanh and Nguyen Ai Viet 2015 *Journal of Physics: Conference Series* **627** 012022

Journal of Physics: Conference Series **726** (2016) 012020 doi:10.1088/1742-6596/726/1/012020

# The study of dynamics heterogeneity and slow down of silica by molecular dynamics simulation

**L T San[1], P K Hung[1], H V Hue[2]**

[1]Department of Computational Physics, Hanoi University of Science and Technology, Vietnam, 1 Dai Co Viet, Hanoi, Viet Nam.

[2]Tay nguyen University, 567 Le Duan-Buon Ma Thuot-Dak Lak, Viet Nam.

E-mail: san.luyenthi@hust.edu.vn

**Abstract.** We have numerically studied the diffusion in silica liquids via the $SiO_x \rightarrow SiO_{x\pm1}$, $OSi_y \rightarrow OSi_{y\pm1}$ reactions and coordination cells (CC). Five models with temperatures from 1000 to 3500 K have been constructed by molecular dynamics simulation. We reveal that the reactions happen not randomly in the space. In addition, the reactions correlated strongly with the mobility of CC atom. Further we examine the clustering of atoms having unbroken bonds and restored bonds. The time evolution of these clusters under temperature is also considered. The simulation shows that both slow down and dynamic heterogeneity (DH) is related not only to the percolation of restored-rigid clusters near glass transition but also to their long lifetime.

## 1. Introduction

The dynamical slowing down in super-cooled liquids when the temperature approaches to their glass transition point $T_g$ is one of challenging problems of physics in condensed matter. There are many theoretical approaches developed to provide useful theoretical guides to understand and analyze various aspects of dynamical heterogeneity and slowing down, such as mode-coupling theory [1], the Adam-Gibbs hypothesis [2], the random first-order transition theory [3], dynamic facilitation [4]... The mode coupling theory predicts a power-law divergence of relaxation times at a critical temperature $T_{MCT} > T_g$ [5]. The Adam-Gibbs study suggests the glass transition phenomenon as resulting from an increase of cooperation of molecular dynamics upon cooling [5]. The simulations have shown that liquid silica behavior differs from strong liquid under wide temperature range. At high temperature, the diffusion violates the Arrhenius law and is well described by a power law $D \sim (T - T_{MCT})^{\gamma}$; here $\gamma$ is close to 2.1 (compared to 1.4 for water); $T_{MCT} \approx 3330$K [6]. This means that at high temperature, liquid silica behaves like a fragile liquid. The heterogeneous dynamics is well established by experiment [7, 8] and simulation [9, 10, 11]. However, experimental studies of strong glasses are rare due to the high $T_g$, and the structural relaxation is not result of simple bond-breaking process at random positions [5]. In previous work [12], we show that the diffusion constant depends not only on the rate of reactions, but also on the spatial distribution of reactions. It is interesting to determine how the spatial correlation of reactions affects the DH and the dynamical slowing down. Moreover, the clusters are predicted as result of specific distribution of reactions in the space.

Journal of Physics: Conference Series **726** (2016) 012020 doi:10.1088/1742-6596/726/1/012020

## 2. Computational procedure

Molecular dynamics (MD) simulation is carried out for the silica model composing 666 silicon and 1332 oxygen particles. We used the van BeestKramervan Santen (BKS) potential which is commonly applied to reproduce structural and dynamical properties of silica liquid and glass [13, 14, 15]. The MD step is equal to 0.47 fs. Initial configuration is generated by randomly placing all particles in the simulation box. This configuration is heated to 5000 K and then cooled down to the desired temperature. After that the obtained sample is relaxed in N-E-V ensemble (constant volume and energy) for long times. The pressure is kept to zero. The time consumed to construct a well-equilibrated model is about $10^7$ MD steps. We have prepared five models at temperatures of 1000, 1400, 1800, 2600 and 3500 K. More details about the procedure of preparing the BKS MD model can be found elsewhere [12]. The dynamics are analyzed through the CC which is determined as a set of central atom and its neighbors. We calculated the coordination cell of Si or O by integrating the Si-O or O-Si radial distribution function up to first minimum (cut off distance), respectively. Here, the bond is created by two atoms Si an O, whose distance is less than the cut off distance. A cluster is defined as a group of particle that resides in the first-neighbor cell of each other [9, 10]. So, atoms form at least a bond with another atom in the cluster. During this time simulation, if a bond remains unchanged from initial point, this bond is called unbroken bond. Atoms linked via unbroken bonds forming the cluster called rigid cluster. The bond may break then restore after some times, however, which is called restore bond. The cluster of atoms linked via these bonds is called restoredrigid cluster. A reaction corresponds to a bond broken or a new bond being created inside a coordination cell, such as $SiO_x \rightarrow SiO_{x'}$ and $OSi_y \rightarrow OSi_{y'}$.

## 3. Result and discussion

Fig.1 shows the number of reactions as a function of MD steps. All points fall well on straight lines. The slopes of these lines are used to determine the rate of reactions ($\xi$). The diffusion constant is calculated via Einstein equation using the mean square displacement of atoms (the values are showed in table 1). When temperature is lower than 1800K, the rate of reactions and diffusion constant are increase with the same proportion. However, $\xi$ increases much slower than diffusion constant as temperature is higher than 1800K. It means that $\gamma$ is just one of factors affecting the diffusion process. It suggests that the temperature changes, how reactions happen in the considered samples is not by the same way. The diffusion is a result of the reactions at CC, but it is one of the keys to effect to the diffusion.

**Table 1.** The diffusion constant and the rate of reactions

| Temperature, K | The rate of reactions ($\xi$) | $D_{\mathrm{Si}}$, cm$^2$/s | $D_{\mathrm{O}}$, cm$^2$/s |
|---|---|---|---|
| 1000 | 0.1599 | 2.5E-08 | 2.6E-08 |
| 1400 | 0.3307 | 4.0E-08 | 4.4E-08 |
| 1800 | 0.3992 | 4.4E-08 | 5.0E-08 |
| 2600 | 0.8540 | 2.1E-07 | 2.6E-07 |
| 3500 | 2.0209 | 3.4E-06 | 4.7E-06 |

Here $m_{cc}$ is the number of coordination cells happening reactions; $m_{unbr}$ is the number of unbroken bonds; $m_b$ is the number of bonds inside sample including unbroken bonds and restored bonds. The reaction processes can analyze through $m_{cc}$, $m_{unbr}$, $m_b$. We can see that the reactions spread on more coordination cells when increasing the number of steps. In the case of 3500K,

Journal of Physics: Conference Series **726** (2016) 012020 doi:10.1088/1742-6596/726/1/012020

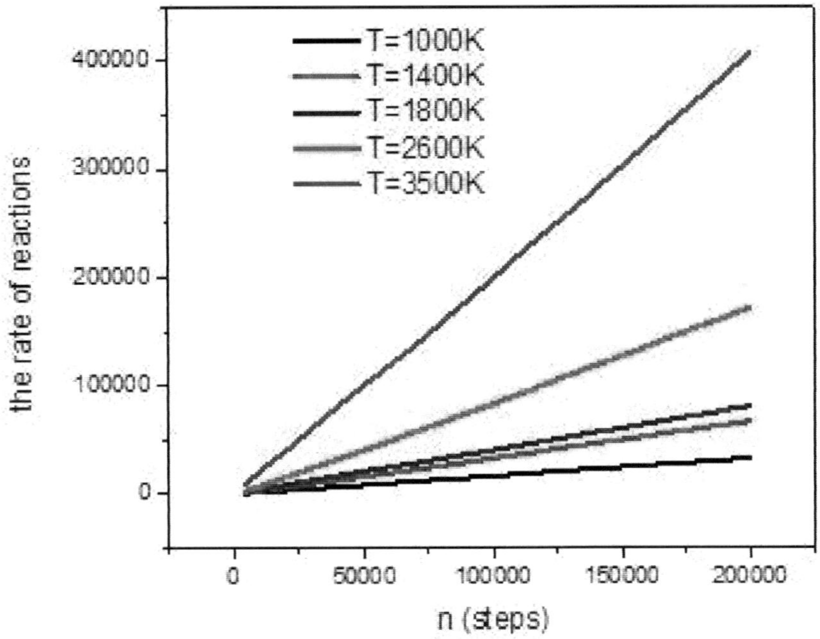


**Figure 1.** The number of reactions as function of steps.

the reactions expand to the entire sample approximately 60000 steps. At the same number of the steps, the temperature is higher, the number of reactions is bigger. But, it is not clearly the indication of spatio-temporal distributions. In fig.3, reactions can been showed via the process of breaking and restoring bonds. These processes occur simultaneously at any temperature. However, we can look at the different picture at high and low temperature. The reactions happen more frequently with some coordination cells or bonds, $m_{cc}$ becomes smaller and $m_{unbr}$ bigger than for the case when the distribution of reactions among coordination cells or bonds is ununiformed. Therefore, the spatial distribution of reactions becomes more heterogeneously upon lowering the temperature.

Atoms are linked with neighbors via unbroken, restored or new bonds. We have considered the clusters which are formed through unbroken bonds (rigid-cluster) and their role for dynamic heterogeneity and slowing down. The image of rigid cluster composing 33 atoms is presented in fig.4. The mean square displacement (MSD) per atom is approximately 1.4315 $Å^2$, is smaller than MSD per atom in the entire sample (about 1.8803 $Å^2$). Their shape changes slightly. The number of atoms of the biggest cluster and the number of rigid cluster at three moments are showed in table 2. These results tell us that these rigid clusters tend to decay when increasing time simulation. However, in table 3, as temperature is continually cooled down, the largest rigid cluster is expanded. It proves that rigid cluster exits in the liquid for long time and percolates over the sample at low temperature. Because of the rigidity of the atoms of rigid cluster, the dynamics decreases dramatically. So, the existence of percolation rigid cluster can be a result of slowing down at near glass transition.

Next, we describe results of our study on the restored-rigid cluster, which include atoms linked through unbroken bonds and restored bonds. We consider a group of Si and O having

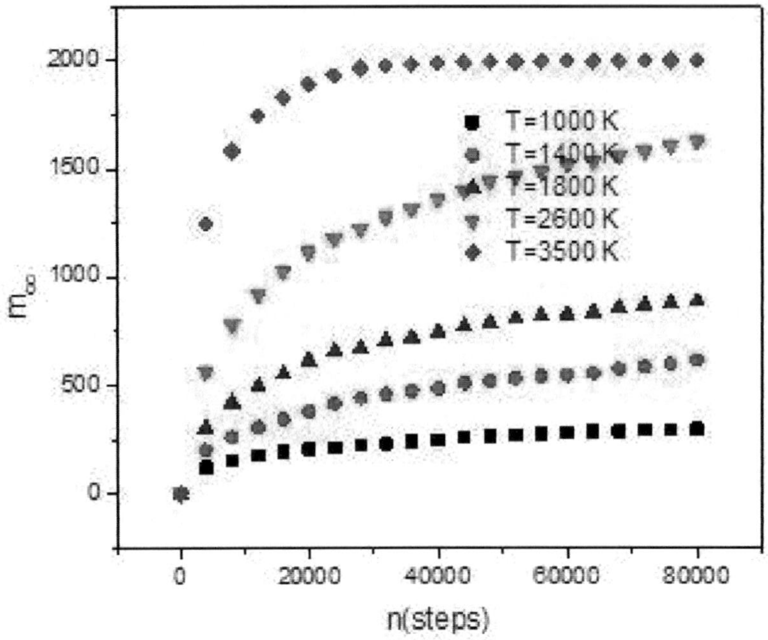


**Figure 2.** The dependence of $m_{cc}$ (a) as a function of number of steps.

4 or 2 bonds like initial time, respectively. For case of 2600 K and $2x10^5$ steps, we found a set containing 1513 atoms, with MSD is about 1.2394 $Å^2$. The number of atoms in largest cluster detected at 11 moments (from 0 steps to $2x10^5$ steps) changed from 1507 atoms to 1513 atoms. It means that during $2x10^5$ steps, there are lager cluster of immobile atoms which percolates over entire sample.

**Table 2.** The main characteristics of atoms of rigid clusters for the sample at 2600 K and three interval times

| Steps | sdam | mtam | $\langle d^2 \rangle, Å^2$ | $\langle dd^2 \rangle, Å^2$ |
|---|---|---|---|---|
| 40000 | 18 | 1852 | 0.6998 | 0.8685 |
| 80000 | 46 | 1672 | 0.749 | 0.9168 |
| 200000 | 158 | 268 | 1.0698 | 1.8803 |

sdam: the number of rigid clusters; mtam: the number of atoms of the largest rigid cluster; $\langle d^2 \rangle$: the MSD per atom in the largest rigid cluster; $\langle dd^2 \rangle$: the MSD per atom in entire sample.

In table 4, we considered Si and O atoms having k bonds (including unbroken bonds and restored bonds) at time simulation. We can see that the MDS and the number of reactions decreases when k increase. At final time, we can choose 180 atoms having largest MSD, monitor the cluster creating process and track the evolution of these clusters at 11 moments during $2x10^5$ steps. The size of biggest cluster varies from 13 to 19 atoms, with the MSD of biggest

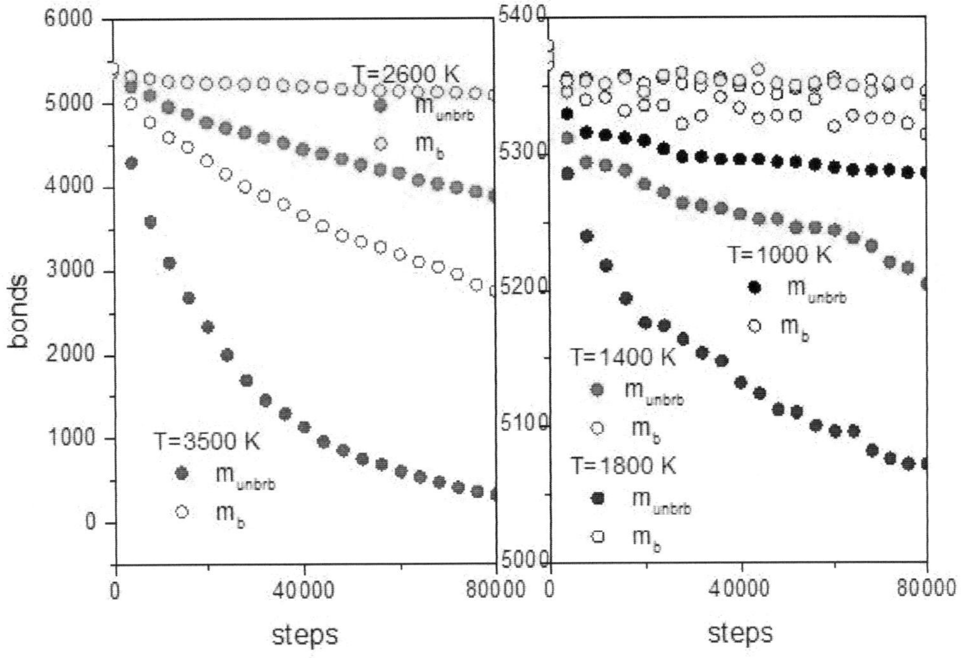


**Figure 3.** The dependence of $m_{unbr}$ and $m_b$ as a function of number of steps.

**Table 3.** The main characteristics of atoms of rigid clusters for the sample at three temperatures, n=2x10$^5$ steps

| Temperature | sdam | mtam | $\langle d^2 \rangle$, Å$^2$ | $\langle dd^2 \rangle$, Å$^2$ |
|---|---|---|---|---|
| 2600K | 158 | 268 | 1.0698 | 1.8803 |
| 1800K | 2 | 1967 | 0.5370 | 0.5370 |
| 1400K | 2 | 1981 | 0.4612 | 0.4804 |

cluster increase monotonically from 0 to 12.7653 Å$^2$. So, in the entire simulation time, the atoms of clusters tend to move together. These clusters of most mobile atoms and rigid clusters are convincing evidence of heterogeneity in silica liquid. At the same time, the trajectories distinguish widely from one atom to another atom in the sample. Some particle can undergo many jumps, even, link together to move; while some particles are nearly immobile over time simulation, they only vibrate around well-defined locations and link to another atoms by lasting bonds called rigid bonds. The most mobile atoms can be answer of the problem what is origin of diffusion beside the reaction of CC, especially, their trajectories. It is suggested that these atoms are bridge between these rigid clusters. When temperature is decrease near glass transition, the bonds exist a longer and longer time. The reactions in private CC is said that being only cause of the diffusion. We stress that the percolation in whole sample of the rigid cluster is associated with slowing down.

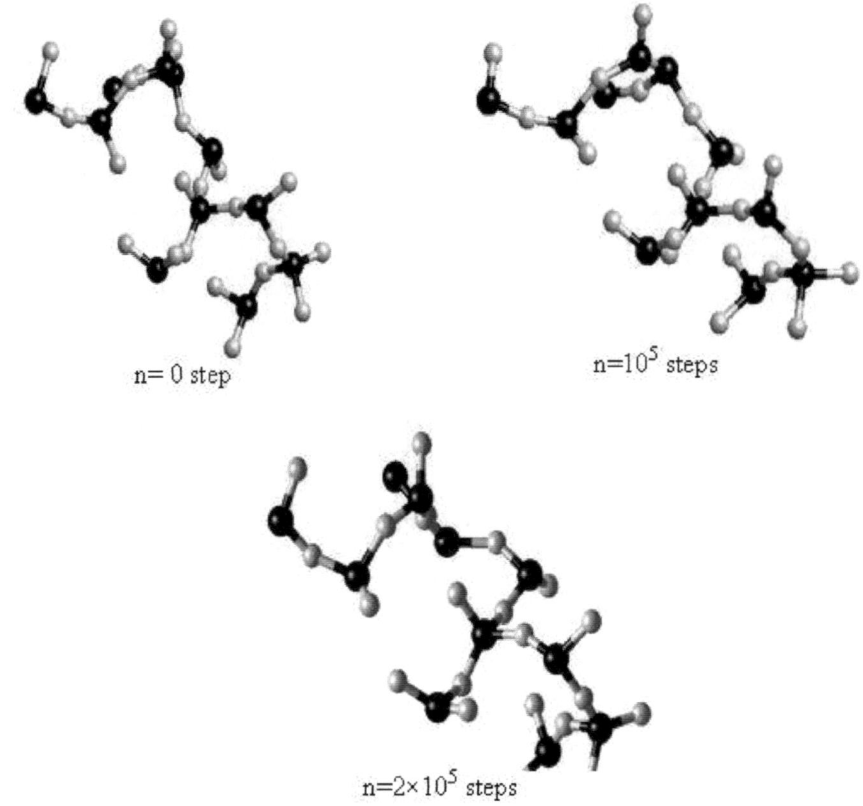


**Figure 4.** The image of five rigid clusters obtained at three moments within n = 2x10$^5$ steps; the temperature is 2600 K; the black and yellow spheres are the atom Si and O, respectively.

**Table 4.** The main characteristics of atoms in rigid and restored bond groups for the sample at 2600 K and n= 2x10$^5$ steps;

| Group(k) | Si | | | O | | |
|---|---|---|---|---|---|---|
| | $N_{group}$ | $\langle d^2 \rangle$, Å$^2$ | $\langle m_{reaction} \rangle$ | $N_{group}$ | $\langle d^2 \rangle$, Å$^2$ | $\langle m_{reaction} \rangle$ |
| 0 | 1 | 11.5018 | 448 | 19 | 14.6135 | 250.3158 |
| 1 | 5 | 9.3522 | 453 | 190 | 4.79 | 132.2842 |
| 2 | 40 | 3.7553 | 303.225 | 1122 | 1.351 | 49.016 |
| 3 | 128 | 2.4358 | 212.875 | 1 | 0.3392 | 620 |
| 4 | 492 | 1.0826 | 88.2114 | | | |

$N_{group}$ is the number of atoms having k bonds (including rigid bonds and restored bonds) at final time, which is the same with initial time.

## 4. Conclusion

The reactions affect strongly the diffusion, but diffusion is not a result of a simple bond breaking process occurring at random positions. As the temperature decreases, the rate of reactions

Journal of Physics: Conference Series **726** (2016) 012020 doi:10.1088/1742-6596/726/1/012020

decreases slower than the diffusion constant. As the temperature is lower, the rigid cluster spans a longer and longer period of time and a lager restored-rigid cluster percolated over whole system making diffusion can no longer be detected. It is mean that they are origin of dynamics slowing down near glass transition. The DH can be seen from the sets of atoms belonging to rigid clusters or restoredrigid clusters or most mobile atoms. We conclude that DH is a result of none-uniform distribution of reactions in the space.

## Acknowledgments

This research is funded by the Vietnam National Foundation for Science and Technology Development (NAFOSTED) under Grant No.: 103.05-2014.40

## References

[1] Gotze W and Sjogren L 1992 *Reports on progress in Physics* **55** 241
[2] Adam G and Gibbs J H 1965 *The Journal of Chemical Physics* **43** 139
[3] Lubchenko V and Wolynes P G 2007 *Annual Review of Physical Chemistry* **58** 235–266
[4] Elmatad Y S, Jack R L, Chandler D and Garrahan J P 2010 *Proceedings of the National Academy of Sciences* **107** 12793–12798
[5] Vogel M and Glotzer S C 2004 *Phys. Rev. Lett.* **92** 255901
[6] Horbach J and Kob W 1999 *Phys. Rev. B* **60** 3169–3181
[7] Tracht U, Wilhelm M, Heuer A, Feng H, Schmidt-Rohr K and Spiess H W 1998 *Phys. Rev. Lett.* **81** 2727–2730
[8] Sillescu H 1999 *Journal of Non-Crystalline Solids* **243** 81–108
[9] Gebremichael Y, Vogel M and Glotzer S 2004 *The Journal of chemical physics* **120** 4415–4427
[10] Giovambattista N, Buldyrev S V, Starr F W and Stanley H E 2003 *Phys. Rev. Lett.* **90** 085506
[11] Aichele M, Gebremichael Y, Starr F W, Baschnagel J and Glotzer S C 2003 *The Journal of Chemical Physics* **119** 5290
[12] Hung P and Vinh L 2006 *Journal of Non-Crystalline Solids* **352** 5531–5540
[13] Barrat J L, Badro J and Gillet P 1997 *Molecular Simulation* **20** 17–25
[14] Horbach J and Kob W 2001 *Physical Review E* **64** 041503
[15] Saika-Voivod I, Poole P H and Sciortino F 2001 *Nature* **412** 514–517

Journal of Physics: Conference Series **726** (2016) 012021 doi:10.1088/1742-6596/726/1/012021

# Dynamic Simulation of Backward Diffusion Based on Random Walk Theory

**Vu Ba Dung, Bui Huu Nguyen**
Hanoi University of Mining and Geology, Duc Thang, Bac Tu Liem, Hanoi, Vietnam.

E-mail: vubazung305@gmail.com

**Abstract.** Results of diffusion study in silicon showed that diffusion of the self-interstitial and vacancy could be backward diffusion and their diffusivity could be negative [1]. The backward diffusion process and negative diffusivity is contrary to the fundamental laws of diffusion such as the law of Fick law, namely the diffusive flux of backward diffusion goes from regions of low concentration to regions of high concentration. The backward diffusion process have been explained [2]. In this paper, the backward diffusion process is simulated. Results is corresponding to theory and show that when thermal velocity of the low concentration area is greater than thermal velocity of the high concentration area, the backward diffusion can be occurred.

## 1. Introduction

One of the most fundamental processes in nature is the diffusion. The diffusion theory of Fick has shown that the diffusion flux J is proportional to the diffusivity and along the reducing direction of the concentration. Ficks theory also showed that diffusion coefficient (diffusivity) is always positive. Backward diffusion processes (diffusion flux J is along the increasing direction of the concentration) and negative diffusivity can taken palace? Backward diffusion and negative diffusivity could be occurred [1]. Based on the thermodynamics theory the backward diffusion and negative diffusion coefficient is presented [1, 2]. In this paper the dynamics simulation of the backward diffusion process is studied and presented.

## 2. The equation of general and backward diffusion

Based on thermodynamics the more general definition of mass flux has the form [3, 4]

$$J = uC, \tag{1}$$

where $J$ is density of mass flux $(moles/m^2s)$, $C$ is molecules concentration $(moles/m^2)$, u is thermal velocity of molecules. Consider the two thickness slices of illustrated in fig.1. In slice I, the concentration of molecules is $C_1$ and thermal agitation velocity of molecules is $u_1$. In slice II, the concentration of molecules is $C_2$ and thermal velocity of molecules is $u_2$. The molecule flux is [4, 5]

$$J_1 = \frac{1}{6}\left(u_1 C_1 - u_2 C_2\right), \tag{2}$$

Assume that thermal velocity of molecules in high concentration slice are lower than in low concentration slice $C_1 < C_2, u_1 = u + nu_2$ and $u_2 = u$. Using expression [2]

$$\Delta C = C_2 - C_1 = \lambda \frac{\partial C}{\partial x}, \tag{3}$$

Journal of Physics: Conference Series **726** (2016) 012021 doi:10.1088/1742-6596/726/1/012021

Where $\lambda$ is characteristic. The diffusion flux is [1, 2, 6]

$$J = -D_{gen}\frac{\partial C}{\partial x}, \tag{4}$$

and

$$\frac{\partial C}{\partial t} = -D_{gen}\frac{\partial^2 C}{\partial x^2}, \tag{5}$$

with the general diffusivity

$$D_{gen} = \frac{2-n}{12}u\lambda, \tag{6}$$

Equations (4) and (5) are called general diffusion equation because that can described every diffusion process (forward, backward and Ficks diffusion). The diffusion (6) showed that the thermal velocity of molecules in low concentration is greater than two times diffusivity is negative and there is a backward diffusion process, namely that the diffusion flows in the direction of increasing concentration. When $n > 2$ equation (4) and (5) is the equations of backward diffusion with negative diffusivity (6) $D < 0$.

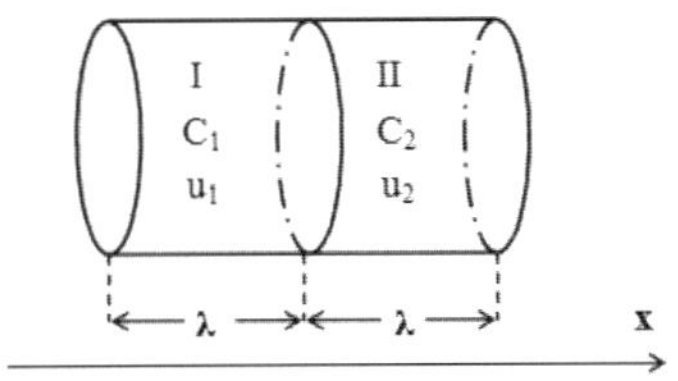


Figure 1: Modeling of the backward diffusion

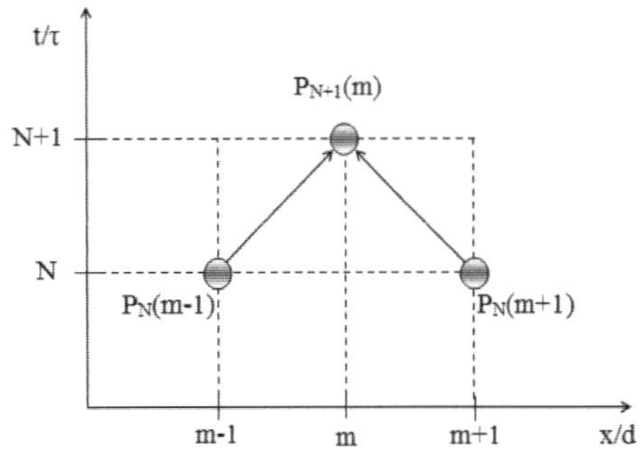


Figure 2: Diagram of a random walk

## 3. Random walk theory and diffusion

From the atomistic point of view, diffusion is considered as result of the random walk of diffusing particles. One important method of generating the fundamental solution introduced earlier is to treat the problem of a random walk. Assume that a molecule (as a drunken sailor) start at $x = 0$ initially. Furthermore, that each time the particle can move one step to the right probability $p$, or one step to the left with probability $q(q = 1-p)$. Suppose that $N$ steps have occurred, each of length $L$. What is probability that the particle is now at position $x = mL$, where -N $\leqslant$ m $\leqslant$ N. If the particle takes nR steps to the right and $n_L$ steps to the left, then $N = n_R + n_L$ and $m = n_R - n_L$. So $n_R$ and $n_L$ can be written [7-10]

$$n_R = \frac{N+m}{2} \;\; and \;\; n_L = \frac{N-m}{2}, \tag{7}$$

The probability of taking a particular sequence of steps with $n_R$ steps to the right and $n_L$ steps to the left is

$$P = p^{n_R}q^{n_L}, \tag{8}$$

The particular sequence of the right steps and the left is not different. Therefore, adding all possible combinations of sequences of $n_R$ steps to the right and $n_L$ steps to the left. This number is given by the usual counting formula

$$\frac{N!}{n_R!n_L!}, \tag{9}$$

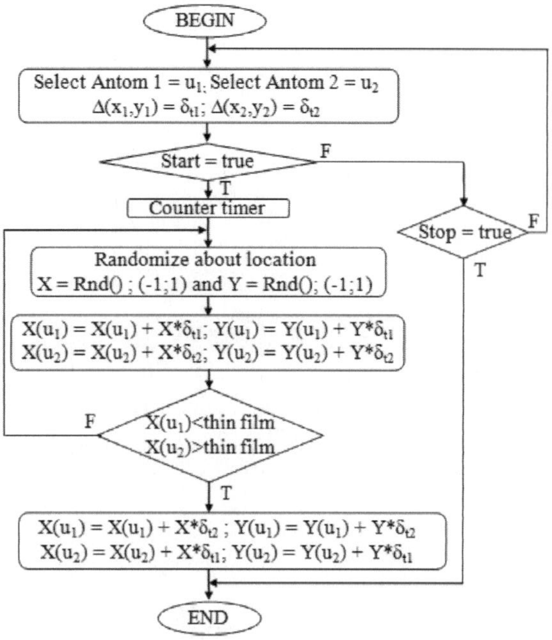


Figure 3: Diagram of the simulation program

The probability of observing the particle at position $x = mL$ is

$$P_N(m) = \frac{N!}{\left(\frac{N+m}{2}\right)!\left(\frac{N-m}{2}\right)!}p^{n_R}q^{n_L}, \tag{10}$$

Assuming that $p = q = 0.5$, $N >> 1$ and $N >> m$. Using Sterlings approximation

$$lnx! = NlnN - N + \frac{1}{2}lnN + \frac{1}{2}ln2\pi, \tag{11}$$

The equation (11) can be applied to (10) and dropping the term of order $1/N^2$, the probability (10) can be written [7-10]

$$P_N(m) = \frac{1}{\sqrt{\pi N}}exp\left(-\frac{m^2}{2N}\right), \tag{12}$$

$P_N(m)$ in equation (12) yields the typical bell-shaped curve of diffusion processes from a point source.

## 4. Applying to the backward diffusion

Based on thermodynamics the backward diffusion process cold be explained and the equation of the backward diffusion have been presented. The next, we simulate the backward diffusion process based on random walk theory. A random walk is a Markov process. Let $j$ and $k$ be states and let $P(j \to k)$ be the probability for a transition from $j$ to $k$, then the transition probabilities are independent of time and only depend on the states $j$ and $k$. In addition probabilities obey the expression

$$\sum_k P(j \to k) = 1, \tag{13}$$

The probability $P_N(m)$ to find a particle at position $x = md$ and at $t = N\tau$ satisfies the stochastic difference equation

$$P_{N+1}(m) = p^{P_N(m-1)} + q^{P_N(m+1)}, \tag{14}$$

Journal of Physics: Conference Series **726** (2016) 012021 doi:10.1088/1742-6596/726/1/012021

in which $p$ is probability for a step to the right and $q$ is the probability for a step to the left ($p = q = 0.5$). So equation (14) can be written

$$P_{N+1}(m) = \frac{1}{2}P_N(m-1) + \frac{1}{2}P_N(m+1), \tag{15}$$

Subtract $P_N(m)$ on both sides of equation (15)

$$P_{N+1}(m) - P_N(m) = \frac{1}{2}\left[P_N(m-1) - 2P_N(m) + P_N(m+1)\right], \tag{16}$$

in the limit of large $N$ differences become differentials

$$\frac{\partial P}{\partial t} = \frac{d^2}{2\tau}\frac{\partial^2 P}{\partial x^2}, \tag{17}$$

Solve the equation (17) with boundary and initial condition

$$P(\pm\infty, t) = 0, \tag{18}$$

and

$$P(x, 0) = \delta(x), \tag{19}$$

where $\delta(x)$ is delta equation. The solution of equation (17) is determined following

$$P(x,t) = \frac{1}{\sqrt{4\pi Dt}} exp\left(-\frac{x^2}{4Dt}\right), \tag{20}$$

where

$$D = \frac{d^2}{2\tau}, \tag{21}$$

Mean square displacement in one dimension

$$\langle x^2 \rangle = \int_{-\infty}^{+\infty} x^2 P(x)dx = 2Dt, \tag{22}$$

Applying to simulate the backward diffusion with assume the thermal velocity of molecules in two slices is different $u_1 > u_2$ (fig.1). If the time of a random walk in two slices are the same ($\tau_1 = \tau_2 = \tau$), the length of a random walk in slice I is greater than in slice II ($d_1 > d_2$). At the time of t the mean displacement of molecules in two slices are

$$\overline{x_1} = \sqrt{\langle x_1^2 \rangle} = \sqrt{2D_1 t} = d_1\sqrt{\frac{t}{\tau}}, \tag{23}$$

$$\overline{x_2} = \sqrt{\langle x_2^2 \rangle} = \sqrt{2D_2 t} = d_2\sqrt{\frac{t}{\tau}}, \tag{24}$$

where $\overline{x_1}$ and $\overline{x_2}$ is the mean displacement of molecules in slice $I$ and slice $II$.

## 5. Program and results of dynamics simulation of the backward diffusion

The purpose of computer simulations is to show the presence of backward diffusion process. We simulated random walk of individual particles. Simulations of the particle migration were executed in the two-dimensions coordinate axes and in different two parts $I$ and $II$. The walk probabilities of every particles $P^{jk}$ from the occupied $j_{th}$ site to the nearest-neighbor free site $k^{th}$ are the same and equal to 0.25 in both the parts. The time of step random jumps of particles is the same and which could be changed. However, the length of a random walk are chosen differently in part $I$ and part $II$ and which can be changed. In the first instance ($t = 0$), there are 22 and 44 particles in the part $I$ and part $II$. The place of all particles in two parts are determined randomly. We chose the thermal velocity of the particles in lower concentration (part $I$) are greeter than four times the thermal velocity of the part $II$ (high concentration part). To do this, we have chosen:

- The minimum length of the particle random walk step is $d_0 = 0.2$ mm.
- The time of the both part I and part II are the same ($\tau_1 = \tau_2 = \tau = 1ms$).
- The length of the random walk steps can be changed ($d_1 = n_1 d_0$ $and$ $d_2 = n_2 d_0$).

The program of simulation is written by Visual Basic language on PC. Fig.3 is diagram of the simulation program. Result is presented by the motion pictures on the monitor

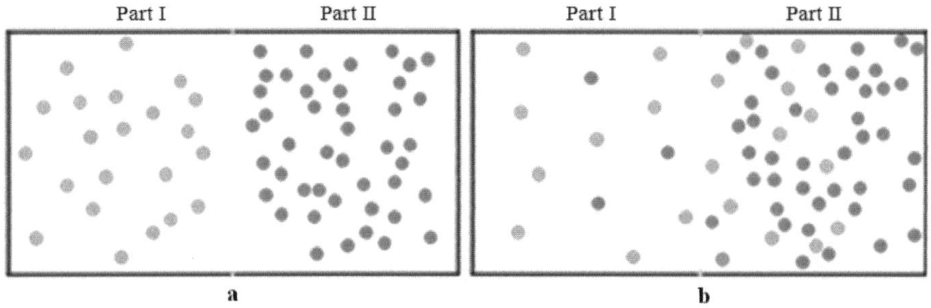


Figure 4: Result of backward diffusion simulation

of PC. Fig.4a shows the positions of particles in part I and part II at the initial time $t = 0$. Fig.4b presents the pictures of the positions of particles after the ten minutes, which shows that particles of part I moves to part II and particles of part II also moves to part I. However, the numbers of particles of the part I moves to part II is greater than the numbers of particles of the part II moves to part I. As result that there is a mass flux goes from low concentration area (part I) to high concentration area (part II). This mass flux is the backward diffusion flux.

## 6. Conclusions

The program of the simulation have demonstrated that when the thermal velocity of low concentration is greater than the thermal velocity of high concentration, backward diffusion process can be occurred, it mean there is a mass flux goes from a low concentration area to a high concentration area. Although the backward diffussion processes are contrary to the Fick law, its can be occurred and are compatible with the molecular dynamics theory. This simulation is great significance in explaining the backward diffusion and the other diffusion types.

## References

[1] Vu Ba Dung and Dinh Van Thien 2014 *Journal of Physics: Conference Series* **537** 012011.
[2] Vu Ba Dung 2015 *Far East Journal of Dynamical Systems* **27** 79.
[3] Olah K 2005 *Periodica Polytecnica Se. Chem. Eng.* **49** 91.
[4] Perrot P 1998 *A to Z of Thermodynamic* Oxford University Press.
[5] Garbaczewski P 2008 *Acta. Phys. Pol.* **E 39** 1087.
[6] Vu Ba Dung, Dinh Van Thien, Ta Thi Dung and Hoang Dinh Chieu 2012 *Proc. Natl. Conf. Theo. Phys.* **37** 67.
[7] Rudnick J and Gaspari G 2004 *Elements of the Random Walk* (Cambridge University press, Cambridge, UK).
[8] Allnatt A R and Lidiard A B 2009 *Random-walk theories of atomic diffusion* (Cambridge University press, Cambridge, UK) pp 337-379.
[9] Muoz E, Selvan M, Xiong R, Ojha M, Keffer D, Nicholson D and Egami T 2011 *Phys. Rev.* **E 83** 011120.
[10] Oliver C 2013 *Elements of Random Walk and Diffusion Processes* (John Wiley and Sons, Inc.) pp 160-164.

# Density functional theory based tight binding study on theoretical prediction of low-density nanoporous phases ZnO semiconductor materials

**Vu Ngoc Tuoc*[1], Tran Doan Huan[1,2], Nguyen Viet Minh[1], Nguyen Thi Thao[1,3]**
[1] Institute of Engineering Physics, Hanoi University of Science and Technology, 1 Dai Co Viet Road, Hanoi, Vietnam
[2] Institute of Materials Science, University of Connecticut, Storrs, CT 06269-3136 USA
[3] Hong Duc University, 307 Le Lai, Thanh Hoa city, Vietnam

E-mail: tuoc.vungoc@hust.edu.vn

**Abstract.** Polymorphs or phases - different inorganic solids structures of the same composition usually have widely differing properties and applications, thereby synthesizing or predicting new classes of polymorphs for a certain compound is of great significance and has been gaining considerable interest. Herein, we perform a density functional theory based tight binding (DFTB) study on theoretical prediction of several new phases series of II-VI semiconductor material ZnO nanoporous phases from their bottom-up building blocks. Among these, three phases are reported for the first time, which could greatly expand the family of II-VI compound nanoporous phases. We also show that all these generally can be categorized similarly to the aluminosilicate zeolites inorganic open-framework materials. The hollow cage structure of the corresponding building block $Zn_kO_k$ (k= 9, 12, 16) is well preserved in all of them, which leads to their low-density nanoporous and high flexibility. Additionally the electronic wide-energy gap of the individual $Zn_kO_k$ is also retained. Our study reveals that they are all semiconductor materials with a large band gap. Further, this study is likely to be the common for II-VI semiconductor compounds and will be helpful for extending their range of properties and applications

Keywords: Nanoporous phases, DFT, II-VI semiconductor.

## 1. Introduction

Zinc oxide (ZnO) is a semiconducting material with more interesting and important properties than many other materials, due to the vast number of scientific publications on ZnO, particularly since 2000 [1]. The reason for this is that ZnO can be realized very easily in many different nanostructures, its excellent luminescent properties and wide possibility of bandgap engineering, e.g. core-shell nanoheterostructures, its piezoelectric found an application as nanogenerator, its bio-compatible in chemical-sensing effects, its high transparency and room-temperature ferromagnetism in opto-electronics and spintronics [2]. To date the seek to synthesize or predict new classes of structures.

* To whom any correspondence should be addressed.

polymorphs or phase for this compound is still of great significance and has been gaining considerable interest. Beside the extensive experiments to find new phases, purely theoretical crystal structure prediction from first principles have also been widely reported [3-6 ].
Since the discovery of the excellent properties and applications of aluminosilicate zeolites in the areas of ion-exchange, separations, and catalysis, the area of inorganic open-framework materials has become one of the intense research activities. One of the major goals of the research over the open-framework materials is to find materials possessing channels and other features that make them porous or nanoporous. Porous materials play an important role in many applications related to energy and sustainability, e.g. catalysis, gas separation, water purification, and batteries. Further in the current search for new and interesting nanoporous open-frameworks, the predictability of their framework design and dimensionality are essential [7] . The International Zeolite Association (IZA) database shows that the number of structural types of unique microporous frameworks has been growing rapidly, from 27 in 1970, to 38 in 1978, to 64 in 1988, to 98 in 1996 and to 174 in 2007 [8]. The properties of cages/hollow structures decide their suitability for different applications, are intimately connected with their structures. Therefore synthesize a specific architecture is important and for most cases remained challenging. Experiments are most often used a " top-down" approach involving the disassembly of a parent 'zeolite' structure to come to the targeted topologies [9]. In the theoretical side, Bromley et al. [10] has recently proposed that it is possible to stabilize new ultralow-density phases of metal oxide materials with low-density by taking smallest nanocages as building blocks of the possible new phases, and proving that there is no barrier to stop the nanocages from coalescing to form low-density nanoporous phases, e.g. SOD, LTA, and FAU [†]. Subsequently, another $Zn_{12}O_{12}$-assembled low-density phase reported by Song et al.[11] and $Zn_{12}O_{12}$ and $Zn_{16}O_{16}$-cub by Z.F. Liu [4]. On the basis of this approach " from the bottom up" that provides a viable way from free clusters to cluster-assembled solid phases, we report our results of density functional theory based prediction starting from other ZnO clusters of $Zn_9O_9$ and $Zn_{12}O_{12}$ of three newly found ZnO-hollow structures.

## 2. Bottom-up structure prediction approach

In this section we briefly desribe the "bottom up" approach for theoretical crystal howllow predictions. The possibility of the formation of different low-density framework materials based on highly stable $Zn_kO_k$ (with n=12, 16) clusters is systematically investigated [4-11]. Generally, the preferred cluster building blocks are chosen as the highest symmetries with large HOMO–LUMO gaps, that are supposed to be most energetically stable. Via the coalescence or linking scheme these cluster building blocks, many kinds of low-density framework of ZnO materials of varying porosity are thus proposed.

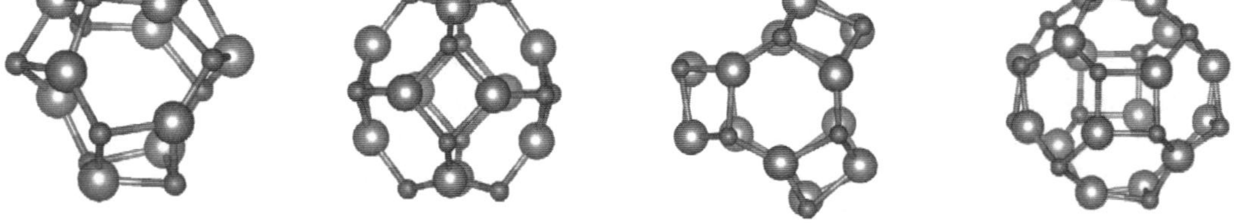

**Fig1.** The relaxed structures of the magic clusters used as the building blocks in this work from the left to right $Zn_9O_9$ (a), $Zn_{12}O_{12}$ (octahedron structures- SOD) (b), another $Zn_{12}O_{12}$ high symmetry multimer (to built a GME structure) (c) and $Zn_{16}O_{16}$-cube-structure (d). Small (red) balls are O atoms, big (gray) ones are Zn. All graphics illustrations in this work is prepared by VESTA [10].

[†] these acronyms are zeolite framework type codes according to the nomenclature of the International Zeolite Association (IZA)

Journal of Physics: Conference Series **726** (2016) 012022 doi:10.1088/1742-6596/726/1/012022

Starting from binary tetrahedral compounds of ZnO structures that is stable at room temperature composed of three-coordinated atoms, sets of which create the even-number of atom rings face side due to the equal role of Zn and O in tetrahedral symmetry, i.e. each Zn/O atom is connected by the three O/Zn atoms with ionic bonding. It is commonly accepted that the higher-symmetry configurations of the cluster are typically found to be more stable and have a greater stability than clusters of same and/or other sizes. Therefore for high-symmetry cluster structures are only possible for certain sizes, e.g. 9, 12, 16, 18, 24 ..., which often called the *magic* clusters (see Fig.1).

As is mentioned above, the magic cluster $Zn_{12}O_{12}$ have already served as the building blocks of the constructed low-density phases, i.e., SOD, LTA, FAU [10]. The growth paths of cluster–cluster coalescence proposed by Bromley [10] and Song et al. [11] are the phases constructed by the member-ring faces linkages. For instance, the case of Sodalite cage (SOD) know as the truncated octahedron structure (see Fig. 2a first line) - the structure comes with eight six-membered rings (6-MR) and six four-membered rings (4-MR) obeying the isolated tetragon rule, withe the two types of Zn–O bonds as the longer one in (4-MR) and the shorter (RSB) connecting two neighbor 4-MR ring s. The SOD linking together with 12 square (4-MR facing 4-MR ring), see details description in [4]. Where as the FAU with four hexa-links (6-MR facing 6-MR), see second line in Fig.2a , LTA with six double-square links (cubic links), see third line in Fig.2a, EMT with four double-hexa links (see the fourth line in Fig.2a).

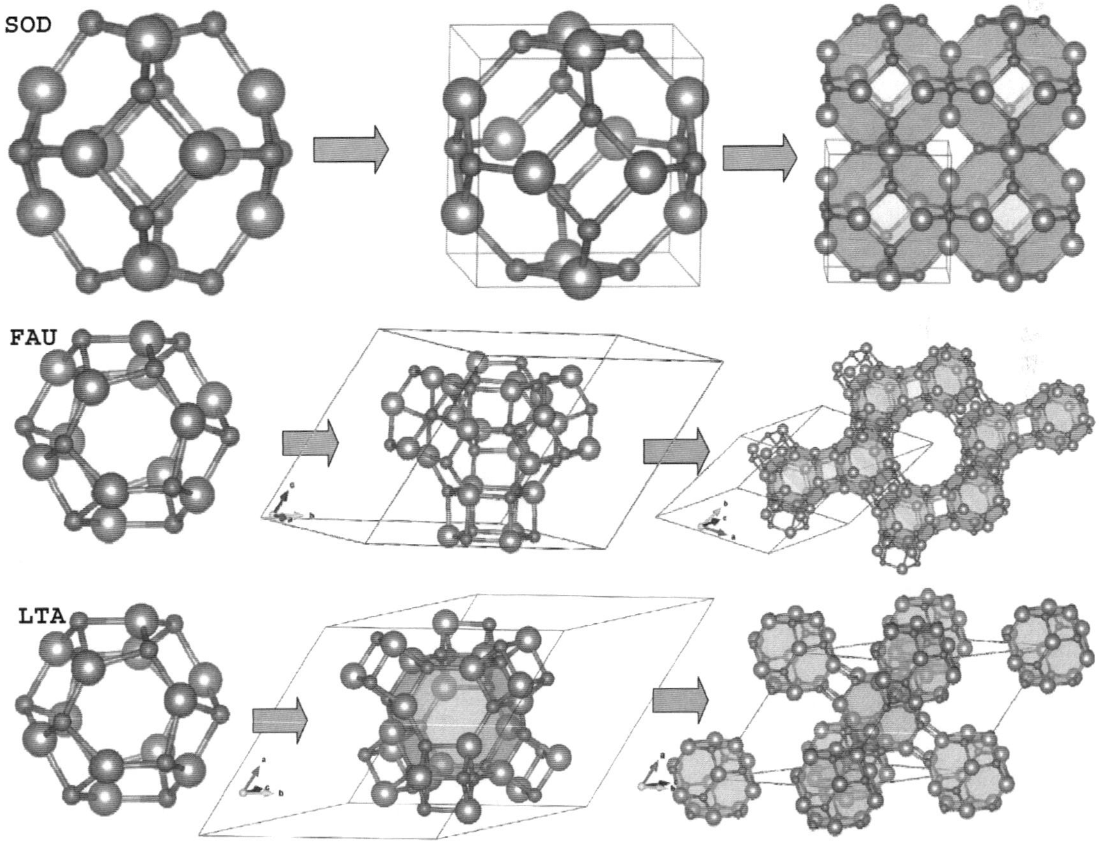

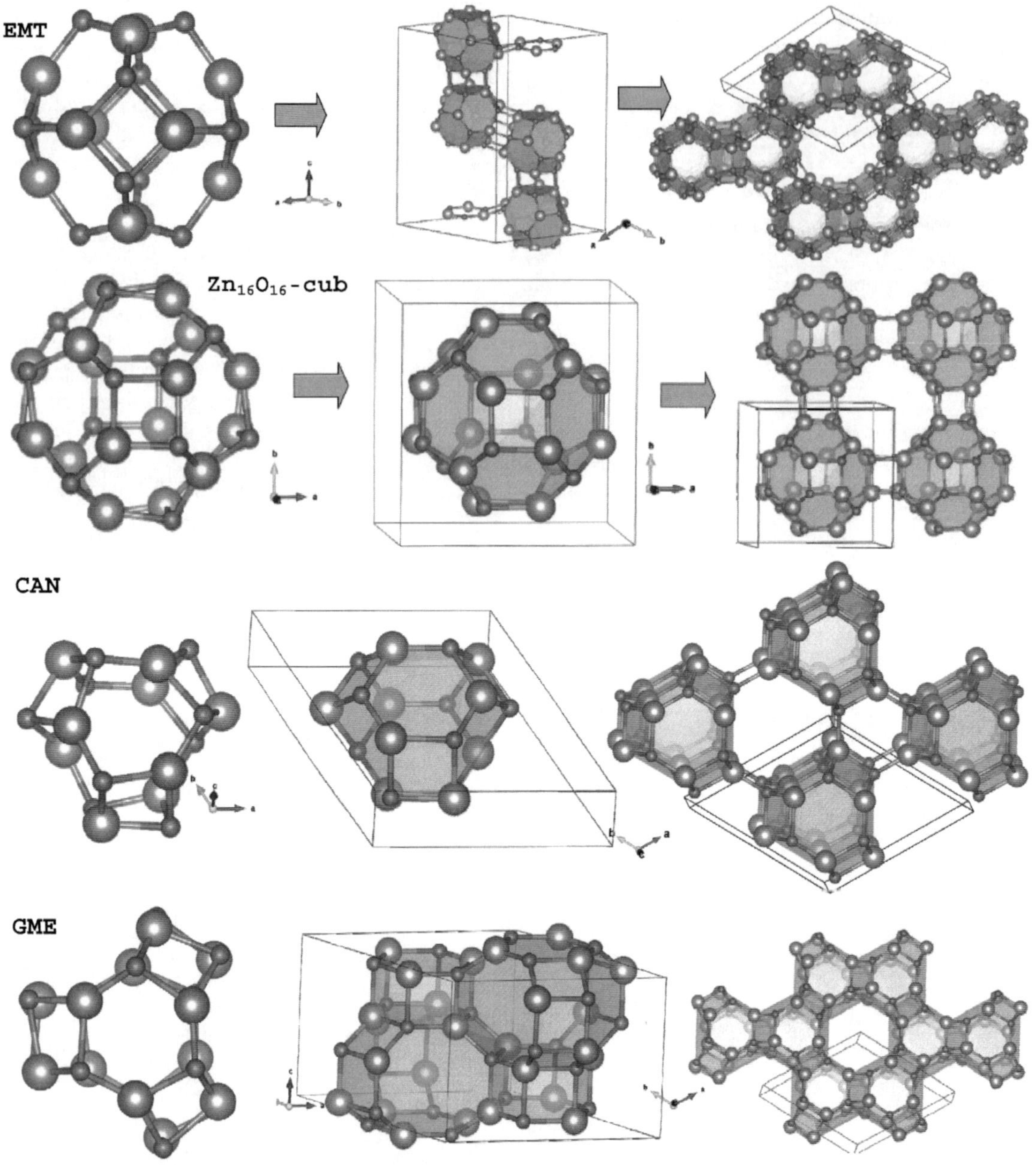


**Fig2a.** From left to right the relaxed cluster - optimized crystal lattice primitive cell – hollow-bulk structure of the investigated structures SOD, FAU, LTA by Bromley et. al. [11], EMT, $Zn_{16}O_{16}$-cube by Z. Liu et.al [4] and CAN, GME by Bromley et. al. [12].

As well as the $Zn_{16}O_{16}$-cubic, the LTA, comes with six double-square links (cubic links) (see the fifth line in Fig.2a). The other two structures CAN and GME found in [12], where CAN is resulted from $Zn_9O_9$ cluster and comes with two hexa-links and 12 square-links closed packed, and GME come with two double hexa-links and six square-links.

Inspired by the approach of creating assembled phases *from the bottom up*, we extend the search and have carried out electronic calculation for the next possible $Zn_9O_9$ and another type of cluster multimer $Zn_{12}O_{12}$ from now on denote as $Zn_{12}O_{12}$-GME (see fig1.c). We found the new three structure which we categorized by its symmetry acording to the IZA classification as AST, SAT and SBT. Herein, we describe three new phases found by us in Fig.2b. The most specific is the AST structure (see the first line in Fig.2b) that comes with 6 double-square (cubic) and 8 hexa-links that is appear to be more packed than the previous SOD structure (see Table. 1 for comparing). Next, SAT and SBS structures, boths low-density hollow strcutures, originated from $Zn_9O_9$ come with two double-hexa links (see the second and the third line in Fig.2b). The detail structure files of these newly found phases are listed in the Appendix A.

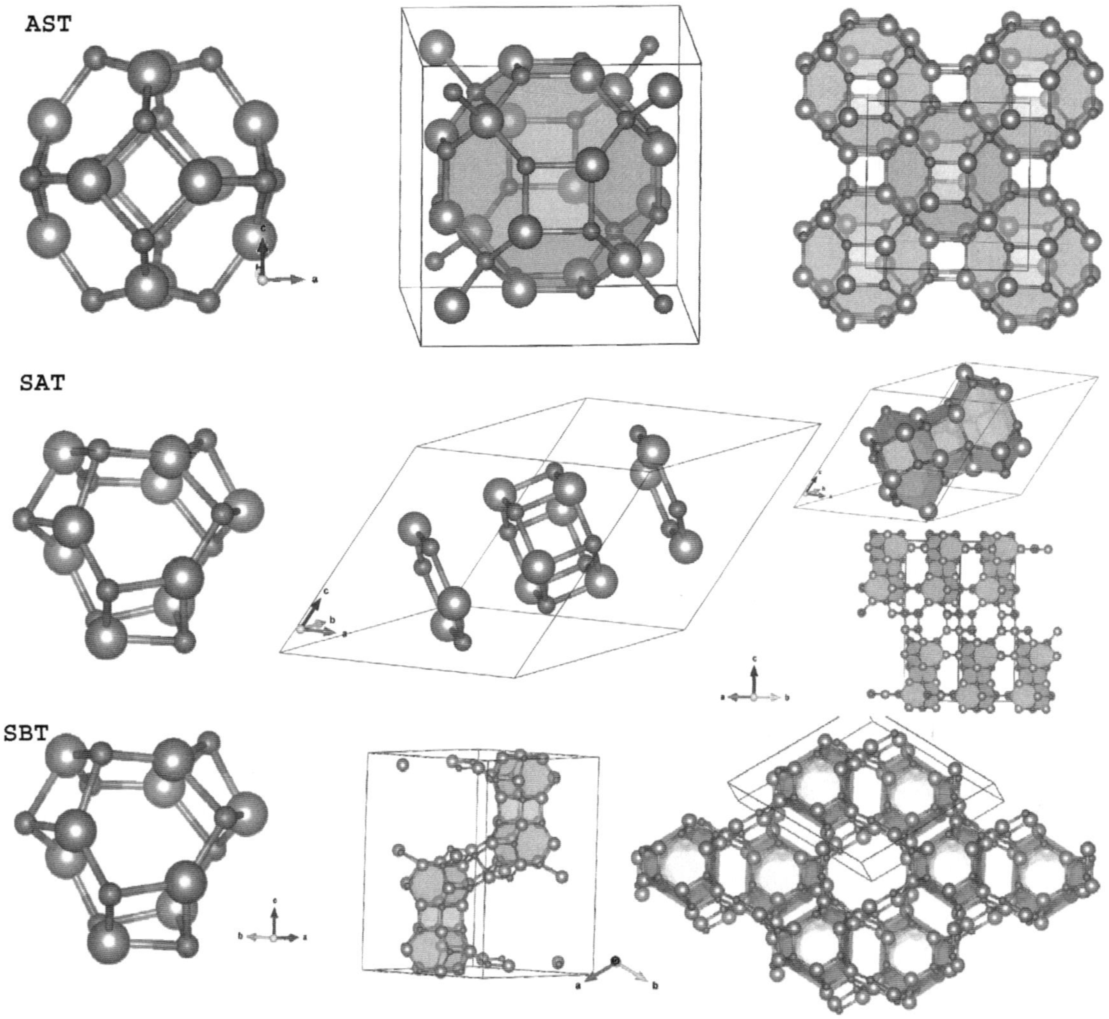


**Fig2b.** From left to right the relaxed cluster - optimized crystal lattice primitive cell – hollow-bulk structure of this work's structures AST, SAT and SBT.

## 3. Computational details

### *3.1. Density-Functional-Based Tight-Binding Plus (DFTB+) method*

The spin-polarized, charge self-consistent, DFTB approach is based on a second-order expansion of the spin-dependent Kohn-Sham total energy functional with respect to a given reference charge and magnetization density. The method has been extensively discussed elsewhere [13-16] and briefly outlined here as: 1/ Expand the orbitals as a linear combination of Slater type orbitals (LCSTO):

$$\psi_n(r) = \sum_{\text{atomic site}\, i} \sum_{\text{orbital}\, \nu} C_{i\nu} \phi_{i\nu}(\mathbf{r}-\mathbf{R}_i) \,. \tag{1}$$

The basis functions $\phi_{i\alpha}(\mathbf{r}-\mathbf{R}_i)$ centered on the atomic nucleus $i$, with position $\mathbf{R}_i$, are themselves a linear combination of single Slater orbitals $\phi_\nu(\mathbf{r}) = (\sum_{j=1} (\sum_{n=0} a_{jn}\mathbf{r}^{l_\nu+n}) e^{-\alpha_j r}) Y_{l_\nu m_\nu}$ The angular and magnetic quantum number are indicated with $l_\nu$ and $m_\nu$. $Y_{l_\nu m_\nu}$ is the corresponding real spherical harmonic.

2/ Tight-binding expansion of the wave functions (calculation of the matrix elements in the two-centers approximation).

$$\sum_{\text{atomic site}\, i} \sum_{\text{orbital}\, \nu} | H_{i\nu,j\mu} - ES_{i\nu,j\mu} | C_{i\nu,j\mu} = 0 \,, \tag{2}$$

with

$$H_{i\nu,j\mu} = \left\langle \phi_{i\nu} | \hat{H} | \phi_{j\mu} \right\rangle \text{ and } S_{i\nu,j\mu} = \left\langle \phi_{i\nu} | \phi_{j\mu} \right\rangle \,. \tag{3}$$

3/ Second order-expansion of Kohn-Sham energy functional (self-consistency in the charge density - SCC-DFTB):

$$E_{tot} = \sum_i^{occ.} n_i \langle \psi_i | H_0 | \psi_i \rangle + \frac{1}{2} \sum_{\mu,\nu} \gamma_{\mu\nu} \Delta q_\mu \Delta q_\nu + E^{rep} \,, \tag{4}$$

where $\Delta q_\mu$ - charge fluctuation decomposed into atomic contribution (Mulliken charge), $\gamma_{\mu\nu}$ - some integral coefficient, $E^{rep}$ - repulsive term (see Ref. 4-6).

With all matrix elements and orbitals are derived from Density Functional calculation, the advantage of DFTB method relies on the use of small basic set of atomic orbitals (in order to reduce the matrix dimension for diagonalization speed-up) and the restriction to two center nonorthogonal Hamiltonian (allowing extensive use of look-up table). What it distinguishes from semi-empirical method is the explicite calculation of the basic wave function which allow deeper physics insight and better control of the approximation used. The method solved Kohn-Sham equation self-consistently using Mulliken charge projection. This approach have proved to give transferable and accurate interaction potential as well as numerical efficiency allowing Molecular Dynamic (MD) simulation of supercell containing several hundreds to a thousand atoms. Thus this is particular suitable to study the electronics proper and dynamics of large mesoscopic system and organic molecule such as Nano-wire, CNT's, DNA stands or absorbate on surface, semiconductor hetero-structure etc. see review in [14-15] and references there in. The advantage of DFTB parameterization is that only few, possibly well chosen systems are needed to create the parameters, i.e. in DFTB fit systems can also be purely ideal systems, if they are chemically acceptable and can be described carefully with an ab-initio approach. Next this well tested parameter, e.g. in an attention for solid state systems and for defect physics as in current case, can be used for much larger system due to its transferability. In our calculation the parameter and its transferability have been successfully applied in several DFTB works [16-18].

The electronic DFTB parameters (i.e., Hubbard parameters, $H^0_{\mu\nu}$ and $S_{\mu\nu}$ matrix elements) were derived directly from DFT calculations, performed within the generalized gradient approximation (GGA) and using the Perdew, Burke, and Ernzerhof (PBE) exchange-correlation functional.

Journal of Physics: Conference Series **726** (2016) 012022 doi:10.1088/1742-6596/726/1/012022

*3.2. Corhersive energy, phase stability and Equation of State (EOS).*

To compare the stability of these nanoporous frameworks with that of traditional ZnO phases, we have calculated total energy per ZnO-pair vs. volume for every studied structures together with well-known phases wurtzite ZnO and zinclende ZnO. The results are presented in Figure 3, in which the solid curves derive from third-order Birch– Murnaghan EOS fit.

On the basis of these fitted curves, we may draw some conclusions on their stabilities. As shown in the Figure 3 and table2, one can note that the SOD phase has the lowest equilibrium energy, indicating that it is the most energetically stable phase among the studied phases. Following next is the CAN, AST. The LTA and SAT have similar stability. Following by GME, $Zn_{16}O_{16}$-cub and finally FAU and SBT. Note that most of the calculated ZnO polytypes, exhibit essentially similar E(V) curve as wurtzite modification at this level of calculation and are indistinguishable on the scale of the figure.

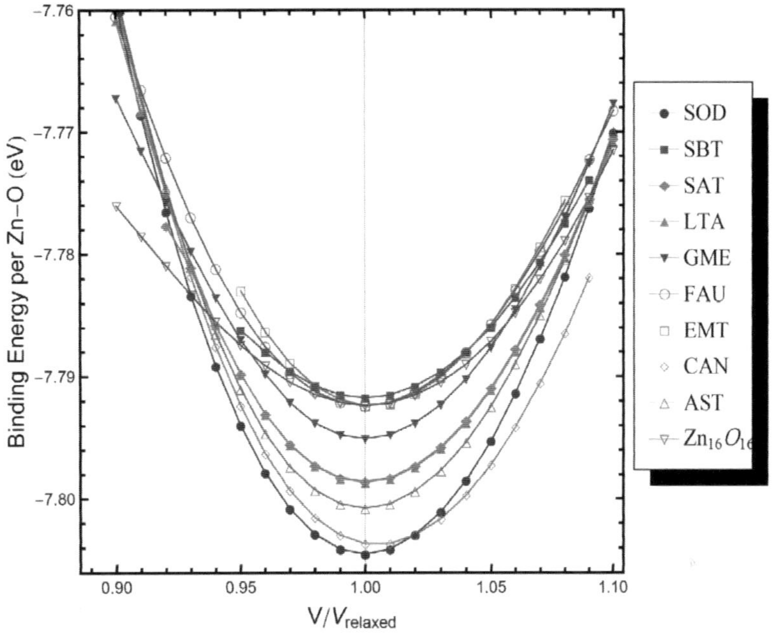


**Fig3.** Total energy (eV/Zn-O) vs. relative volume for the considered phases of ZnO.

*3.3. Electronic band structure*

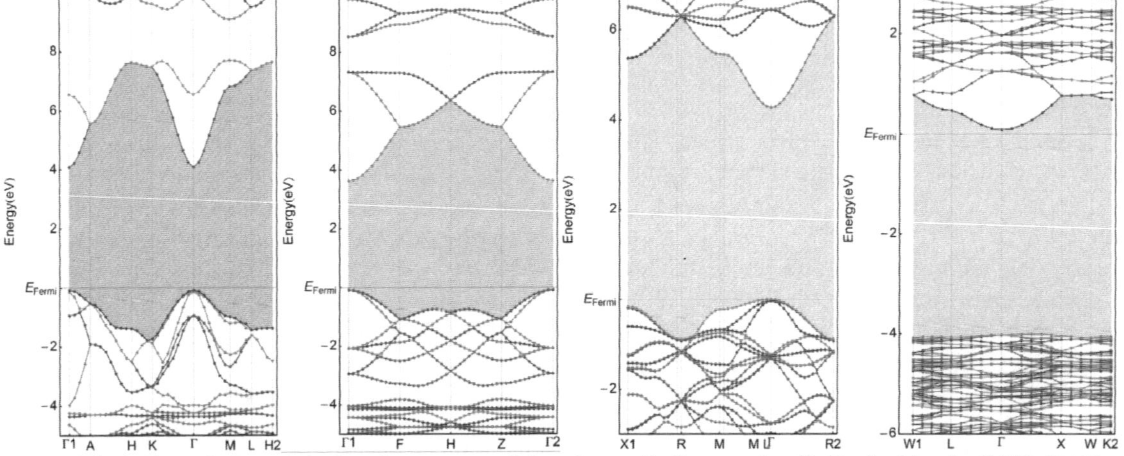


**Fig4.** From left to right band structure of the phases ZnO-wurtzite, ZnO-zincblende, SOD, FAU

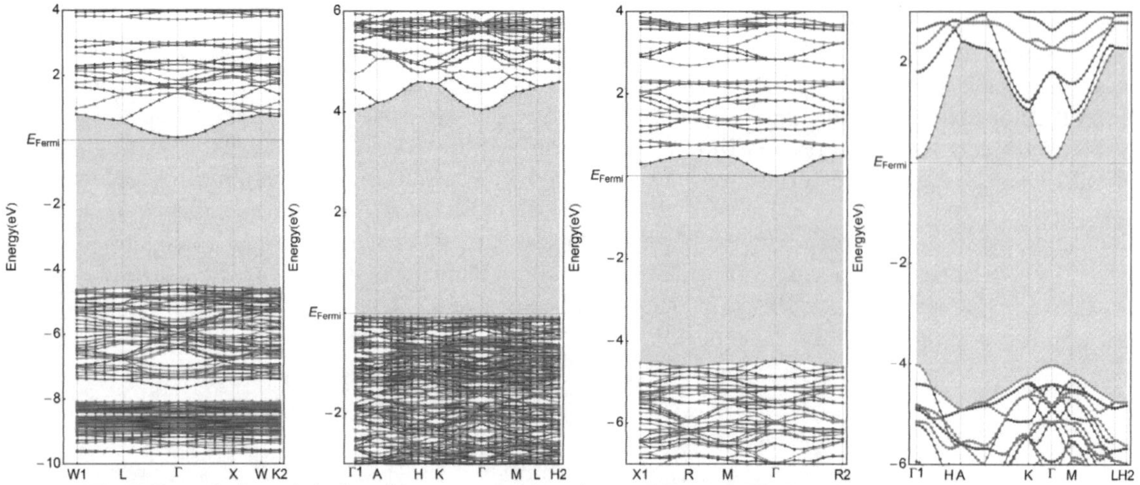


**Fig5.** From left to right band structure of the phases LTA, EMT, $Zn_{16}O_{16}$-cub and CAN

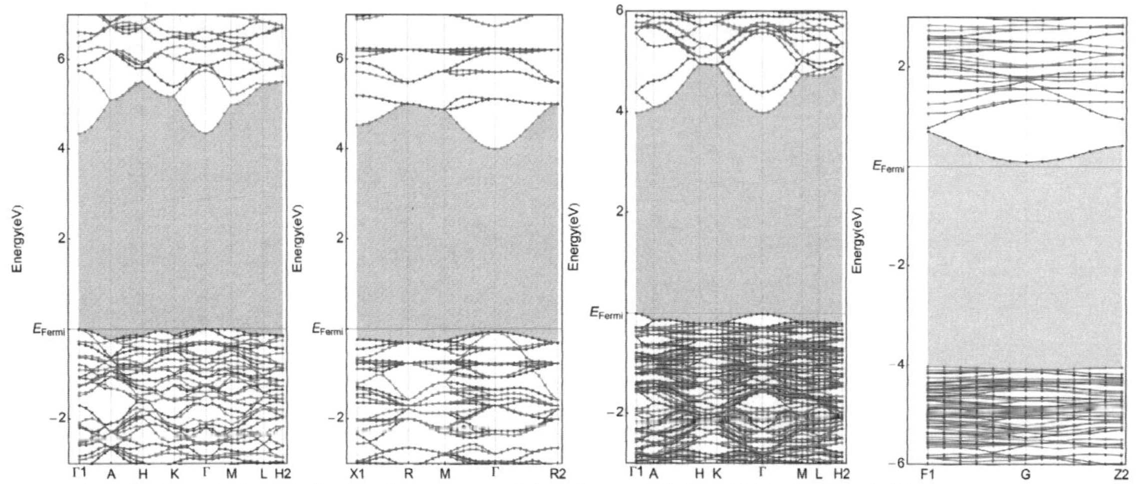


**Fig6.** From left to right band structure of the phases GME, AST, SAT and SBT

Concerning the effects of polytypism on the electronic structure, i.e. bandstructures and gaps. Our calculations show that the band gaps among the polytypes ranging from 4.0 eV for SAT to 4.57 eV for LTA. It is noted that within our DFTB parametrization the ZnO wurtzite's bandgap is the order of 3.73eV and ZnO zincblende is equal to 4.16eV, which are close enough to the experimental values. So that all the new phases are still wide bandgap semiconducting materials.

## 4. Results and discussion

The stability of studied structures evident that, it is reasonable to believe that these cluster-assembled phases would be the attractive synthesis targets. In the $Zn_kO_k$-assembled phases, the coordination number for every phase is all four (except some small dangling in phase $Zn_{16}O_{16}$-cub), i.e. each Zn (O) atom has four O (Zn) neighbors forming distortional $sp^3$-type hybridization for almost all phases. Thus preserving the valuable properties of the ZnO materials, such as semicoducting, piezoelectric and optically transparent.

As is shown in characteristics Table 1, the volume per atom of ZnO hollow phases higher than the WZ and ZB phase, respectively, from 29.0% up to 85% larger than that of WZ. Naturally, it leads to higher flexibility and compressibility (lower bulk moduli) of the new hollow phases. Therefore, these new nanoporous phases if they are could be synthesized, will be one of the most promising candidates for replacing the expensive and mechanically fragile atomic or molecular selective materials.

Journal of Physics: Conference Series **726** (2016) 012022 doi:10.1088/1742-6596/726/1/012022

Hereby, we also show the estimations of pore volume as well as specific surface area of the optimized primitive unit cell for these nanoporous phases (see Table 1), which does reflect the porosity of their framework to a certain extent.

The other essential structural parameters, such as the lattice contents, particle density, mass density, crystall structure together with the symmetry space group, which shows that all the phases are high symmetry. i.e. come with high stabilities, are also listed in characteristic Table 1.

**Table 1.** Calculated characteristics of all studied phase and most common phase of ZnO.

| Structures | **WZ** | **ZB** | **SOD** | **LTA** | **FAU** | **$Zn_{16}O_{16}$cube** |
|---|---|---|---|---|---|---|
| Mass Density (gcm$^{-3}$) | 5.68 | 5.4 | 4.306 | 3.563 | 3.081 | 3.28 |
| V/at Å$^3$ | 11.9 | 12.6 | 15.69 | 18.967 | 21.93 | 20.6 |
| Part.density ($10^{23}$cm$^{-3}$) | 8.8 | 4.84 | 12.64 | 48.53 | 48.46 | 32.48 |
| Coord. number | 4 | 4 | 4 | 4 | 4 | 3.75 |
| Crystall structure | Wurt (hex) | Cub | Cub | FCC | Cub | Cub |
| Symmetry groups | P63M186 | F43216 | PM3N 223 | FM3C 226 | FD-3 203 | P43-M 215 |
| Unit cell (atoms) | 8 | 4 | 12 | 48 | 48 | 32 |
| Lattice param. Å (a-c) | 3.249-5.205 | 4.63 | 5.732 | 10.89 | 16.15 | 8.704 |
| Average bond Å | 1.977 | 2.015 | 2.027 | 2.032 | 2.043 | 2.074 |
| Average angle | | | | | | |
| Zn-O-Zn | 109.5 | 109.5 | 110 | 107.3 | 109.2 | 108.18 |
| O-Zn-O | 109.5 | 109.5 | 110 | 107.6 | 108.7 | 108.88 |
| Band gap eV | 3.73 | 4.16 | 4.218 | 4.572 | 4.108 | 4.49 |
| UC specific surface area Å$^2$ | 0 | 0 | 80.63 | 377.6 | 1464 | 268.74 |
| UC Pore Volume Å$^3$ | 0 | 0 | 248.3 | 1201 | 2096 | 659.5 |
| | **EMT** | **GME** | **CAN** | **AST** | **SAT** | **SBT** |
| Mass Density (gcm$^{-3}$) | 3.078 | 3.702 | 4.404 | 4.089 | 4.073 | 3.242 |
| V/at Å$^3$ | 21.953 | 18.254 | 15.35 | 16.53 | 16.592 | 20.84 |
| Part.density ($10^{23}$cm$^{-3}$) | 96.46 | 24.55 | 12.65 | 40.60 | 72.6 | 48.48 |
| Coordination number | 4 | 4 | 4 | 4 | 4 | 4 |
| Crystall structure | Trigonal | Trigonal | Hex | Cub | Triclinic | Hex |
| Symmetry groups | P31C 163 | P31C 163 | P63M 176 | F43M 216 | R3 148 | R3 148 |
| Unit cell (atoms) | 96 | 24 | 12 | 40 | 24 | 48 |
| Lattice param. Å (a-c) | 11.43-18.64 | 8.85-6.47 | 8.047-3.288 | 8.711 | 7.88 | 11.32-27.0 |
| Average bond Å | 2.043 | 2.037 | 2.037 | 2.031 | 2.036 | 2.043 |
| Average angle | | | | | | |
| Zn-O-Zn | 108.7 | 108.32 | 108.3 | 108.0 | 108.2 | 107.6 |
| O-Zn-O | 109.1 | 108.41 | 108.07 | 107.9 | 108.2 | |
| Band gap eV | 4.109 | 4.327 | 4.245 | 4.054 | 4.0 | 4.136 |
| UC specific surface area Å$^2$ | 3159 | 188.3 | 71.87 | 273 | 396.6 | 324.4 |
| UC Pore Volume Å$^3$ | 736.7 | 608.6 | 228.9 | 188 | 120.9 | 1000.5 |

From the above table 1, the symmetry of the hollow phases also comfirm that in various structures different stacking and/or linking orders, resulted from mixing the hcp- and the fcc-type stacking, would produce different polytypes.

### 5. Conclusions

Our theoretical prediction on the three interesting ZnO candidates for ZnO hollow phases follow by the bottom up approach from diffirent types of ZnO magic cluster which exhibit similar stacking orders to the many investigated ZnO polytypes. Comparing with the phases found recently by others, it is clearly expected that, the hollow phases could survive in periodic systems without structural collapse, which leads to their nanoporous and low-density features. Furthermore structurally, we can relate these polytype structures to those of other II-VI semiconducting materials, such as ZnS, CdSe, CdTe where different polytypes might also be observed.

### Acknowledgments

This work was supported by Vietnam National Foundation for Science and Technology Development (NAFOSTED) under grant number 103.01-2014.25.

### References

[1] Willander M 2014 *Zinc Oxide Nanostructures Advance and Applications* ed. Magnus Willander (Pan Stanford Publisher) 1st edition
[2] Kołodziejczak-Radzimska A and Jesionowski M 2014 *Materials* **7** 2833
[3] Woodley M and Catlow R 2008 *R. Nat. Mater.* **7** 937
[4] Zhifeng Liu, Xinqiang Wang, Jiangtao Cai, Gaobin Liu, Ping Zhou, Kan Wang and Hengjiang Zhu 2013 *J. Phys. Chem. C* **117** 17633
[5] Sokol A A, Farrow M R, Buckeridge J, Logsdail A J, Catlow C R A, Scanlonab D O and Woodley S M 2014 *Phys. Chem. Chem. Phys.* **16** 21098
[6] Yongliang Yong, Xiaohong Li, Xiping Hao, Jingxiao Cao and Tongwei Li 2014 *RSC Adv.* **4** 37333
[7] Peng Guo, Jiho Shin, Alex Greenaway G, Jung Gi Min, Jie Su, Hyun June Choi, Leifeng Liu, Paul Cox A, Suk Bong Hong, Paul Wright A and Xiaodong Zou 2015 *Nature* **524** 74
[8] Xu R, Pang W, Yu J, Huo Q and Cheng J 2007 *Chemistry of Zeolites and Related Porous Materials: Synthesis and Structure* (John Wiley & Sons Asia) Singapore
[9] Roth W J, Nachtigall P, Morris R E, Wheatley P S, Seymour V R, Ashbrook S E, Chlubná P, Grajciar L, Položij M, Zukal A, Shvets O and Čejka J 2013 *Nature Chemistry* **5** 628
[10] Carrasco J, Illas F and Bromley S T 2007 *Phys. Rev. Lett.* **99** 235502
[11] Yong Y, Song B and He P 2011 *J. Phys. Chem. C* **115** 6455
[12] Demiroglu I, Tosoni S, Illasa F and Bromley S T 2014 *Nanoscale* **6** 1181
[13] Elstner M, Porezag D, Jungnickel G, Elsner J, Haugk M, Frauenheim Th, Suhai S and Seifert G 1998 *Phys. Rev. B* **58** 7260
[14] Kohler C, Seifert G and Frauenheim Th 2005 *Chem. Phys.* **309** 23
[15] Elstner M, Porezag D, Jungnickel G, Elsner J, Haugk M, Aradi B, Hourahine B and Frauenheim Th 2007 *J. Phys. Chem. A* **111** 5678
[16] simulation code dftb+ at http://www.dftb-plus.info/
[17] Moreira N H, Dolgonos G, Aradi B, Da Rosa A L and Frauenheim Th 2009 *J. Chem. Theory Comput.* **5** 605
[18] Vu Ngoc Tuoc 2010 *Comp. Mat. Sci.* **49** S161
[19] Momma K and Izumi F 2008 *J. Appl. Crystallogr.* **41** 653

Journal of Physics: Conference Series **726** (2016) 012022 doi:10.1088/1742-6596/726/1/012022

**Appendix A Structures files of the new hollow structures found by this work**

1/ structure: AST
cell parameters
8.7111 8.7111 8.7111 90.0000 90.0000 90.0000
fractional coordinates

| | | | | |
|---|---|---|---|---|
| Zn1 | Zn | 0.75000 | 0.25000 | 0.25000 |
| Zn2 | Zn | 0.25000 | 0.75000 | 0.25000 |
| Zn3 | Zn | 0.25000 | 0.25000 | 0.75000 |
| Zn4 | Zn | 0.75000 | 0.75000 | 0.75000 |
| O1 | O | 0.25000 | 0.25000 | 0.25000 |
| O2 | O | 0.75000 | 0.75000 | 0.25000 |
| O3 | O | 0.75000 | 0.25000 | 0.75000 |
| O4 | O | 0.25000 | 0.75000 | 0.75000 |
| Zn5 | Zn | 0.11880 | 0.11880 | 0.11880 |
| Zn6 | Zn | 0.11880 | 0.61880 | 0.61880 |
| Zn7 | Zn | 0.61880 | 0.11880 | 0.61880 |
| Zn8 | Zn | 0.61880 | 0.61880 | 0.11880 |
| Zn9 | Zn | 0.88121 | 0.88121 | 0.11880 |
| Zn10 | Zn | 0.88121 | 0.38121 | 0.61880 |
| Zn11 | Zn | 0.38121 | 0.88121 | 0.61880 |
| Zn12 | Zn | 0.38121 | 0.38121 | 0.11880 |
| Zn13 | Zn | 0.88121 | 0.11880 | 0.88121 |
| Zn14 | Zn | 0.88121 | 0.61880 | 0.38121 |
| Zn15 | Zn | 0.38121 | 0.11880 | 0.38121 |
| Zn16 | Zn | 0.38121 | 0.61880 | 0.88121 |
| Zn17 | Zn | 0.11880 | 0.88121 | 0.88121 |
| Zn18 | Zn | 0.11880 | 0.38121 | 0.38121 |
| Zn19 | Zn | 0.61880 | 0.88121 | 0.38121 |
| Zn20 | Zn | 0.61880 | 0.38121 | 0.88121 |
| O5 | O | 0.88235 | 0.11765 | 0.11765 |
| O6 | O | 0.88235 | 0.61765 | 0.61765 |
| O7 | O | 0.38235 | 0.11765 | 0.61765 |
| O8 | O | 0.38235 | 0.61765 | 0.11765 |
| O9 | O | 0.11765 | 0.88235 | 0.11765 |
| O10 | O | 0.11765 | 0.38235 | 0.61765 |
| O11 | O | 0.61765 | 0.88235 | 0.61765 |
| O12 | O | 0.61765 | 0.38235 | 0.11765 |
| O13 | O | 0.11765 | 0.11765 | 0.88235 |
| O14 | O | 0.11765 | 0.61765 | 0.38235 |
| O15 | O | 0.61765 | 0.11765 | 0.38235 |
| O16 | O | 0.61765 | 0.61765 | 0.88235 |
| O17 | O | 0.88235 | 0.88235 | 0.88235 |
| O18 | O | 0.88235 | 0.38235 | 0.38235 |
| O19 | O | 0.38235 | 0.88235 | 0.38235 |
| O20 | O | 0.38235 | 0.38235 | 0.88235 |

2/ structure: SAT
cell parameters
8.2486 8.2486 8.2487 60.0000 60.0000 60.0000
fractional coordinates

| | | | | |
|---|---|---|---|---|
| Zn1 | Zn | 0.05402 | 0.28158 | 0.54414 |
| Zn2 | Zn | 0.78945 | 0.56183 | 0.30996 |
| O1 | O | 0.45639 | 0.70881 | 0.95079 |
| O2 | O | 0.70489 | 0.44874 | 0.19866 |
| Zn3 | Zn | 0.54412 | 0.05404 | 0.28157 |
| Zn4 | Zn | 0.30995 | 0.78943 | 0.56184 |
| O3 | O | 0.95076 | 0.45643 | 0.70879 |
| O4 | O | 0.19868 | 0.70488 | 0.44871 |
| Zn5 | Zn | 0.28159 | 0.54410 | 0.05404 |
| Zn6 | Zn | 0.56183 | 0.30989 | 0.78949 |
| O5 | O | 0.70879 | 0.95080 | 0.45639 |
| O6 | O | 0.44873 | 0.19869 | 0.70483 |
| Zn7 | Zn | 0.94597 | 0.71843 | 0.45585 |
| Zn8 | Zn | 0.21051 | 0.43818 | 0.69006 |
| O7 | O | 0.54361 | 0.29118 | 0.04922 |
| O8 | O | 0.29512 | 0.55125 | 0.80135 |
| Zn9 | Zn | 0.45589 | 0.94596 | 0.71840 |
| Zn10 | Zn | 0.69007 | 0.21056 | 0.43815 |
| O9 | O | 0.04922 | 0.54358 | 0.29121 |
| O10 | O | 0.80131 | 0.29513 | 0.55130 |
| Zn11 | Zn | 0.71844 | 0.45586 | 0.94599 |
| Zn12 | Zn | 0.43816 | 0.69011 | 0.21051 |
| O11 | O | 0.29120 | 0.04920 | 0.54360 |
| O12 | O | 0.55128 | 0.80133 | 0.29511 |

3/ structure: SBT
cell parameters
11.2501 11.2501 11.2501 59.7560 59.7560 59.7560
fractional coordinates

| | | | | |
|---|---|---|---|---|
| Zn1 | Zn | 0.36488 | 0.53552 | 0.71822 |
| Zn2 | Zn | 0.20688 | 0.66907 | 0.02736 |
| Zn3 | Zn | 0.84451 | 0.64093 | 0.30812 |
| Zn4 | Zn | 0.51336 | 0.34297 | 0.16431 |
| O5 | O | 0.64933 | 0.27398 | 0.46790 |
| O6 | O | 0.79499 | 0.96347 | 0.33379 |
| O7 | O | 0.16897 | 0.70157 | 0.34119 |
| O8 | O | 0.48722 | 0.84403 | 0.65581 |
| Zn9 | Zn | 0.71822 | 0.36488 | 0.53552 |
| Zn10 | Zn | 0.02736 | 0.20688 | 0.66907 |
| Zn11 | Zn | 0.30812 | 0.84451 | 0.64093 |
| Zn12 | Zn | 0.16431 | 0.51336 | 0.34297 |
| O13 | O | 0.46790 | 0.64933 | 0.27398 |

| | | | | |
|---|---|---|---|---|
| O14 | O | 0.33379 | 0.79499 | 0.96347 |
| O15 | O | 0.34119 | 0.16897 | 0.70157 |
| O16 | O | 0.65581 | 0.48722 | 0.84403 |
| Zn17 | Zn | 0.53552 | 0.71822 | 0.36488 |
| Zn18 | Zn | 0.66907 | 0.02736 | 0.20688 |
| Zn19 | Zn | 0.64093 | 0.30812 | 0.84451 |
| Zn20 | Zn | 0.34297 | 0.16431 | 0.51336 |
| O21 | O | 0.27398 | 0.46790 | 0.64933 |
| O22 | O | 0.96347 | 0.33379 | 0.79499 |
| O23 | O | 0.70157 | 0.34119 | 0.16897 |
| O24 | O | 0.84403 | 0.65581 | 0.48722 |
| Zn25 | Zn | 0.63512 | 0.46448 | 0.28178 |
| Zn26 | Zn | 0.79312 | 0.33093 | 0.97264 |
| Zn27 | Zn | 0.15549 | 0.35907 | 0.69188 |
| Zn28 | Zn | 0.48664 | 0.65703 | 0.83569 |
| O29 | O | 0.35067 | 0.72602 | 0.53210 |
| O30 | O | 0.20501 | 0.03653 | 0.66621 |
| O31 | O | 0.83103 | 0.29843 | 0.65881 |
| O32 | O | 0.51278 | 0.15597 | 0.34419 |
| Zn33 | Zn | 0.28178 | 0.63512 | 0.46448 |
| Zn34 | Zn | 0.97264 | 0.79312 | 0.33093 |
| Zn35 | Zn | 0.69188 | 0.15549 | 0.35907 |
| Zn36 | Zn | 0.83569 | 0.48664 | 0.65703 |
| O37 | O | 0.53210 | 0.35067 | 0.72602 |
| O38 | O | 0.66621 | 0.20501 | 0.03653 |
| O39 | O | 0.65881 | 0.83103 | 0.29843 |
| O40 | O | 0.34419 | 0.51278 | 0.15597 |
| Zn41 | Zn | 0.46448 | 0.28178 | 0.63512 |
| Zn42 | Zn | 0.33093 | 0.97264 | 0.79312 |
| Zn43 | Zn | 0.35907 | 0.69188 | 0.15549 |
| Zn44 | Zn | 0.65703 | 0.83569 | 0.48664 |
| O45 | O | 0.72602 | 0.53210 | 0.35067 |
| O46 | O | 0.03653 | 0.66621 | 0.20501 |
| O47 | O | 0.29843 | 0.65881 | 0.83103 |
| O48 | O | 0.15597 | 0.34419 | 0.51278 |

# On the fluctuation induced mass enhancement

**Nguyen Van Hoa[1], Vu Ngoc Tuan[2], Le Van Xuan[3], Nguyen Tri Lan[4] and Nguyen Ai Viet[4]**

[1]Hochiminh Pedagogical University, 270 An Duong Vuong, Hochiminh City, Viet Nam

[2]Military Institute of Science and Technology, 17 Hoang Sam, Cau Giay, Ha Noi, Viet Nam

[3]Hanoi College of Technology and Economics, Soc Son, Ha Noi, Viet Nam

[4]Institute of Physics, Vietnam Academy of Science and Technology, 10 Dao Tan, Ha Noi, Viet Nam

E-mail: xuanlv.hn@gmail.com

**Abstract.** The effective mass induced by the background fluctuation on particles is considered. The analytical results show that the effective mass depends only on the properties of fluctuation, and takes non–zero value when and only when fluctuation mean value is non—zero. The possible applications of the obtained results to complex systems such as biology and ecology where environmental factors lead to the changes of the information exchange ranges from long to short one are discussed, i.e. the possibility of using physical modeling techniques to investigate macroscopic behaviors of some complex systems under consideration.

## 1. Introduction

In conventional understanding of modern physics, motion of a particle is characterized by its dispersion law $\mathcal{E}(p)$ where the dependence of $\mathcal{E}(p)$ on $p$ reflects whether particle is massive or massless.

In one hand, the massless particle has dispersion law in form

$$\mathcal{E}(p) = cp, \tag{1}$$

and in other hand, the massive particle $m_0$ has the dispersion law written in the relativistic form[1, 2] as

$$\mathcal{E}(p) = \sqrt{(mc^2)^2 + (cp)^2}, \tag{2}$$

which tends to additive law of energy

$$\mathcal{E}(p) = mc^2 + \frac{1}{2m}p^2 + \underbrace{V(p)}_{o(p^4)}, \tag{3}$$

when momentum $p$ approaches to zero, where $c$ is some maximal velocity of the massless motion in given media. It is important to think of a possibility that the coordinate representation of function $V(p)$

$$V(r) = \frac{1}{\sqrt{V_\Omega}} \int_\Omega \mathrm{d}p \; \mathrm{e}^{ipr} V(p) , \tag{4}$$

would contain some physical insights of potential affecting on the motion of a massive particle in the medium, or in other words, interaction between massive particle and surrounding environment.

Following the standard mathematical routine found in textbooks, the mass of a moving particle with given dispersion law $\mathcal{E}(p)$ is determined by second order derivative of $\mathcal{E}(p)$ with respect to $p$ [3, 4] as

$$\frac{1}{m} = \frac{\mathrm{d}^2}{\mathrm{d}p^2} \mathcal{E}(p) \bigg|_{p=0} . \tag{5}$$

When a particle moves under influence of some fluctuations which changes the dispersion law of the particle, energy of massive particle reads[5, 6, 7]

$$\mathcal{E}(p, \epsilon) = \sqrt{(mc^2)^2 + c^2 p^2 + \epsilon^2} , \tag{6}$$

then the energy of moving particle is changed, and obtained by including all contributions of fluctuations via integrating out with fluctuation distribution function $\mathcal{D}(\epsilon)$

$$\mathcal{E}[cp; \mathcal{D}] = \int_\Omega \mathrm{d}\epsilon \mathcal{D}(\epsilon) \sqrt{(mc^2)^2 + c^2 p^2 + \epsilon^2}. \tag{7}$$

The effective mass of moving particle will be determined by the definition (5).

## 2. The Effective Mass

To determine the fluctuation induced mass [3, 4] of initial particle, the second order derivative of dispersion law $\mathcal{E}(cp; \mathcal{D})$ with respect to momentum should be calculated, and the result reads

$$\begin{aligned} \frac{1}{m^*} &= \frac{\mathrm{d}^2}{\mathrm{d}p^2} \mathcal{E}[cp; \mathcal{D}] \bigg|_{p=0} \\ &= c^2 \int_\Omega \mathrm{d}\epsilon \mathcal{D}(\epsilon) \frac{1}{\sqrt{(mc^2)^2 + \epsilon^2}} . \end{aligned} \tag{8}$$

The impression enlightened by second derivative of the integral is that fluctuations of the medium would cause a change of mass of particle moving in it. However, to see more insights, some simple type of fluctuation will be investigated.

### *2.1. Simplest single mode fluctuation – Dirac distribution*

In single mode fluctuation, the distribution function $\mathcal{D}(\epsilon)$ is nothing but the Dirac $\delta$–function[10]

$$\mathcal{D}(\epsilon) = \delta(\epsilon - \epsilon_0) . \tag{9}$$

Inserting fluctuation distribution function into the integral and performing integrating over random variable $\epsilon$, the result reads

$$\begin{aligned} m^* &= \left( \frac{\mathrm{d}^2}{\mathrm{d}p^2} \mathcal{E}\left[cp; \mathcal{D}\right]\Big|_{p=0} \right)^{-1} \\ &= \frac{1}{c^2}\sqrt{\left(mc^2\right)^2 + \epsilon_0^2}. \end{aligned} \tag{10}$$

By introducing new mass denoted by $m_\epsilon = \frac{\epsilon_0}{c^2}$ characterizing the fluctuations, the effective mass $m^*$ of a moving particle will be rewritten as follows

$$m^* = \sqrt{m^2 + m_\epsilon^2}. \tag{11}$$

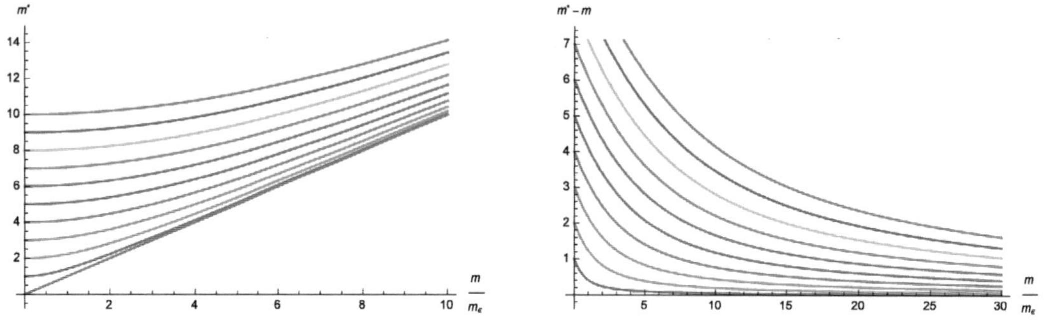


**Figure 1.** The graph of the effective mass under the influence of fluctuations in single mode fluctuation.

### *2.2. Boltzmann fluctuation*

In the Boltzmann mode of fluctuation, the Boltzmann distribution function[9]

$$\mathcal{D}\left(\epsilon\right) = \frac{1}{2\lambda} \exp\left\{ -\frac{\left|\epsilon - \epsilon_0\right|}{\lambda} \right\}, \tag{12}$$

will be used as distribution function of media fluctuations. The corresponding effective mass takes the form

$$\begin{aligned} m^* &= \left( \frac{\mathrm{d}^2}{\mathrm{d}p^2} \mathcal{E}\left[cp; \mathcal{D}\right]\Big|_{p=0} \right)^{-1} \\ &= \frac{1}{c^2}\left( \frac{\pi\left(\mathbf{H}_0\left(\frac{mc^2}{\lambda}\right) - Y_0\left(\frac{mc^2}{\lambda}\right)\right)}{2\lambda} \right)^{-1}, \end{aligned} \tag{13}$$

where $\mathbf{H}_n\left(z\right)$ is the Struve function, and $Y_n\left(z\right)$ is the Bessel function of the second kind.

### *2.3. Gaussian fluctuation*

The media with white noise provides the Gaussian fluctuation which is expressed by normal distribution function [8, 9, 11]

$$\mathcal{D}\left(\epsilon\right) = \frac{1}{\sqrt{2\pi}\sigma} \exp\left\{ -\frac{\left(\epsilon - \epsilon_0\right)^2}{2\sigma^2} \right\}. \tag{14}$$

Moving in this kind of media, the effective mass of a particle is

Journal of Physics: Conference Series **726** (2016) 012023 doi:10.1088/1742-6596/726/1/012023

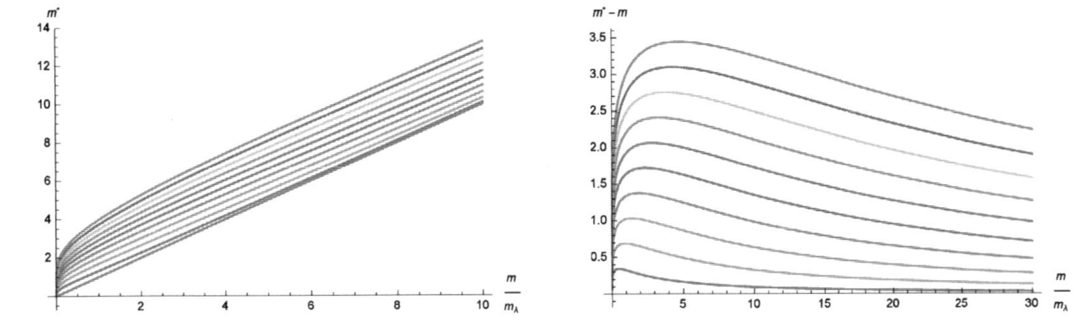


**Figure 2.** The graph of the effective mass under the influence of fluctuations in Boltzmann fluctuation, where $m_\lambda = \frac{\lambda}{c^2}$ is characteristic mass of Boltzmann fluctuation.

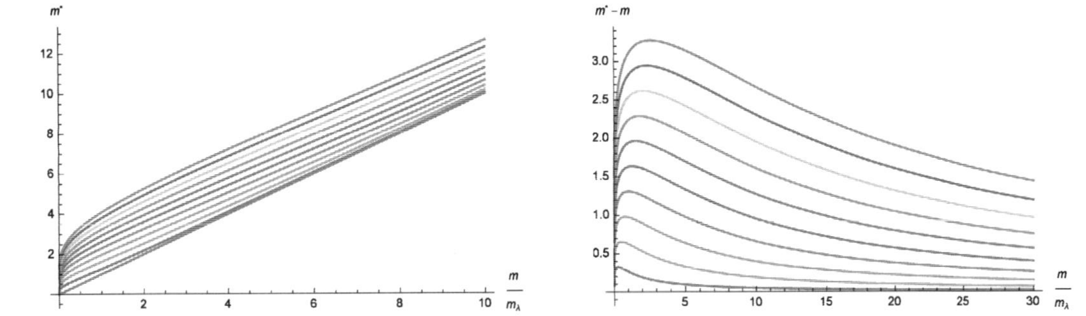


**Figure 3.** The graphs of the effective mass under the influence of fluctuations in Gaussian fluctuation, where $m_\sigma = \frac{\sigma}{c^2}$ is characteristic mass of Gaussian fluctuation.

$$\begin{aligned} m^* &= \left( \frac{\mathrm{d}^2}{\mathrm{d}p^2} \mathcal{E}\left[cp; \mathcal{D}\right] \Big|_{p=0} \right)^{-1} \\ &= \frac{1}{c^2} \exp\left\{ -\frac{c^4 m^2}{4\sigma^2} \right\} \left( \frac{K_0\left(\frac{c^4 m^2}{4\sigma^2}\right)}{\sqrt{2\pi}\sigma} \right)^{-1}, \end{aligned} \tag{15}$$

where $K_n(z)$ is the modified Bessel function of the second kind.

### *2.4. Discussion*

When a particle moves in the fluctuations of an environment, the environmental impact will produce the change of particle mass which is not linear. Furthermore, a light particle, $\frac{m}{m_\epsilon} \ll 1$, becomes much heavier within fluctuation of the media, but less influence of fluctuation is observed for heavy particles, $\frac{m}{m_\epsilon} \gg 1$.

For all kinds of fluctuations under consideration the effective mass is heavier than the mass of particle, so that the effect would be called the mass enhancement under influence of media fluctuations. As a result of the effect, an observer will see slower movement of initial particle, or in other words, the kinetic energy of the particle decreases when it gets moving in any environmental fluctuations.

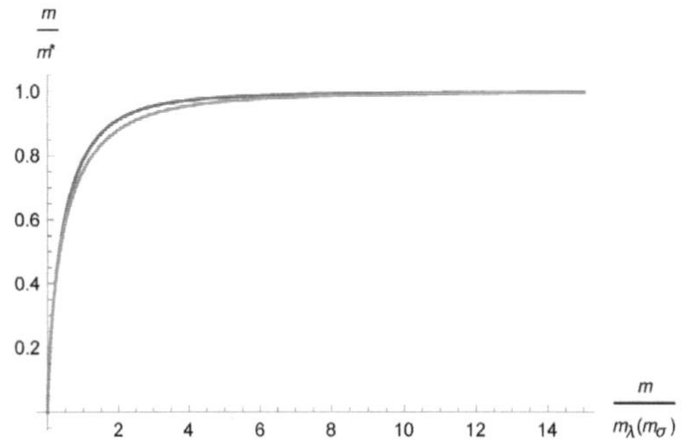


**Figure 4.** The graph of the ratio of effective mass and mass under the influence of fluctuations in Gaussian and Boltzmann fluctuations.

One of the most important issues appeared in the investigation is the mass of massless particle moving in noisy background. This problem can be formulated in following theorem

**Theorem** The fluctuation induced mass on a massless particle

$$m^* = \left( c^2 \int\limits_{\Omega} d\epsilon \frac{\mathcal{D}(\epsilon)}{|\epsilon|} \right)^{-1}, \tag{16}$$

depends only on the nature of the fluctuations, and is non–zero when and only when fluctuation mean value gets non–zero value.

However, the proof of the theorem will be left to future research.

## 3. Conclusion

In conclusion, by considering different probability density functions of background fluctuation, the change of effective mass of a moving massive and massless particles has been studied. In the case of massless particle, the effective mass gets non–zero value when and only when the expectation value of background fluctuation gets non–zero. In the case of massive particle, the dependence of effective mass on the ratio of mass and characteristic mass of background fluctuations, $m_\sigma$ and $m_\lambda$, respectively has been analytically found.

The obtained results are possibly applicable to study many collective behaviors of complex systems, where environmental factors would change the information exchange ranges from long to short one.

## Acknowledgments

This research is funded by Vietnam National Foundation for Science and Technology Development (NAFOSTED) under grant number 103.01-2015.42.

## References

[1] Einstein A 1905 Zur Elektrodynamik bewegter Körper *Annalen der Physik* (in German) **322**(10) 891–921
[2] Einstein A 1906 Über eine Methode zur Bestimmung des Verhältnisses der transversalen und longitudinalen Masse des Elektrons *Annalen der Physik* (in German) **21**(13) 583–586
[3] John Roche 2005 What is mass? *European Journal of Physics 2005* **26**(2) 225
[4] Okun L B 1989 The Concept of Mass *Physics Today* **42**(6) 31–36
[5] Landau L D 1946 On the vibration of the electronic plasma *J. Phys. USSR* **10** 25
[6] Landau L D and Lifshitz E M 1984 *Electrodynamics of Continuous Media* (Pergamon Press)

[7] Andreev A A 2000 *An Introduction to Hot Laser Plasma Physics* (Huntington, New York: Nova Science Publishers, Inc.)
[8] Lyon A 2014 Why are Normal Distributions Normal? *The British Journal for the Philosophy of Science* **65**(3) 621–649
[9] Park Sung Y and Bera Anil K 2009 Maximum entropy autoregressive conditional heteroskedasticity model *Journal of Econometrics* **150** 219–230
[10] Dirac P 1958 *The Principles of Quantum Mechanics* 4th ed. (Oxford at the Clarendon Press)
[11] Gauss C F 1809 *Theory of the Motion of the Heavenly Bodies Moving about the Sun in Conic Sections*

Journal of Physics: Conference Series **726** (2016) 012024 doi:10.1088/1742-6596/726/1/012024

# On application of non–extensive statistical mechanics to studying ecological diversity

**Le Van Xuan[1], Nguyen Tri Lan[2], and Nguyen Ai Viet[2]**

[1]Hanoi College of Technology and Economics, Soc Son, Ha Noi, Viet Nam

[2]Institute of Physics,Vietnam Academy of Science and Technology, 10 Dao Tan, Ha Noi, Viet Nam

E-mail: xuanlv.hn@gmail.com

**Abstract.** The concept of Tsallis entropy provides an extension of thermodynamics and statistical physics. In the ecology, Tsallis entropy is proposed to be a new class of diversity indices $S_q$ which covers many common diversity indices found in ecological literature. As a new statistical model for the Whittaker plots describing species abundance distribution, the truncated exponential distribution is used to calculate the diversity and evenness indices. The obtained results in new model are graphically compared with those in previous publication in the same field of interests, and shows a good agreement. A further development of a thermodynamic theory of ecological systems that is consistent with entropic approach of statistical physics is motivated.

## 1. Introduction

In the history of science, the applications of thermodynamics and statistical physics, and their methods on research of complex systems, especially biological and ecological systems is not a new trend in interdisciplinary science [1–12]. The basic of these branches of physics brings new methodologies, new ideas and new understanding into ecological science [2, 3, 8–19]. One of the main interests in research of ecological systems is biological diversity, co–called biodiversity [20–30], which is used informally to refer to diversity at all levels of taxonomic organization, particularly the level of species. In order to make a quantitative knowledge of conceptual biodiversity [4, 31–37], many scientific ideas have been developed to provide possible approaches to an explicit definition of biodiversity. However, due to the complexity of the ecological systems, it is hardly to find a single and simple definition of biodiversity that would be accepted by all scientific communities who have different scientific backgrounds, even they still have ecological modelling and thermodynamics as a common platform.

In spite of scientifically conceptual controversy, some methods of measuring biological diversity are widely accepted to be used in ecological research such as Shannon–Wiener index (H) [38–40] and Simpson index (D) [31] which are derived from information and probability theories, respectively. The justification of Shannon–Wiener and Simpson indices in ecological investigation and their applicable conditions are temporally not the subject of this presentation, but some basic important discussion of the theme would found in literature. These indices, are acceptable for the communities because of its simplicity and ability to provide a single numerical value, a partial description of species richness (S) and species evenness (E) [32,41–48]for a given ecological system.

In their mathematical nature, Shannon–Wiener and Simpson indices are the functionals of some given distributions and have the mathematical expressions of different entropies – the most important physical quantity in thermodynamics and statistical physics which has been generalized and extended to provide its applicability not only in ecological research [31, 38, 39, 49, 50] , but also in many other fields of interests such as economics, sociology, etc. It has to notice that the entropy–like indices are not entropy in original meaning, they have their meaning in their certain context. The advantage of entropy formalism outside of physics is that the back connections to thermodynamic and statistical theories of collective behaviors in science of complexity supporting the evaluation of complex aspects that are not capable of measuring in complex systems.

In the framework of the mathematical similarity of biodiversity and entropy, a entropic approach is proposed to consider biological systems. As the very early attempt in the approach, a model for a rank abundance curve or Whittaker plot [51–54]which plays the role of a distribution function of a ecological system is the subject of consideration and the consistency of the model is evaluated in the context of the $q$–generalized entropy – Tsallis entropy ($S_q$) [55–58]. In general, the model in its mathematical form provides a possibility to analytically obtain all relevant indices needed to explore ecological characteristics of general ecological systems. The derivation of a thermodynamic theory for ecological systems will be left to the further investigation.

## 2. Diversity indices and entropy

To understand what a diversity index is, it could be referred to the simple definition in Wikipedia [59] as "A diversity index is a quantitative measure that reflects how many different types, such as species of animals or of plants, there are in a dataset, and at the same time take into account how evenly the basic entities such as individuals are distributed among those types. The value of a diversity index increases both when the number of types increases and when evenness increases. For a given number of types, the value of a diversity index is maximized when all types are equally abundant."

The Wikipedia's formulation of a diversity index supports for a large number of possibilities to develop a mathematical definition of diversity index based on so–called a rank abundance curve or Whittaker plot [52] within the satisfactory of the principle of maximization of desired quantity. In point of view of physicists, the most preferable quantity for the definition of a diversity index is the entropy [7, 15, 19, 53, 60–65], but mathematicians think of any functionals depending on probability that satisfy a set of the given conditions. Having different minds for a long time, the mathematicians and the physicists agree about some common mathematical definitions of diversity index as the same as the mathematical formulation of the entropy. In last decades, it is easy to find in any university textbook [11, 66] or internet the mathematical definition of well–known diversity indices and other relevant indices such as species richness and species evenness. To avoid conceptual misunderstanding, it would be worth to emphasize that entropies are reasonable indices of diversity, but this is no reason to claim that entropy is diversity. [67] In one hand, the similarity of diversity index and entropy provides the basic for thermodynamic and statistical physical theories of ecology, in other hand, the interpretation of physical equivalent quantities in those theories has to be realized with very careful and fruitful knowledge of ecological science.

### *2.1. True Diversity*

In ecology, the numerical value of true diversity, or the effective number of types [10, 14, 23, 24, 33, 59, 68], is obtained by first taking the weighted generalized mean $M_{q-1}$

$$M_{q-1} = \left(\sum_{i=1}^{C} p_i^q\right)^{q-1}, \tag{1}$$

of the proportional abundances of the types in the dataset, and then taking the reciprocal of $M_{q-1}$ [32, 36, 37, 69, 70] as

$$\begin{aligned} {}^qD &= (M_{q-1})^{-1} \\ &= \left(\sum_{i=1}^{C} p_i^q\right)^{1/1-q}, \end{aligned} \tag{2}$$

where $C$ is the total number of types in the dataset, and $p_i$ is the proportional abundance of the $i$th type. In the expression (2), the parameter $q$ is a positive number and often called as the order of the diversity.

From the equations (1) and (2), it is obvious to recognize that increasing the value of $q$ increases the effective weight given to the most abundant species. This leads to obtaining a larger value of $M_{q-1}$ and a smaller value of true diversity (${}^qD$) with increasing $q$. Furthermore, true diversity (2) approaches to the exponential of "the so–called Shannon entropy" in limit of $q = 1$

$$\lim_{q\to 0} {}^qD = \exp\left\{-\sum_{i=1}^{C} p_i \ln p_i\right\}. \tag{3}$$

General form of true diversity would be the generating functional of many well–known diversity indices found in several literature of quantitative ecology [36].

### 2.2. *Diversity Indices*

Among the most common diversity indices, the Shannon–Wiener index (H) and Simpson index (D) are repeatedly recalled in this presentation because of their popularity not only in ecology, but also in physics, information theory, economics and so on. These indices have proved themselves as the powerful tool in interdisciplinary sciences reflecting the primary connection between physics and other sciences.

The Shannon–Wiener index, or the Shannon entropy,

$$\mathrm{H} = -\sum_{i=1}^{C} p_i \ln p_i, \tag{4}$$

was originally proposed by C. E. Shannon to quantify the information content in the information theory [38, 39] and has been a popular diversity index in the ecological literature. By bringing the information theory's meaning of Shannon entropy into ecology, the Shannon–Wiener index quantifies the uncertainty in predicting the species identity of an individual that is taken at random from the dataset.

As it was mentioned in the general expression of true diversity (3), Shannon–Wiener index is the logarithm of true diversity ${}^qD$ when $q$ tends to 1

$$\mathrm{H} = \ln {}^1D. \tag{5}$$

The other common index being also very popular in ecology literature is the Simpson index, or the Gini–Simpson index

Journal of Physics: Conference Series **726** (2016) 012024 doi:10.1088/1742-6596/726/1/012024

$$\begin{aligned} \mathrm{D} &= 1 - \sum_{i=1}^{C} p_i^2 \\ &= 1 - \left({}^2D\right)^{-1}. \end{aligned} \tag{6}$$

which was first introduced in 1949 by E. H. Simpson to measure the degree of concentration when individuals are classified into types. The same type of measure as this index can be also found in sociology, economics and management science which is known as the Blau index.

Two above common indices are obvious not the same, and have advantages and disadvantages in different scenarios of ecological calculations [14,24,27,29,33,71]. For example, in experiences of ecologists, for a measure of dominance the Simpson index is quite good, and it is a poor indicator of richness, however the Shannon–Wiener index provides an immediate measure for both aspects of diversity [33,35]. In mathematical impression, both Shannon–Wiener and Simpson indices are very similar to same definitions of entropy which is a very important quantity of thermodynamics and statistical physics where a new definition of entropy that extends statistical physics to new fields of interests is always the subject of consideration. It is surprised that the so–called Tsallis entropy [55–58]

$$S_q = \frac{1}{q-1}\left(1 - \sum_{i=1}^{C} p_i^q\right), \tag{7}$$

might be a good generalization of the Shannon–Wiener and Simpson indices because they are nothing but particular limiting cases of the Tsallis entropy with $q = 1$

$$\begin{aligned} \lim_{q\to 1} S_q &= S_1 \\ &= -\sum_{i=1}^{C} p_i \ln p_i \\ &= \mathrm{H}, \end{aligned} \tag{8}$$

and $q = 2$

$$\begin{aligned} \lim_{q\to 2} S_q &= S_2 \\ &= 1 - \sum_{i=1}^{C} p_i^2 \\ &= \mathrm{D}, \end{aligned} \tag{9}$$

respectively. In the expression (7) and its derivatives, $C$ is the number of states, $p_i$ is the probability of the $i$ state, and $q$ is a real parameter.

Within the methodology of analogical thinking [9], it is no doubt of applying the concept of Tsallis entropy to the concept of diversity to transform bulky and cumbersome calculations for each given diversity index into the single and unified platform with the help of Tsallis entropy [35]. This approach also offers an opportunity of taking into account the order of the diversity $q$ in all possible calculating processes and revising this parameter in ordinary definition (2). It has to be noticed that the order of the diversity $q$ in (2) and statistical generalization parameter $q$ in (7) are generally not the same, but in only some particular cases both these parameters match each other. Further study of interchange between two interesting parameters is of long–term investigation in the future.

## 3. A rank abundance curve and exponential distribution

Before the diversity indices are calculated, the observable dataset must be processed and displayed as species abundance data. In practice, there are many different ways to visualize the species abundance distribution. Among them, the best known and most informative method is the rank (abundance) plot or dominance (diversity) curve termed Whittaker plots in celebration of their inventor [33, 52, 68, 72]. A typical visualization of species abundance data – rank abundance curve or Whittaker plot could be found in several textbooks and research on ecology [33, 52, 68, 72]. The more detailed information explaining why the Whittaker plot is the most preferable in study of diversity could be found in a number of university textbooks and references therein.

In theoretical view, it might be agreed that a simple numerical plot brings inside not so much useful contents and does not support a good generalization of the research. A number of statistical models was proposed to empirically fit a observed dataset with a given distribution function providing the possibility for comparing obtained results among different datasets. In the last eighty–year period, many statistical models have been proposed and developed to maximize the information extracted from large observed datasets in different ecological projects. Several arguments have been listed by investigators to prove or to support how and why one model is chosen, and to validate the results of the chosen model. The development makes calculations of diversity more complicated and controversial in all aspects [33, 68].

To avoid complexity caused by furnishing well–known statistical models, another statistical model based on the knowledge of Boltzmann – Gaussian transition [73], the well–known phenomenon in complexity science, is proposed. The distribution function of newly proposed model is simply an exponential function. Hence the number of different species in a dataset could not be infinite, the rank abundance curve of the given dataset has to be fitted by a truncated exponential distribution

$$D_C(x_i, \lambda) = p_i = \begin{cases} \frac{\exp\{\frac{C}{\lambda}\}}{\lambda\left(\exp\{\frac{C}{\lambda}\}-1\right)} \exp\left\{-\frac{x_i}{\lambda}\right\} & \text{for } 0 \le x_i \le C \\ 0 & \text{otherwise,} \end{cases} \tag{10}$$

where $C$ is maximum value of species rank and $\lambda$ is some rate parameter of the rank abundance curve under consideration. The model parameters $C$ and $\lambda$ are directly obtained from observed dataset.

The distribution function (10) allows to perform an analytical and exact calculation for Tsallis entropy. This property of the truncated exponential distribution can be made as one of all reasons why it is selected to be the distribution function of the statistical model under consideration. The resultant entropy reads

$$S_q = \frac{1}{q-1} \frac{1-\lambda\left(e^{\frac{Cq}{\lambda}}-1\right)\left(\lambda\left(e^{\frac{C}{\lambda}}-1\right)\right)^{-q}}{q}. \tag{11}$$

The Tsallis entropy $S_q$ (11) is mathematical exact and supporting a construction of a formal thermodynamic and statistical theory for a class of assemblages which can be described by distribution function (10). The question on the accuracy of the newly proposed statistical model in comparison within a particular dataset should be the subject of ongoing consideration. Regardless of this question, the Tsallis entropy (11) will be used as root expression in calculating the diversity indices in below.

## 4. Results and discussions

As a primary attempt, in this section the proposed model is used to calculate some indices of diversity within the concept of the Tsallis entropy. Due to the lack of real data, the calculation of indices, in this work, is limited only with abstract and symbolic set of parameters $C$ and $\lambda$, and the obtained results are justified with those of other works [24, 27, 33, 35, 71] by means of graphical comparison.

It is almost obvious to confirm that the Tsallis entropy as function of parameter $q$ is one of the possible interpolations of the Shannon–Wiener and Simpson indices. Decreasing $q$ elevates monotonically the graphics of entropy approaching the Shannon–Wiener and Simpson indices when $q = 1$ and $q = 2$, respectively. This reconfirms the initial assumption of similarity between the order of diversity and non–extensive parameter $q$ and encourages the idea of using Tsallis entropy as a diversity index in ecological research.

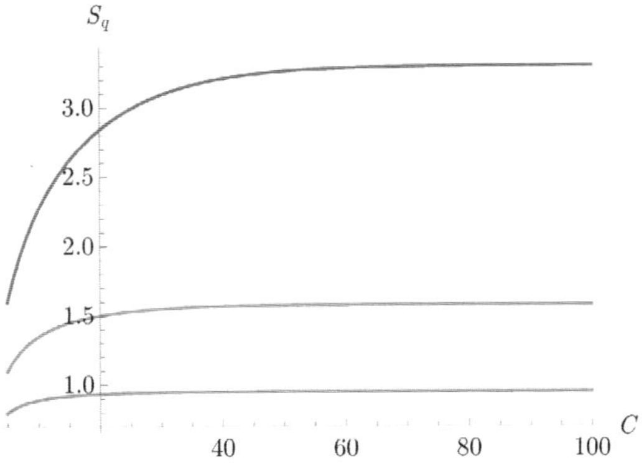


**Figure 1.** The Tsallis entropy (11) is the function of the total number of types in the dataset $C$ with different values of non–extensive parameter $q$. The Shannon–Wiener index H and Simpson index D are the top and bottom lines corresponding to $q = 1$ and $q = 2$, respectively.

Treating the Tsallis entropy as a function of non–extensive parameter $q$ with different fixed values of $C$ shows the change of diversity index within variation of the total number of types in the dataset. There is no clear reason to unify parameter $C$ and a hidden parameter of environment such as latitude, altitude or underground water ..., however in some particular measurement, when several environmental factors are kept unchanged, the parameter $C$ could be interpreted as one of those [33, 35].

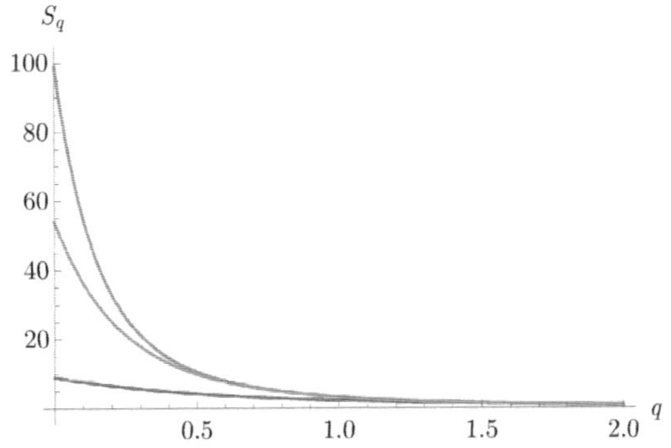


**Figure 2.** The Tsallis entropy (11) is the function of non–extensive parameter $q$ with different values of the total number of types in the dataset $C$ . Based on the abundance, the region with $q \leq 1$ emphasizes the richness whereas the region $q \geq 1$ emphasizes species dominance.

Another evaluation taking into account the role of the rate parameter $\lambda$ supports an additional information to allow characterizing the global conditions as the single approximative one.

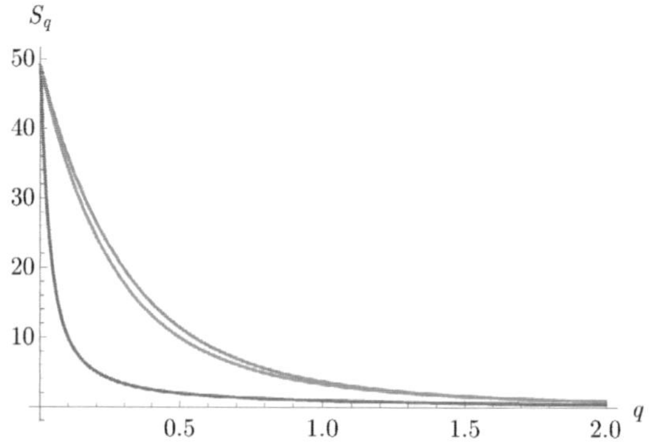


**Figure 3.** The Tsallis entropy (11) is the function of non–extensive parameter $q$ with different values of the rate parameter $\lambda$.

Various mathematical behaviors of Tsallis entropy (11) are evaluated to show out the applicability of truncated exponential distribution and the new family of diversity indices in modelling the observed datasets and in searching the possible ecological meanings of each appearing parameter. In one hand, the fitting parameters $C$ and $\lambda$ might contain an effective information of ecological systems, the non–extensive parameter $q$ would be new well–defined order of diversity. The first relates strongly to the incompleteness of datasets hiding the uncertainty of the systems under consideration and observations, and the latter provides an alternative of reducing the influence of uncertainty in macroscopic results.

In consistence with that happens to the diversity indies, evenness indices are also considered as derivatives that have to be calculated for each dataset. In common sense, the evenness index reflects how close in numbers each species in a observed dataset is. Mathematically the evenness index is determined through a diversity index such as Shannon–Wiener index or Simpson index, a measure of biodiversity which quantifies how equal the community is numerically. In the context of Tsallis entropy, the evenness reads

$$E_q = \frac{S_q}{S_q^{max}} , \tag{12}$$

where $S_q^{max}$ is the maximum value of $S_q$. It is not difficult to find out the analytical and exact expression of the evenness index in current situation as

$$S_q = \frac{1}{1-C^{1-q}} \frac{1-\lambda\left(e^{\frac{Cq}{\lambda}}-1\right)\left(\lambda\left(e^{\frac{C}{\lambda}}-1\right)\right)^{-q}}{q}. \tag{13}$$

Performing same manipulations as it was done with the diversity indies, allows to look mathematical behaviors of the evenness indices in more details. It is no doubt that the evenness index has a local minimum for each dataset. The existence of the local minimum in the current formalism is quite clear from the mathematical view, and actually agrees with the results found in [35].

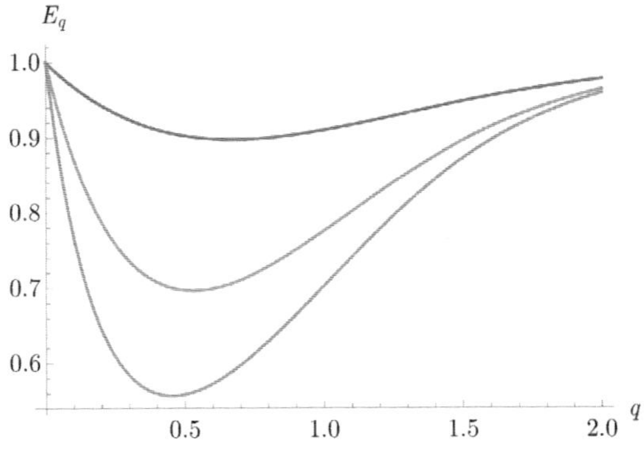


**Figure 4.** The evenness index (13) is the function of versus non–extensive parameter $q$ with different values of the total number of types in the dataset $C$.

It is also clear that the values where the evenness indices $E_q$ reach minimum increases for decreasing the total number of species $C$. The same impressions are in view of the rate parameter $\lambda$.

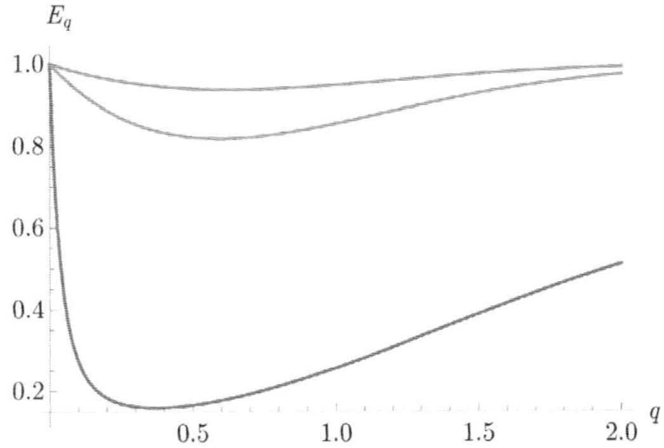


**Figure 5.** The evenness index (13) is the function of versus non–extensive parameter $q$ with different values of of the rate parameter $\lambda$.

## 5. Conclusions

What has been done in this work is to propose a new statistical model based on truncated exponential distribution for the rank abundance curve of observed datasets. The model is simple and provides analytical calculations for diversity indices. Within the context of Tsallis entropy, which is an extension of statistical physics to other fields of interests in complexity, the diversity and evenness indices are exactly analytically obtained, and numerically evaluated to verify the applicability of the model in ecological science and to look for ecological meaning of parameters appeared in the model and in the Tsallis entropy.

Investigating mathematical behaviors of diversity and evenness indices shows that the non–extensive parameter $q$ would be considered as the order of diversity while the total number of species and its rate parameter reflect an effective influence of environmental factors in statistics of ecology.

The model also motivates an idea to construct a consistent thermodynamic and statistical physical theory for ecological systems. This and many other open problems that could not be fully listed here, are waiting for further investigation.

## Acknowledgments

This research is funded by Vietnam National Foundation for Science and Technology Development (NAFOSTED) under grant number 103.01-2015.42.

## References

[1] Anand M, Gonzalez A, Guichard F, Kolasa J and Parrott L 2010 *Diversity* **2** 395
[2] Jørgensen S E and Svirezhev Y M 2004 *Towards a Thermodynamic Theory for Ecological Systems* (Oxford: Pergamon)
[3] Svirezhev Y M 2000 *Ecological Modelling* **132** 11 – 22
[4] Ricotta C 2000 *Ecological Modelling* **125** 245 – 253
[5] Young L J and Young J H 2002 *Statistical ecology : a population perspective* (Springer Science+Business Media, LLC)
[6] Austin M P 2002 *Ecological Modelling* **157** 101 – 118
[7] Banavar J R, Maritan A and Volkov I 2010 *Journal of Physics: Condensed Matter* **22** 063101
[8] Colinvaux P A 1973 *Introduction to Ecology* (Wiley)
[9] Colyvan M and Ginzburg L R 2010 *The Quarterly Review of Biology* **85** 171–182 (*Preprint* `http://dx.doi.org/10.1086/652321`)
[10] Dewar R C and Porté A 2008 *Journal of Theoretical Biology* **251** 389 – 403
[11] Fox G A, Negrete-Yankelevich S and Sosa V J (eds) 2015 *Ecological Statistics: Contemporary Theory and Application* (Oxford University Press)
[12] Goerner S J 1994 *World Futures* **42** 193–214 (*Preprint* `http://dx.doi.org/10.1080/02604027.1994.9972515`)
[13] Haynie D T 2008 *Biological Thermodynamics* 2nd ed (Cambridge University Press)
[14] Colwell R K 2009 *The Princeton Guide to Ecology* (Princeton University Press) chap Biodiversity: Concepts, Patterns and Measurement, pp 257–263
[15] Martyushev L and Seleznev V 2006 *Physics Reports* **426** 1 – 45
[16] Garcia-Morales V and Pellicer J 2003 *ArXiv Mathematical Physics e-prints* (*Preprint* `math-ph/0304013`)
[17] Würtz P and Annila A 2010 *Biosystems* **100** 70 – 78
[18] Futuyma D J, Shaffer H B and Simberloff D (eds) 2009 *Annual review of ecology, evolution, and systematics.* (Palo Alto, Calif.: Annual Reviews)
[19] Harte J, Zillio T, Conlisk E and Smith A B 2008 *Ecology* **89** 2700–2711
[20] What is biodiversity? Tech. rep. United Nations Environment Programme, World Conservation Monitoring Centre
[21] Cardinale B J 2011 *American Journal of Botany* **98** 572–592
[22] Cardinale B 2012 *Nature* **486** 59–67
[23] Gaston K J 2000 *Nature* **405** 220–227
[24] Gaston K J and Spicer J I 2004 *Biodiversity - An Introduction* 2nd ed (Blackwell Publishing)
[25] Gitay H, Suárez A, Dokken D J and Watson R T 2005 Climate change and biodiversity Tech. rep. Intergovernmental Panel on Climate Change (WMO & UNEP)
[26] Harrison I, Laverty M and Sterling E Species diversity
[27] Hawksworth D L 1996 *Biodiversity: measurement and estimation* (Springer)
[28] Hubbell S 2001 *The unified neutral theory of biodiversity and biogeography* (Princeton University Press, Princeton, N.J.)
[29] Benton M J 2001 *Geological Journal* **36** 211–230
[30] Jones-Walters L and Mulder I 2009 *Journal for Nature Conservation* **17** 245–247
[31] Simpson E H 1949 *Nature* **163** 688–688
[32] Hill M O 1973 *Ecology* **54** 427–432
[33] Magurran A E 2005 *Measuring Biological Diversity* (Blackwell Publishing)
[34] Tuomisto H 2010 *Oecologia* **164** 853–860
[35] Mendes R S, Evangelista L R, Thomaz S M, Agostinho A A and Gomes L C 2008 *Ecography* **31** 450–456
[36] Jost L 2006 *Oikos* **113** 363–375
[37] Tuomisto H 2010 *Oecologia* **164** 853–860

[38] Shannon C E 1948 *The Bell System Technical Journal* **27** 379–423
[39] Shannon C E 1948 *The Bell System Technical Journal* **27** 623–656
[40] Chao A and Shen T J 2003 *Environmental and Ecological Statistics* **10** 429–443
[41] Mulder C P H, Bazeley-White E, Dimitrakopoulos P G, Hector A, Scherer-Lorenzen M and Schmid B 2004 *Oikos* **107** 50–63
[42] Rabosky D L 2009 *Ecology Letters* **12** 735–743
[43] Stirling G and Wilsey B 2001 *The American Naturalist* **158** 286–299 pMID: 18707325 (*Preprint* `http://dx.doi.org/10.1086/321317`)
[44] Waide R B, Willig M R, Steiner C F, Mittelbach G, Gough L, Dodson S I, Juday G P and Parmenter R 1999 *Annual Review of Ecology and Systematics* **30** 257–300 (*Preprint* `http://dx.doi.org/10.1146/annurev.ecolsys.30.1.257`)
[45] Field R 2009 *Journal of Biogeography* **36** 132–147
[46] Hill M O 1973 *Ecology* **54** 427–432
[47] Jost L 2010 *Diversity* **2** 207
[48] Heip C H and Soetaert P M H K 1998 *Océanis* **24** 61–78
[49] Tsallis C 2011 *Entropy* **13** 1765
[50] Isa H, Ito M, Kamei E, Tohyama H and Watanabe M 2013 *Linear Algebra and its Applications* **439** 3148 – 3155
[51] Andrewartha HG B L 1954 *The Distribution and Abundance of Animals* (The University of Chicago Press, Chicago, Illinois)
[52] Whittaker R H 1965 *Science* **147** 250–260 (*Preprint* `http://science.sciencemag.org/content/147/3655/250.full.pdf`)
[53] Harte J 2011 *Maximum Entropy and Ecology: A Theory of Abundance, Distribution, and Energetics* Oxford Series in Ecology and Evolution (Oxford University Press)
[54] Tokeshi M 1990 *Journal of Animal Ecology* **59** 1129–1146
[55] Tsallis C 1988 *Journal of Statistical Physics* **52** 479–487
[56] Tsallis C 2002 *Chaos, Solitons & Fractals* **13** 371 – 391 classical and Quantum Complexity and Non-Extensive Thermodynamics
[57] Tsallis C 2009 *Brazilian Journal of Physics* **39** 337–356
[58] Tsallis C 2009 *Introduction to Nonextensive Statistical Mechanics: Approaching a Complex World* (Springer–Verlag)
[59] Diversity index (wikipedia)
[60] Harte J and Newman E A 2014 *Trends in Ecology & Evolution* **29** 384 – 389
[61] Martínez-Kahn M and Martínez-Castilla L 2010 *Ecological Psychology* **22** 69–87 (*Preprint* `http://dx.doi.org/10.1080/10407410903493160`)
[62] Martyushev L M and Seleznev V D 2014 *Physica A: Statistical Mechanics and its Applications* **410** 17 – 21
[63] Plastino A and Curado E M F 2005 *Phys. Rev. E* **72**(4) 047103
[64] Uffink J 1995 *Studies in History and Philosophy of Science Part B: Studies in History and Philosophy of Modern Physics* **26** 223 – 261
[65] Vakarin E V and Badiali J P 2011 *Modern Physics Letters B* **25** 1821–1828 (*Preprint* `http://www.worldscientific.com/doi/pdf/10.1142/S0217984911027054`)
[66] Gore A and Paranjpe S 2001 *A Course in Mathematical and Statistical Ecology* THEORY AND DECISION LIBRARY (SERIES B: MATHEMATICAL AND STATISTICAL METHODS) (Springer-Science+Business Media, B.Y.)
[67] Jost L 2006 *Oikos* **113** 363–375
[68] Hill D, Fasham M, Tucker G, Shewry M and Shaw P (eds) 2005 *Handbook of Biodiversity Methods: Survey, Evaluation and Monitoring* (Cambridge University Press)
[69] Tuomisto H 2010 *Ecography* **33** 2–22
[70] Tuomisto H 2010 *Ecography* **33** 23–45
[71] Leveque C J M 2003 *Biodiversity* (New York: John Wiley)
[72] Colwell R K 2009 *The Princeton Guide to Ecology* (Princeton University Press) chap III.1, pp 257–283
[73] Anh C T, Lan N T and Viet N A 2014 *Journal of Physics: Conference Series* **537** 012005

Journal of Physics: Conference Series **726** (2016) 012025 doi:10.1088/1742-6596/726/1/012025

# Second quantization model for surface plasmon polariton in metallic nano wires

**Nguyen Thi Phuong Lan[1], Do Thi Nga[2] and Nguyen Ai Viet[2]**

[1] Hanoi Pedagogical University 2, Nguyen Van Linh Street, Xuan Hoa Ward, Phuc Yen Town, Vinh Phuc Province, Vietnam

[2] Institute of Physics, 10 Dao Tan, Thu Le, Ba Dinh, Ha Noi, Vietnam

E-mail: lanntphuong@gmail.com

**Abstract.** A model of effective Hamiltonian is proposed in second quantization representation for system of surface plasmons and photon (polariton) in metallic nano wires. The dispersion relation curves of surface plasmon polariton was calculated by mean of the Bogoliubov diagonalization method. The surface plasmon photon vertexes are considered. The conditions for excitation surface plasmon, existence plasmon radiate modes, and a possible application of metallic nano wires were also discussed.

## 1. Introduction

In the last tens years, the new field of plasmon has been developed, which discussed the interesting features of metals with nano structure, exciting with an electromagnetic radiation: the plasmon resonance.

The plasmon is a quasiparticle resulting from the quantization of plasma oscillations. Plasmons are collective oscillations of the free electron gas density. A Plasmon can couple with a photon to create another quasiparticle called a plasmon polariton. Surface plasmons are plasmons which are restricted in surfaces and interact strongly with light resulting in a polariton. They occur in the interface of a vacuum and material with a small positive imaginary and large negative real dielectric constant. It is usually a metal or doped dielectric.

The result of this combination is the interesting optical properties of absorption and scatter in the Vis - IR region and can be exploited for many applications in modern technology. They also play an important role in new interesting physics phenomena such as Surface Enhanced Raman Spectroscopy (SERS), Surface Plasmon Energy Transfer (SET), and Forster Resonance Energy Transfer (FRET) and have numerous other applications in plasmonics and nanotechnology and medical treatments [1-16].

Since for all these surface plasmon polariton (SPP) systems the downscaling of the cross section is not limited by the light wavelength, they represent a promising alternative to dielectric optical waveguides. Thus, there is a substantial interest in the fundamental properties of SPP propagation in nanoscale structured matter, which is determined by the respective dispersion relations.

Nowadays, the experimental determination of the dispersion relation in the confined metallic films with nanoscopic cross sections nanowires is a challenging task, as the methods such as total internal reflective attentuated spectroscopy cannot be applied easily because of the small size of the nanowires. An alternate experimental technique, conventional spectroscopic extinction, was proposed in [5], allowing the measurement of plasmon resonances in metallic nanowires with finite length. Absorption and scattering in a nanowire give rise to an extinction band, the maximum of which is used to define the frequency of the resonance of the SPP mode.

In the article [17], the experimental determination of the dispersion relation for Ag and Au nanowires was reported. Interestingly, it was shown at the multipolar plasmon resonances of metal nanowires can be described in terms of standing plasmon waves, allowing one to deduce the dispersion relation from optical extinction measurements. The proposed model is supported by additional investigations with a modified experimental setup.

The focus was taken into the metallic nanowire systems which have already great interests in the SPP waveguides, and could produce a strong evanescence field near their surface. It is also known as an intrinsic problem restricting the applications of surface plasmons, which is the energy loss. To overcome this obstacle, hybrid surface plasmon modes supported by a composite waveguide of metal, spacer and dielectric, have been introduced.

In the previous work [19 ], a simple second quantization Hamiltonian for SPP with planar geometry was proposed.

In this research, we studied the quantum theory of surface plasmons, and plasmon polaritons with metallic nanowires. The Drude model was reviewed first and then applied for the metallic nanowires with cylindrical symmetry. Using the Mie theory for small radius metallic nanowires, the dispersion relation of surface plasmon polaritons was found. Based on that results, a simple two branch surface plasmon polariton model was considered for metallic nanowires with a definition of plasmon-photon effectiveness correlation constant. The dispersion relation SPP was found by applying the Bogoliubov diagonalization technique to the the second quantization model Hamiltonian for system of surface plasmon and photon in metallic nano wires. The comparison of the theoretical results obtained from our model and experimental data give good agreement.

## 2. Drude model for single mode surface plasmon polaritons

In this part we presented a simple semi-classical model, like Dude model for single mode surface plasmon polariton in metallic nano particle.

In the semi-classical theory, the dielectric response of surface palsmon is characterized by the Drude formula, which is the generalized Lorentz model for metals

$$\epsilon_m(\omega) = \epsilon_\infty - \frac{\omega_{SP}^2}{[\omega(\omega + i\gamma_P)]}, \tag{1}$$

where $\gamma_P$ is the damping constant, $\omega_P = \eta\omega_{SP}$ is the surface plasmon frequency with the parameter $\eta$ defined by the dimension and geometry properties of metallic nano particles, $\omega_P = \sqrt{ne^2/\epsilon_0 m^*}$ is the bulk plasmon frequency, $\epsilon_\infty$ and $\epsilon_0$ are the high frequency and static dielectric constants, $n$ is the density and $m^*$ is the effective mass of electrons in the metal.

Considering $\gamma_P << \omega_{SP}$ - the damping constant is much smaller than surface plamon frequency, the standard Drude model is

$$\epsilon_m = \epsilon_\infty - \frac{\omega_P^2}{\omega^2}. \tag{2}$$

Like in the case of Lorentz model, we can separate the real and imaginary parts of the dielectric constant by analyzing the function around the value of $\omega \simeq \omega_P$

$$\epsilon_m(\omega) \simeq \epsilon_\infty - \frac{\omega_{SP}^2}{(\omega - \omega_{SP})^2 + \gamma_P^2} + i\frac{\omega_{SP}^2\gamma_P}{\omega\left[(\omega - \omega_{SP})^2 + \gamma_P^2\right]}. \tag{3}$$

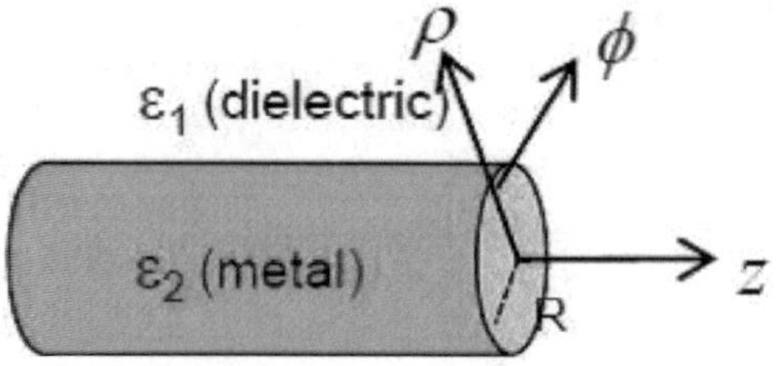


**Figure 1.** Model metallic nano wire

The real part $Re$ and imaginary part $Im$ of plasmon dielectric constant $\epsilon_m(\omega)$ now have the Lorentz-like forms, i.e. have a dispersive

$$\epsilon_{1m}(\omega) = Re\,\epsilon_m(\omega) = \epsilon_\infty - \frac{\omega_{SP}^2}{(\omega-\omega_{SP})^2+\gamma_P^2} = A_{dis}, \tag{4}$$

and for absorption part

$$\epsilon_{2m}(\omega) = Im\,\epsilon_m(\omega) = \frac{\omega_{SP}^2\gamma_P}{\omega\left[(\omega-\omega_{SP})^2+\gamma_P^2\right]} = A_{abs}. \tag{5}$$

It is noted here that the asymmetric absorption peak around $\omega_{SP}$ is the common behavior of plasmon absorption. For the case of metallic half space with planar geometry, we have $\eta_{planar} = 1/\sqrt{1+\epsilon_d}$ or $\eta_{planar} = 1/\sqrt{2}$ in the vacuum $\epsilon_d = \epsilon_0 = 1$.

The dispersion laws of surface plasmon polariton $\Omega(k)$ are defined from the boundary conditions.

## 3. Drude like model for surface plasmon polariton in nano wires

To calculate the group velocity $d\omega = dk$ as a function of a metallic nano wire radius $R$, a model consisting of a metallic cylinder with a dielectric constant $\epsilon_m$ surrounded by a dielectric medium of dielectric constant $\epsilon_d$ was used (see the figure 1).

For the special case with TM mode ($H_z = 0$) and fundamental mode with no winding $m = 0$, continuity of the remaining tangential components $E_z$ and $H_\phi$ at the boundary lead to the equation for SPP dispersion relations. In this case the surface plasmon propagation is governed by the dispersion relation for the fundamental transverse magnetic modes, which is given by [16]

$$\frac{k_2^2}{k_{2\perp}}\frac{J_0'(k_{2\perp}R)}{J_0(k_{2\perp}R)} - \frac{k_1^2}{k_{1\perp}}\frac{H_0'(k_{1\perp}R)}{H0(k_{1\perp}R)} = 0, \tag{6}$$

where $k_i = \sqrt{\epsilon_i}\omega/c = \sqrt{k_{i\perp}^2 + k_{i\|}^2}$ and $J_m$, $H_m$ are Bessel and Hankel functions of the first kind, respectively. By numerically solving the above equation (6), we can calculte the group velocity as a function of $R$ with frequency $\omega$ and given $\epsilon_d$ and $\epsilon_m$. In the limit of $k_{i\perp} = \sqrt{k_i^2 - k_{i\|}^2} \simeq ik_\|$ we have

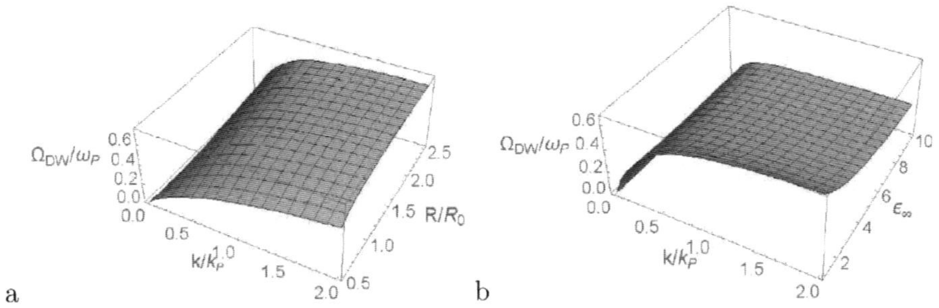


**Figure 2.** SPP dispersion relation for metallic nano wire in the Drude like model

$$\frac{\epsilon_m}{\epsilon_d} = \frac{K_0'\left(k_{\parallel}R\right)\, I_0\left(k_{\parallel}R\right)}{K_0\left(k_{\parallel}R\right)\, I_0'\left(k_{\parallel}R\right)}, \tag{7}$$

where $K_m$, $I_m$ are modified Bessel and Hankel functions, respectively.

Taking the case of metal/vacuum interface $\epsilon_d = 1$, from the Drude model the frequency $\omega_k$ equals

$$\omega^2 = \omega_P^2\left(\epsilon_\infty - \epsilon_m\right) \tag{8}$$

In the Drude like model, the SPP dispersion relation for metallic nano wires is

$$\omega_{DW}\left(k_{\parallel}, R\right) = \omega_P \left\{\epsilon_\infty - \frac{K_0'\left(k_{\parallel}R\right)\, I_0\left(k_{\parallel}R\right)}{K_0\left(k_{\parallel}R\right)\, I_0'\left(k_{\parallel}R\right)}\right\}^{-1/2}, \tag{9}$$

or using the properties of modified Bessel and Hankel functions, it can be rewritten as follow

$$\omega_{DW}\left(k_{\parallel}, R\right) = \omega_P \left\{\epsilon_\infty - \frac{K_1\left(k_{\parallel}R\right)\, I_0\left(k_{\parallel}R\right)}{K_0\left(k_{\parallel}R\right)\, I_1\left(k_{\parallel}R\right)}\right\}^{-1/2}, \tag{10}$$

and is plotted in the figure for the cases $\epsilon_\infty = 1$, and $\epsilon_\infty = 3.7$ (Ag).

For almost metals $\lfloor \epsilon_m/\epsilon_d \rfloor \gg 1$ that $\left\lfloor k_{\parallel}R \right\rfloor \ll 1$

$$\frac{\epsilon_m}{\epsilon_d} \simeq \frac{2}{\left[a + ln\left(k_{\parallel}R\right)\right]\left(k_{\parallel}R\right)^2}, \tag{11}$$

where $a = \gamma_E - ln2$ , $\gamma_E = 0.577$ is the Euler's constant. Denote $k_{\parallel} = nk_0$ we got

$$n\left(R\right) \simeq \frac{C\left(\epsilon_d, \epsilon_m\right)}{k_0 R}, \tag{12}$$

where the function $C$ is

$$C\left(\epsilon_d, \epsilon_m\right) = \frac{\sqrt{-2\epsilon_d/\epsilon_m}}{ln\sqrt{-4\epsilon_m/\epsilon_d} - \gamma_E}. \tag{13}$$

For $k_0 R \to 0$, the phase velocity $v_{ph} = c/n \to 0$, and the group velocity $v_{gr} = c/\left[d\left(n\omega\right)/d\omega\right] \to 0$.

In the case of small radius nanowires $k_{\parallel}R \leq 1$, we obtain the dispersion relation for SPP

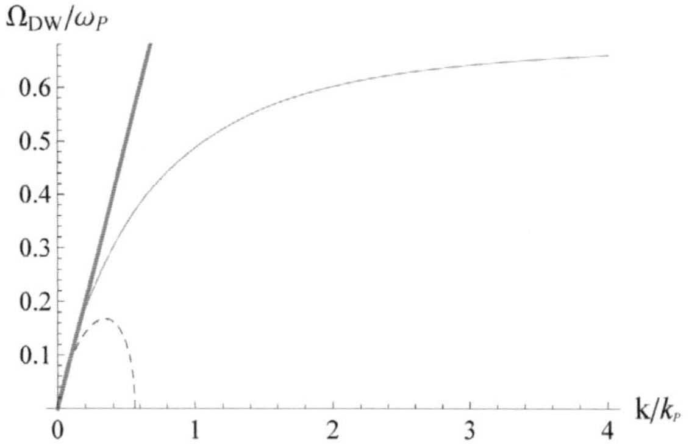


**Figure 3.** The dispersion relations for SPP obtained by equation (10) curve, and (14) dashed line

$$\omega_{DW}\left(k_{\|}, R\right) \simeq \omega_P \left\{ \epsilon_\infty - \frac{2\epsilon_d}{\left[a + ln\left(k_{\|}R\right)\right]\left(k_{\|}R\right)^2} \right\}^{-1/2} . \tag{14}$$

In the figure 3, we plot the dispersion relations for SPP obtained by equation (10) curve, and (14) dashed line. For guiding, the photon line (thick) also is plotted. The dispersion relation from (14) valid only for small nanowires.

## 4. Second quantization quantum Hamiltonian model for surface plasmon polaritons in metallic nano wires

In the similar cases of exciton plariton and phonon polariton, we consider a model Hamiltonian in 2nd quantization form for surface plasmon polariton in metallic nano wires with single mode

$$H = \sum_k H_k = \sum_k \left\{ \omega_{\gamma k} a_k^+ a_k + \omega_{P1} b_k^+ b_k + g_k \left( a_k^+ b_k + b_k^+ a_k \right) \right\}, \tag{15}$$

where $a_k(a_k^+)$ and $b_k(b_k^+)$ are the annihilation (and creation) photon and plasmon operators correspondingly with momentum $k$, $\omega_{P1}$ is the surface surface plasmon energy with $m = 1$, and $\omega_{\gamma k} = ck/\sqrt{\epsilon_d}$ is the photon dispersion law. We denote $g_k$ the plasmon-photon transition vertex (or coupling constant), this vertex is absence in traditional plasmon theory $g_{Bk} = 0$ because the bulk plasmon is longitudinal excitation, while photon is transverse excitation. For the case of surface plasmons due to the existence of the boundary conditions that electromagnetic waves must be satisfied at the interface, the plasmon-photon transition vertex might not be zero, and being the main parameter of our theory.

## 5. Bogoliubov transformation and dispersion relation

We use the Bogoliubov transformation technique taken from the theory of superconductivity for the plasmon polariton diagonalized Hamiltonian

$$H_k = \sum_k \left\{ \Omega_{Uk} \gamma_{Uk}^+ \gamma_{Uk} + \Omega_{Lk} \gamma_{Lk}^+ \gamma_{Lk} \right\}, \tag{16}$$

where $\gamma_{ik}$ (and$\gamma_{ik}^+$ ) are the annihilation (and creation) operators of the surface plasmon polarition SPP with momentum $k$, and $i$ is the branch number $i = L = 1$ for lower and $i = U = 2$ for upper branch.

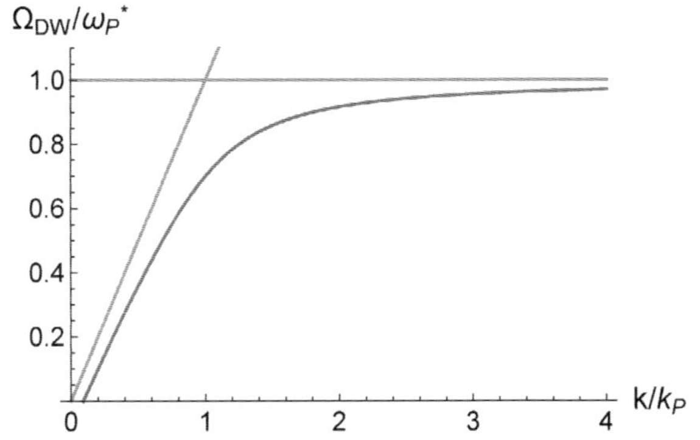


**Figure 4.** Simple two bands SPP model with the effctive $g_k = const$

The transformations with unity condition $u_k^2 + v_k^2 = 1$ are

$$\gamma_{1k} = u_k a_k + v_k b_k,\ \gamma_{2k} = -v_k a_k + u_k b_k. \tag{17}$$

Using the commutative relations for annihilation and creation operators $\left[a_k,\, a_k^+\right] = 1$, $[b_k,\, b+] = 1$, $\left[\gamma_{ik},\, \gamma_{ik}^+\right] = 1$, and equalling to zero in other cases, by standard calculation as in [4] we obtain the surface plasmon polariton SPP dispersion relations for lower branch

$$\Omega_{Lk}\left(R\right) = \frac{1}{2}\left\{\left[\omega_{\gamma k} + \omega_{P1}\left(R\right)\right] - \sqrt{\left[\omega_{\gamma k} - \omega_{P1}\left(R\right)\right]^2 + 4g_k^2}\right\}, \tag{18}$$

and for upper branch

$$\Omega_{Uk}\left(R\right) = \frac{1}{2}\left\{\left[\omega_{\gamma k} + \omega_{P1}\left(R\right)\right] + \sqrt{\left[\omega_{\gamma k} - \omega_{P1}\left(R\right)\right]^2 + 4g_k^2}\right\}. \tag{19}$$

Note that in the case of planar boundary geometry, the upper branch is lying in the energy gap where damping is too high so only the lower branch exist, while in the case of metallic nano spherical geometry, both branches could be exist. The surface plasmon polariton dispersion relation $\Omega_k$ depends on wave vector $k$ and coupling constant $g_k = 0.3\omega_P^*$ is presented in the figure 4.

Note that the simple two bands model with the effective $g_k = const$ of surface plasmon polariton might be best in the most important bottom region but may be fare in the long wave limit $k = 0$.

## 6. Coupling constant $g_k$ with k-dependence

As mentioned above, the simple two bands model of surface plasmon polariton might be best in the most important neck region but may be failed in the long wave limit $k = 0$. In this part, we propose to overcome this problem by investigation the k-dependence of the plasmon-photon coupling constant.

Assuming the two dispersion relations of Drude and lower SPP branch with $\omega_{\gamma k} \leq \omega_{P1}$ of our quantum model are equal $\omega_{DW} = \Omega_{Lk}$, and putting $k = k_{\|}$ we got the equation for finding the plasmon-photon coupling constant $g_{Wk}$

$$\omega_{DW}\left(k, R\right) = \frac{1}{2}\left\{\left(\omega_{\gamma k} + \omega_{P1}\right) - \sqrt{\left(\omega_{\gamma k} - \omega_{P1}\right)^2 + 4g_{Wk}^2}\right\}. \tag{20}$$

The solution of this equation is

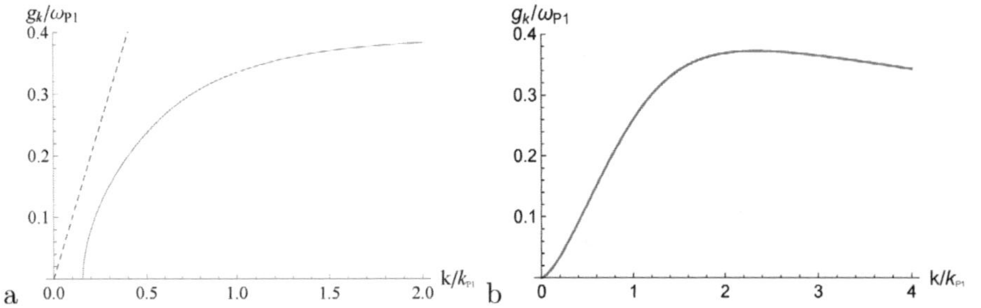


**Figure 5.** The value of the plasmon-photon coupling constant $g_k$ depends on wave vector $k$ , a) calculated from equation (10) (curve), and from optical model [15] (dashed line), b) from comparison with Drude mode

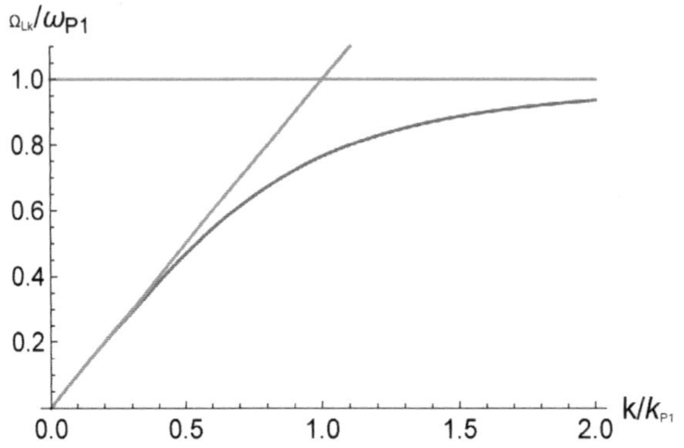


**Figure 6.** The dispersion relation obtained by our model

$$g_{Wk}^2(R) = [\omega_{k+} - \omega_{DW}(k, R)]^2 - \omega_{k-}^2, \tag{21}$$

where

$$\omega_{k\pm} = \frac{1}{2}(\omega_{P1} \pm \omega_{\gamma k}), \tag{22}$$

for the SPP lower branch $k = k$ if $k < k_{p1}$, and $k = k_{P1}$ if $k \geq k_{p1}$.

The value of the plasmon-photon coupling constant $g_k$ depends on wave vector $k$ at the metal/vacuum interface ($\epsilon_d = 1$) is plotted in the figure 5, here , and will be taken as parameter of our model.

With this plasmon-photon coupling constant, the dispersion relation obtained by our model is presented in the figure 6.

Two branches of surface plasmon polariton of metallic nano wires (curve and Dashing [Large]) is plotted in the figure 7 with the proposed coupling constant $g_k$ (DotDashed).

Note that the existence of upper branch. This branch might play an important role in some physics phenomena.

## Discussion

In this work, we reviewed and studied several semiclassical and quantum models of surface plasmons and surface plasmon polariton in the cylindrical symmetry. We proposed a simple two branch model Hamiltonian for surface plasmon polarization in second quantification

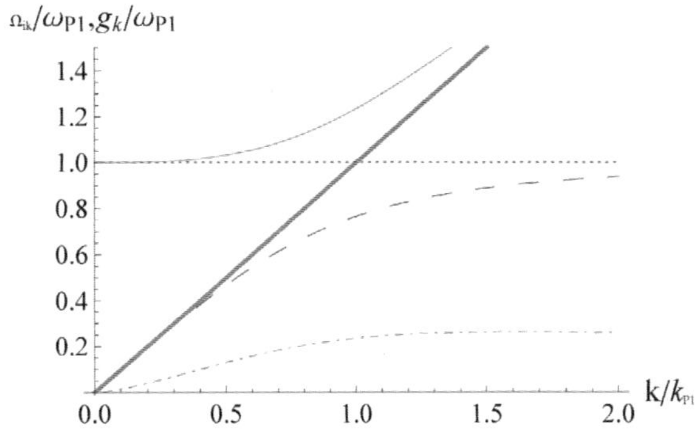


**Figure 7.** Two branch model for surface plasmon polariton of metallic nano wires (Up-curve and Low-Dashing [Large]) with the proposed coupling constant $g_k$ (DotDashed)

representation for metallic nanowires. The main parameter of the model is surface plasmon photon transition vertex (surface plasmon photon effective coupling constant) was obtained. We compared the two cases planar and cylindrical geometries. For the case of planar geometry, the conditions for excitation are needed, while that do not need for the case of cylindrical geometry. The existence of upper surface plasmon polariton branch with the photon-like behavior at high frequency will play an important role in explanation the some experimental results.

## Acknowledgment

This research is funded by Vietnam National Foundation for Science and Technology Development (NAFOSTED) undergrant number 103.01-2015.42.

## References

[1] Garcia M A 2011 Surface plasmons in metallic nanoparticles: Fundamentals and applications *J. Phys.* D **44** 28

[2] Homola J 2006 *Surface plasmon resonance based sensors* (Springer Series on Chemical Sensors and Biosensors vol 4) ed J Homola (Berlin: Springer-Verlag) chapter 1 pp 3–44

[3] Xie W, Qiu P and Mao C 2011 Bio-imaging, detection and analysis by using nanostructures as SERS substrates *J. Mater. Chem.* **21** 5190

[4] An N B, Hieu N V, Thang N T, Thang T L and Viet N A 1981 Polariton theory of scattering of light beams in semiconductors *Ann. Phys. (NY)* **131** 149

[5] Ren-Bin Z, Wei-Hao L, Jun Z and Sheng-Gang L 2012 Surface plasmon wave propagation along single metal wire *Chin. Phys.* B **21** 117303

[6] Schmeits M 1989 Surface-plasmon coupling in cylindrical pores *Phys. Rev.* B **39** 7567

[7] Kottmann J P and Martin O J F 2001 Plasmon resonant coupling in metallic nanowires *Opt. Exp.* **8** 655

[8] Laroche T and Girard C 2006 Near-field optical properties of single plasmonic nanowires *Appl. Phys. Lett.* **89** 233119

[9] Ashley J C and Emerson L C 1974 Dispersion relations for non-radiative surface plasmons on cylinders *Surf. Sci.* **41** 615

[10] Liu L, Han Z and He S 2005 Novel surface plasmon waveguide for high integration *Opt. Express* **13** 6645

[11] Oulton R F, Sorger V J, Genov D A, Pile D F P and Zhang X 2008 A hybrid plasmonic waveguide for subwavelength confinement and long-range propagation *Nat. Photon.* **2** 496

[12] Tian J, Ma Z, Li Q, Song Y, Liu Z, Yang Q, Zha C, Akerman J, Tong L and Qiu M 2010 Nanowaveguides and couplers based on hybrid plasmonic modes *Appl. Phys. Lett.* **97** 231121

[13] Sun S, Hung-Ting C, Wei-Jin Z and Guang-Yu G 2013 Mode conversion of surface plasmon polaritons in silver double-nanowire systems *Opt. Exp.* **21** 14591

[14] Delga A, Feist J, Bravo-Abad J and Garcia-Vidal F J 2014 Quantum emitters near a metal nanoparticle: Strong coupling and quenching *Phys. Rev. Lett.* **112** 253601

[15] Teichroeb J H, Forresta J A, Ngai V and Jones L W 2006 Anomalous thermal denaturing of proteins adsorbed to nanoparticles *Eur. Phys. J. E.* **21** 1924

[16] Chang D E, Sorensen A S, Hemmer P R and Lukin M D 2007 Strong coupling of single emitters to surface plasmons *Phys. Rev.* B **76** 035420

[17] Schider G, Krenn J R, Hohenau A, Ditlbacher H, Leitner A, Aussenegg F R, Schaich W L, Puscasu I, Monacelli B and Boreman G 2003 Plasmon dispersion relation of Au and Ag nanowires *Phys. Rev.* B **68** 155427

[18] Ha C V, Nga D T, Viet N A and Nhung T H 2015 The local field dependent effect of the critical distance of energy transfer between nanoparticles *Opt. Comm.* **353** 49

[19] Nga D T T, Lan N T P and Viet N A 2015 Second quantization model of surface plasmon polariton in metal planar surface *J. Phys.: Conf. Series* **627** 012018

[20] Lan N T P, Nga D T T and Viet N A 2015 Trapping cold atoms using surface plasmons with phase singularities generated by evanescent Bessel beams *J. Phys.: Conf. Series* **627** 012017

[21] Alpeggiani F and Andreani L C 2014 Quantum theory of surface plasmon polaritons: Planar and spherical geometries *Plasmonics* **9** 965

Journal of Physics: Conference Series **726** (2016) 012026 doi:10.1088/1742-6596/726/1/012026

# The pygmy dipole resonance in neutron-rich nuclei

**Nguyen Quang Hung**[1], **Hoang Anh Tuan Kiet**[1], **Huynh Ngoc Duc**[2], **and Nguyen Thi Chuong**[2,3]

[1] Institute of Research and Development, Duy Tan University, K7/25 Quang Trung, Danang city, Vietnam

[2] Graduate School, University of Pedagogy, An Duong Vuong, District 5, Ho Chi Minh City, Vietnam

[3] Le Quy Don High School, Vung Tau city, Ba Ria - Vung Tau Province, Vietnam

E-mail: nqhungdtu@gmail.com

**Abstract.** The pygmy dipole resonance (PDR), which has been observed via the enhancement of the electric dipole strength $E1$ of atomic nuclei, is studied within a microscopic collective model. The latter employs the Hartree-Fock (HF) method with effective nucleon-nucleon interactions of the Skyrme types plus the random-phase approximation (RPA). The results of the calculations obtained for various even-even nuclei such as $^{16-28}$O, $^{40-58}$Ca, $^{100-120}$Sn, and $^{182-218}$Pb show that the PDR is significantly enhanced when the number of neutrons outside the stable core of the nucleus is increased, that is, in the neutron-rich nuclei. As the result, the relative ratio between the energy weighted sum of the strength of the PDR and that of the GDR (giant dipole resonance) does not exceed 4 %. The collectivity of the PDR and GDR states will be also discussed.

## 1. Introduction

The pygmy dipole resonance (PDR) was firstly identified via the enhancement of the electric-dipole strength $E1$ at low energy ($E \leq 10$ MeV) around the particle emission threshold of neutron-rich nuclei [1]. The PDR is commonly represented as the dipole vibration of the excess neutrons against the isospin-symmetry core of protons and neutrons in the nucleus. Nature of the PDR and its properties have been therefore the subjects of various theoretical and experimental studies. From the experimental point of view, the PDR has been investigated systematically from the measurements of dipole response of a large number of stable and unstable nuclei via electromagnetic excitation in heavy-ion collisions [2], photon scattering [3], and Coulomb excitation reactions [4]. From the theoretical aspect, various theoretical models have been proposed to describe the nature of PDR. Some of them are the large-scale shell model calculations [5], relativistic random-phase approximation (RRPA) [6], RRPA plus phonon-coupling model [7], quasiparticle random-phase approximation (QRPA) with Skyrme interaction [8], phonon-damping model (PDM) [9], density functional theory (DFT) [10], ect. One of the major issue in the theoretical study of the PDR is the discrepancy in the predictions of different approaches regarding the strength and collectivity of the PDR. For instant, the RRPA predicts a prominent peak, which is identified as the collective PDR below 10 MeV in $^{120-132}$Sn and $^{122}$Zr [6, 7], whereas the calculations within the QRPA do not expose any collective states in the low-energy region of the $E1$ strength distribution for the same isotope chain [8]. In addition, there has been lack of a systematic calculation within a self-consistent microscopic approach for the PDR in nuclei with the mass numbers ranging from the light to the heavy ones. The aim of present study is to perform such a calculation in order to see the evolution of the PDR in neutron-rich nuclei ranging from the light systems such as $^{16-28}$O to the heavy one such as $^{182-218}$Pb. For that purpose, we employ the self-consistent Hartree-Fock (HF) plus the RPA code with the use of realistic Skyrme-type interactions [11]. Though this code has been successfully used in the study of the nuclear giant resonances such as giant monopole resonance (GMR), giant dipole resonance (GDR), giant quadrupole resonance (GQR), etc., it has not been used to describe the PDR in neutron-rich nuclei.

## 2. The random-phase approximation

As mentioned in the Introduction, present work employs the self-consistent HF + RPA code with realistic Skyrme-type interactions. The formalism of the RPA is presented in details in Refs. [11, 12], so we report here only the main equations. Basically, the RPA equation is derived based on the creation (destruction) operator $B^\dagger_{ph}$ ($B_{ph}$) that creates (annihilates) the particle-hole (p-h) pairs coupled to total angular momentum ($JM$)

$$B^\dagger_{ph}(JM) = \sum_{m_p,m_h} (j_p m_p j_h - m_h|JM) a^\dagger_{j_p m_p} (-1)^{j_h - m_h} a_{j_h - m_h} \ , \qquad B_{ph} = (B^\dagger_{ph})^\dagger \ , \tag{1}$$

where $a^\dagger_{j_p m_p}$ is an operator that creates the particles on the orbital having the angular momentum $j_p$ and projection $m_p$, whereas $a_{j_h - m_h}$ is an operator that annihilates the particles or creates the holes having the angular momentum $j_h$ and projection $-m_h$. Within the RPA, the excited states $|\nu\rangle$ are formed by applying the excitation operators of the p-h pairs $Q^\dagger_\nu$ on the correlated ground state $|\tilde{0}\rangle$, namely

$$|\nu\rangle = Q^\dagger_\nu |\tilde{0}\rangle \ , \tag{2}$$

where $Q^\dagger_\nu$ is expressed as a combination of the creation and destruction operators of the p-h pairs, namely

$$Q^\dagger_\nu = \sum_{ph} \left[ X^\nu_{ph} B^\dagger_{ph}(JM) - Y^\nu_{ph} B_{ph}(\widetilde{JM}) \right] \ , \tag{3}$$

with $\widetilde{JM}$ being the time-reversed state of $JM$, i.e, a state with the same $J$ but opposite $z$-projection ($-M$). The amplitudes $X_{ph}$ and $Y_{ph}$ are the eigenvectors of the so-called RPA matrix having the form as

$$\begin{pmatrix} A & B \\ -B & -A \end{pmatrix} \begin{pmatrix} X^\nu \\ Y^\nu \end{pmatrix} = E_\nu \begin{pmatrix} X^\nu \\ Y^\nu \end{pmatrix} \ , \tag{4}$$

where the submatrices $A$ and $B$ are derived based on the particle-hole configurations

$$A_{ph,p'h'} = (\epsilon_p - \epsilon_h)\delta_{pp'}\delta_{hh'} + \langle \overbrace{p\,h'}^{J}|V_{res}|\underbrace{hp'}_{J}\rangle \ , \tag{5}$$

$$B_{ph,p'h'} = \langle \overbrace{p\,p'}^{J}|V_{res}|\underbrace{hh'}_{J}\rangle \ , \tag{6}$$

where $\epsilon_p$ and $\epsilon_h$ are respectively the energies of the particle and hole levels, whereas the residual interaction $V_{res}$, which is an anti-symmetrized p-h interaction, is obtained within the calculations of the Hartree-Fock plus Skyrme interactions [12]. The reduced transition probabilities from the ground state to the excited states are calculated based on the solutions of the RPA equations

$$B(EJ,\tilde{0}\to\nu)=\left|\sum_{ph}\left(X^{\nu}_{ph}+Y^{\nu}_{ph}\right)\langle p||\hat{F}_J||h\rangle\right|^2 , \tag{7}$$

where $\hat{F}_J$ is a $J$-multipole of the external field because of which the nucleus is excited. $\hat{F}_J$ consists of the isoscalar $\hat{F}^{IS}_J$ and isovector $\hat{F}^{IV}_J$ components, whose explicit form are given as

$$\hat{F}^{IS}_J=\sum_{i=1}^{A}f_J(r_i)Y_{JM}(\hat{r}) , \tag{8}$$

$$\hat{F}^{IV}_J=\sum_{i=1}^{A}f_J(r_i)Y_{JM}(\hat{r})\tau_z(i) , \tag{9}$$

where $\tau_z(i)$ is the $z$-component of the isospin operator, whereas $Y_{JM}(\hat{r}_i)$ is the spherical harmonics functions. The function $f_J(r)$ in Eqs. (8) and (9) is often given in term of a spherical Bessel function for the electromagnetic external fields or it is proportional to $r^J$ in the long wavelength limit. In the case of dipole vibration ($J=1$), in order to remove the center of mass contributions, we employ the electromagnetic operator of the form as

$$\hat{F}_{1M}=\frac{eN}{A}\sum_{i=1}^{Z}r_iY_{1M}(\hat{r}_i)-\frac{eZ}{A}\sum_{i=1}^{N}r_iY_{1M}(\hat{r}_i) , \tag{10}$$

where $\frac{eN}{A}$ and $-\frac{eZ}{A}$ are the effective charge for protons ($Z$) and neutrons ($N$), respectively.

In principle, the distribution of the transition probabilities $B(EJ)$ in Eq. (7) is discrete. To have a continuous distribution, one should use the smoothed strength function $S(E)$, which can be obtained based on the $\delta$-function representation $\delta(x)=\varepsilon/[\pi(x^2+\varepsilon^2)]$ with $\varepsilon$ being the smoothing parameter. The explicit expression of $S(E)$ is given as

$$S(E)=\frac{\varepsilon}{\pi}\sum_{\nu}\frac{B(EJ)}{(E-E_\nu)^2+\varepsilon^2} . \tag{11}$$

## 3. Results

The numerical calculations are carried out within the self-consistent HF + RPA code with Skyrme-type interactions for the isovector (IV) dipole states ($J^\pi=1^-$) of several spherical nuclei ranging from the closed-shell to neutron-rich isotopes such as $^{16,20,24,28}$O, $^{40,48,52,58}$Ca, $^{100,106,114,120}$Sn, and $^{182,194,208,218}$Pb. As for the Skyrme interaction, we employ the realistic SLy5 interaction [13], whose parameter sets are specified and used in the HF calculation. The later is done in a spherical box with radius $R=24$ fm and the cut-off energy $E_c=60$ MeV for the unoccupied states. Shown in Figs. 1 and 2 are the IV reduced transition densities B(E1) as functions of excitation energy $E$. It is clear to see from these figures that beside the GDR region, whose excitation energy $E$ ranges from 10 MeV to 30 MeV, there appears a low energy region ($E\leq 10$ MeV), which is called the PDR region, where the B(E1) values increase with increasing the neutron number, especially in the case of light and medium-mass isotopes such as $^{28}$O [Fig. 1 (d)] and $^{58}$Ca [Fig. 1 (h)]. The more neutron is added to the system, the more pygmy dipole strength is gained. To see the contribution of the pygmy dipole states to the total B(E1) strength, we plot in Fig. 3 the ratio between the energy weighted sum of strength (EWSS)

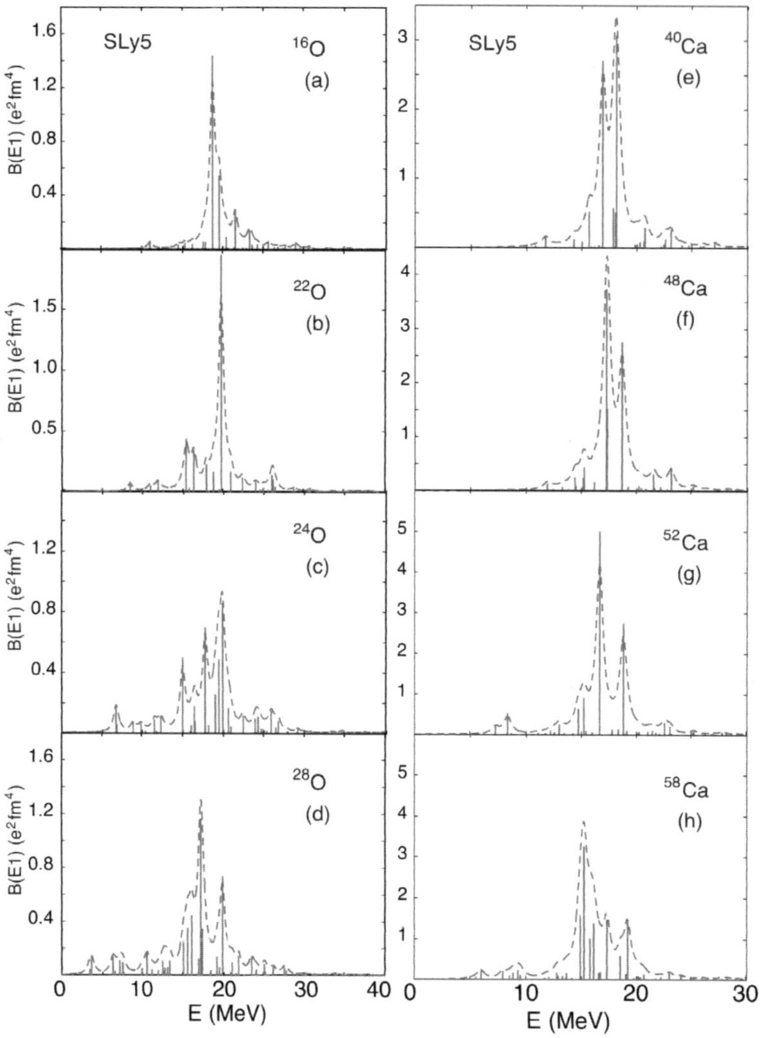


**Figure 1.** Distribution of the reduced transition probabilities $B(E1, \tilde{0} \to 1^-)$ for the isovector dipole states in Oxygen and Calcium isotopes. The dashed lines denote the strength function $S(E)$ with the smoothing parameter $\varepsilon = 0.4$ MeV.

of the PDR and that of the GDR as function of the mass number, namely $r = S_{PDR}/S_{GDR}(\%)$, where $S = \sum_\nu E_\nu B(EJ, E_\nu)$. This Fig. 3 shows that $r$ increases from almost 0% in the closed shell nuclei such as $^{16}$O, $^{40,48}$Ca, $^{100}$Sn, and $^{182}$Pb to about 4 % in very neutron-rich nuclei such as $^{28}$O and $^{58}$Ca, which agrees with the experimental investigations [14, 15] (See e.g., Fig. 2 of Ref. [15]).

Obviously, each of the PDR or GDR states is formed by the combinations of many p-h configurations. As the result, the collectivity of the PDR (GDR) state should be defined based on the number of p-h components, which contribute an appreciable weight to the RPA wave function. The latter is given as

$$A_{ph} = \left(X^\nu_{ph}\right)^2 - \left(Y^\nu_{ph}\right)^2 \; . \tag{12}$$

The normalization of the total weight leads to the condition $\sum_{ph} A_{ph} = 100\%$. The values of

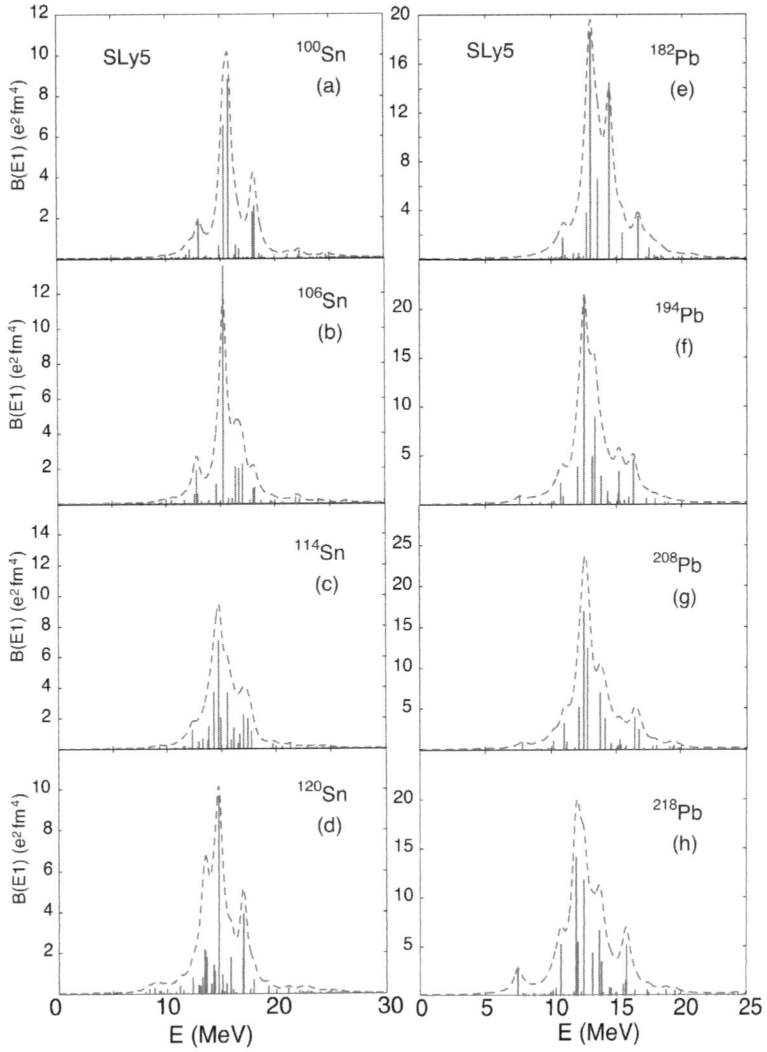


**Figure 2.** Same as in Fig. 1 but for Tin and Lead isotopes.

$A_{ph}$ for the p-h configurations mostly contributed to two selected PDR ($E$ = 3.42 MeV) and GDR ($E$ = 17.19) states are listed in Table 1. One can see easily from this Table that for the GDR state there are 7 p-h configurations which have comparable contributions to it with the largest contribution of only 33.93 % $[(4p_{3/2}, 2s_{1/2})_n]$. It means that this GDR state can be considered as a collective. Meanwhile, for the PDR state only one neutron p-h configuration $(2p_{3/2}, 1d_{3/2})_n$ already contributed about 99.2 % of its total strength. This PDR state therefore has no collective property, which agrees with the QRPA prediction in Ref. [8], whereas it is in contradiction to the RRPA results in Refs. [6] and [7].

## 4. Conclusions

This work studies the evolution of the pygmy dipole resonance in neutron-rich nuclei by employing the self-consistent HF + RPA code with Skyrme interactions. The results obtained show that the PDR states increase significantly with increasing the neutron number.

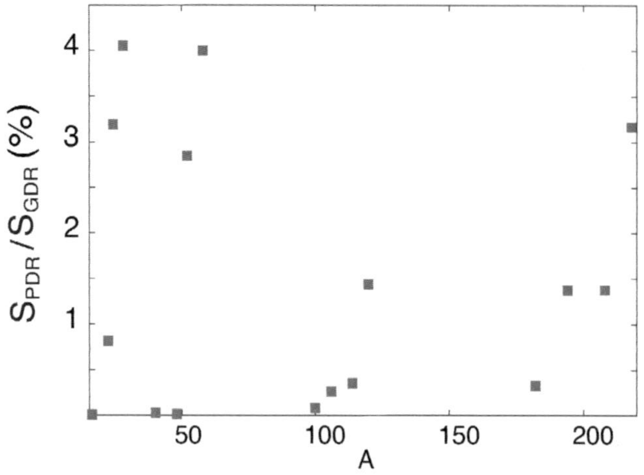


**Figure 3.** Ratio between the energy weighted sum of the strength of the PDR and that of the GDR obtained within the HF + RPA as functions of the mass number.

**Table 1.** The weight $A_{ph}$ of the contribution of the proton ($p$) and neutron ($n$) particle-hole configurations to the PDR and GDR states obtained within the HF + RPA calculations for $^{28}$O nucleus.

| PDR ($E = 3.42$MeV) | | | GDR ($E = 17.19$ MeV) | | |
|---|---|---|---|---|---|
| p-h conf. | $E$ (MeV) | $A_{ph}$(%) | p-h conf. | $E$ (MeV) | $A_{ph}$(%) |
| $(1d_{5/2}, 1p_{3/2})_p$ | 12.65 | 0.10 | $(2d_{5/2}, 1p_{3/2})_p$ | 12.65 | 9.00 |
| $(1d_{3/2}, 1p_{3/2})_p$ | 14.92 | 0.04 | $(1d_{3/2}, 1p_{1/2})_p$ | 14.92 | 7.96 |
| $(1f_{7/2}, 1d_{5/2})_n$ | 11.43 | 0.40 | $(2f_{5/2}, 1d_{5/2})_n$ | 16.69 | 13.99 |
| $(2p_{3/2}, 2s_{1/2})_n$ | 7.31 | 0.01 | $(1f_{7/2}, 1d_{5/2})_n$ | 11.43 | 5.55 |
| $(1p_{1/2}, 1d_{3/2})_n$ | 3.64 | 0.01 | $(2f_{7/2}, 1d_{5/2})_n$ | 13.40 | 4.93 |
| $(2p_{3/2}, 1d_{3/2})_n$ | 3.35 | 99.20 | $(3f_{7/2}, 1d_{5/2})_n$ | 19.43 | 8.63 |
| $(2f_{5/2}, 1d_{3/2})_n$ | 10.13 | 0.01 | $(4p_{3/2}, 2s_{1/2})_n$ | 16.33 | 33.93 |

Consequently, the relative ratio between the energy weighted sum of the strength of the PDR and that of the GDR increases with increasing the mass number and reaches the maximum value of about 4 %, in agreement with the experimental investigation. By analyzing the weight of the contribution of the p-h configurations to the PDR and GDR states, one can conclude that the GDR state can be considered as collective, whereas the PDR has no collectivity. It is worth mentioning that present approach does not take into account the pairing correlation, which is one of the important features of neutron-rich nuclei. In addition, this code also neglects the high-order excitation modes that come from the two particle - two hole (2p-2h) or three particle - three hole (3p-3h) excitations. As the result, the conclusions above are made only in the basis of the present RPA calculations. To have a fully self-consistent microscopic prediction of the PDR, this RPA code should be improved by including the pairing correlation and high-order excitation modes. That is the goal of our oncoming studies. The numerical calculations were carried out using the Integrated Cluster of Clusters (RICC) system at RIKEN. This work is supported by the National Foundation for Science and Technology Development (NAFOSTED) of Vietnam through Grant No.103.04-2013.08.

Journal of Physics: Conference Series **726** (2016) 012026 doi:10.1088/1742-6596/726/1/012026

## References

[1] Bartholomew G A 1961, *Ann. Rev. Nucl. Sci.* **11** 259; Brzosko J. S *et al.*, 1969 *Can. J. Phys.* **47** 2849.
[2] Liestenscheneider *et al.* 2001, *Phys. Rev. Lett.* **86** 5442; Tryggestad E *et al.* 2002, *Phys. Lett.* B **541** 52.
[3] Aumann T *et al.* 2005, *Eur. Phys. J.* A **26** 441; Wieland O and Bracco A (2011), *Prog. Part. Nucl. Phys.* **66** 374.
[4] Gibelin J *et al.* 2007, *Nucl. Phys.* A **788** 153.
[5] Sagawa H and Suzuki T 1999, *Phys. Rev.* C **59** 3116.
[6] Vretenar D, Paar N, Ring P, and Lalazissis G A 2001, *Nucl. Phys.* A **692** 496; Paar N, Niksic T, Vretenar D, and Ring P 2005, *Phys. Lett.* B **606** 88.
[7] Litvinova E, Ring P, and Vretenar D 2007, *Phys. Lett.* B **647** 111.
[8] Sarchi D, Bortignon P F, and Colo G 2004, *Phys. Lett.* B **601** 27; Terazaki J and Engel J 2006, *Phys. Rev.* C **74** 044301; Terazaki T and Engel J 2010, *Phys. Rev.* C **82** 034326.
[9] Dang N D, Suzuki T, and Arima A 2000, *Phys. Rev.* C **61** 064304; Dang N D, Au V K, Suzuki T, and Arima A 2001, *Phys. Rev.* C **63** 044302.
[10] Tsoneva N and Lenske H 2012, *J. Phys.: Conf. Ser.* **366** 012043.
[11] Colo G, Cao L, Giai N V, and Capelli L 2013, *Comp. Phys. Comm.* **184** 142.
[12] Ring P and Schuck P 2004, *The Nuclear Many-Body Problem*, Springer, Heidelberg.
[13] Chabanat E, Bonche P, Haensel P, Meyer J, and Schaeffer R 1998, *Nucl. Phys.* A **635** 231.
[14] Savrana D, Aumann T, Zilgese A 2013, *Prog. Part. Nucl. Phys.* **70** 210.
[15] Zilges A, Derya V, Savran D, Hennig A, Pickstone SG and Spieker M 2015, *J. Phys.: Conf. Ser.* **580** 012052.

Journal of Physics: Conference Series **726** (2016) 012027 doi:10.1088/1742-6596/726/1/012027

# Production and decay of radion and Higgs in $e^+e^-$ and $\mu^+\mu^-$ colliders

**Dang Van Soa [1], Dao Thi Le Thuy[2], Bui Thi Ha Giang[2]**

[1] Hanoi Metropolitan University, 98 Duong Quang Ham, Cau Giay, Hanoi, Vietnam
[2]Hanoi National University of Education, 136 Xuan Thuy, Cau Giay, Hanoi, Vietnam

E-mail: giangbth@hnue.edu.vn

**Abstract.** We analyse the production and decay of the scalar particles of the Randall – Sundrum model. The mixture of the original Higgs and radion causes the interesting physical results. In this work, the production and decay of radion $\phi$ and Higgs $h$ in high energy $e^+e^-$ and $\mu^+\mu^-$ collisions through $\phi, h$ propagator are studied in detail.

## 1. Introduction

In 1999, the Randall – Sundrum (RS) model was conceived to solve the Higgs hierarchy problem [1], which the gravity scale and the weak scale can be naturally generated. The RS setup involves two three – branes bounding a slice of 5D compact anti-de Sitter space taken to be on an $S^1/Z_2$ orbifold. Gravity is localized UV brane or Planck brane, while the Standard Model (SM) fields are supposed to be localized IR brane or TeV brane [2]. The Golberger – Wise mechanism is presented to stabilize the radius of the extra dimension without reintroducing a fine tuning. Fluctuations about the stabilized RS model include both tensor and scalar modes. The fluctuations of the size of the extra dimension, characterized by the scalar component of the metric otherwise known as the radion [2,3].

The radion may turn out to be the lightest new particle in the RS model. The phenomenological similarity and potential mixing of the radion and Higgs boson warrant detailed to distinguish between the radion and Higgs signals at colliders. In our earlier works [3,4], we have calculated the production cross - section of radion in the high energy collisions. In this paper, we study the production and decay of radion $\phi$ and Higgs h in some high energy colliders. This paper is organized as follows. In Section II, we give a review of the RS model. Section III is devoted to the production and decay of radion and Higgs in high energy $e^+e^-$ and $\mu^+\mu^-$ collisions. Numerical evaluation is given in Sec.IV. Finally, we summarize results and make conclusions in Sec.V.

## 2. A review of Randall – Sundrum model

The RS model is based on a 5D spacetime with non – factorizable geometry [1]. The single extradimension is compactified on an $S^1/Z_2$ orbifold of which two fixed points accommodate two three – branes (4D hyper – surfaces): the Planck brane and TeV brane.

The ordinary 4D Poincar invariance is shown to be maintained by the following classical solution to the Einstein equation:

$$ds^2 = e^{-2\sigma(y)}\eta_{\mu\nu}dx^\mu dx^\nu - b_0^2 dy^2, \qquad \sigma(y) = m_0 b_0 |y|, \tag{1}$$

where $x^\mu$ ($\mu = 0, 1, 2, 3$)denote the coordinates on the 4D hyper – surfaces of constant y with metric $\eta_{\mu\nu} = diag(-1, 1, 1, 1)$. The $m_0$ and $b_0$ are the fundamental mass parameter and compactification radius, respectively.
The 4D effective Lagrangian is then [5, 6, 7]

$$\mathcal{L} = -\frac{\phi_0}{\Lambda_\phi}T_\mu^\mu - \frac{1}{\hat{\Lambda}_{\mathrm{W}}}T^{\mu\nu}(x)\sum_{n=1}^{\infty}h_{\mu\nu}^{(n)}(x), \tag{2}$$

where $\Lambda_\phi \equiv \sqrt{6}M_P\Omega_0$ is the VEV of the radion field, and $\hat{\Lambda}_{\mathrm{W}} \equiv \sqrt{2}M_P\Omega_0$ $(M_P \equiv 1/\sqrt{8\pi G_N})$ . The $T^{\mu\nu}$ is the energy – momentum tensor of the TeV brane localized SM fields. The $T_\mu^\mu$ is the trace of the energy – momentum tensor, which is given at the tree level as [7, 8]

$$T_\mu^\mu = \sum_f m_f \bar{f} f - 2m_{\mathrm{W}}^2 \mathrm{W}_\mu^+ \mathrm{W}^{-\mu} - m_Z^2 Z_\mu Z^\mu + (2m_{h_0}^2 h_0^2 - \partial_\mu h_0 \partial^\mu h_0) + .... \tag{3}$$

The gravity – scalar mixing arises at the TeV – brane by [9]

$$S_\xi = -\xi \int d^4x\sqrt{-g_{\mathrm{vis}}}R(g_{\mathrm{vis}})\hat{H}^+\hat{H}, \tag{4}$$

where $R(g_{vis})$ is the Ricci scalar for the induced metric on the visible brane or TeV brane, $g_{vis}^{\mu\nu} = \Omega_b^2(x)\left(\eta^{\mu\nu} + \varepsilon h^{\mu\nu}\right)$. $\hat{H}$ is the Higgs field before re-scaling, i.e. $H_0 = \Omega_0\hat{H}$. The parameter $\xi$ denotes the size of the mixing term. With $\xi \neq 0$, there neither a pure Higgs boson nor pure radion mass eigenstate. This $\xi$ term mixes the $h_0$ and $\phi_0$ fields into the mass eigenstates $h$ and $\phi$ as given by [6, 9]:

$$\begin{pmatrix} h_0 \\ \phi_0 \end{pmatrix} = \begin{pmatrix} 1 & -6\xi\gamma/Z \\ 0 & 1/Z \end{pmatrix}\begin{pmatrix} \cos\theta & \sin\theta \\ -\sin\theta & \cos\theta \end{pmatrix}\begin{pmatrix} h \\ \phi \end{pmatrix} = \begin{pmatrix} d & c \\ b & a \end{pmatrix}\begin{pmatrix} h \\ \phi \end{pmatrix}, \tag{5}$$

where

$$\gamma \equiv v_0/\Lambda_\phi, Z^2 \equiv 1 - 6\xi\gamma^2(1+6\xi) = \beta - 36\xi^2\gamma^2, \beta = 1 - 6\xi\gamma^2, \tag{6}$$

$$a = \frac{\cos\theta}{Z}, b = -\frac{\sin\theta}{Z}, c = \sin\theta - \frac{6\xi\gamma}{Z}\cos\theta, d = \cos\theta + \frac{6\xi\gamma}{Z}\sin\theta. \tag{7}$$

The mixing angle $\theta$ is defined by

$$\tan 2\theta = 12\gamma\xi Z\frac{m_{h_0}^2}{m_{h_0}^2(Z^2 - 36\xi^2\gamma^2) - m_{\phi_0}^2}. \tag{8}$$

The new fields h and $\phi$ are mass eigenstates with masses

$$m_{h,\phi}^2 = \frac{1}{2Z^2}\left[m_{\phi_0}^2 + \beta m_{h_0}^2 \pm \sqrt{(m_{\phi_0}^2 + \beta m_{h_0}^2)^2 - 4Z^2 m_{\phi_0}^2 m_{h_0}^2}\right]. \tag{9}$$

The mixing between the states enable decays of the heavier eigenstate into the lighter eigenstates if kinematically allowed. Overall, the production cross – sections, widths and relative branching fractions can all be affected significantly by the value of the mixing parameter $\xi$ [8, 9, 10]. We

Journal of Physics: Conference Series **726** (2016) 012027 doi:10.1088/1742-6596/726/1/012027

now dicuss the previous estimations on some model parameters. All phenomenological signatures of the RS model including the radion – Higgs mixing are specified by five parameters:

$$\Lambda_\phi,\ \frac{m_0}{M_P},\ m_h,\ m_\phi,\ \xi. \tag{10}$$

For the reliability of the RS solution, the ratio $\frac{m_0}{M_P}$ is usually taken around $0.01 \le \frac{m_0}{M_P} \le 0.1$ to avoid too large bulk curvature. Therefore, we consider the case of $\Lambda_\phi = 5$TeV and $\frac{m_0}{M_P} = 0.1$, where the effect of radion on the oblique parameters is small [11]. In the following, let us choose $\xi = 0,\ \pm 1/6$, in agreement with those in Ref.[9] with $\xi\gamma \ll 1$, $Z^2 \approx 1$.

## 3. The properties of radion $\phi$, Higgs h

### 3.1. The creation of radion $\phi$, Higgs $h$

In this section, we consider the production of radion and Higgs in the high energy $e^+e^-$ and $\mu^+\mu^-$ collisions. The Feynman diagram of the process collision is shown in Fig.1.

We have amplitude squared of this collisions in the following cases:

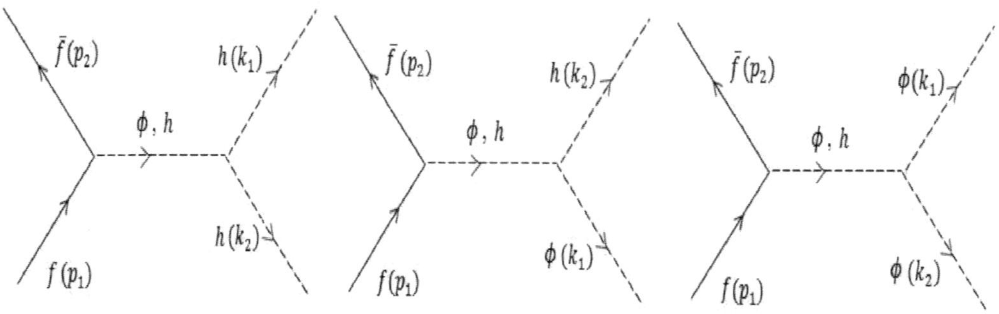


**Figure 1.** Feynman diagrams for $ff$ collision through $\phi, h$ propagator, where f is $e^-, \mu^-$ respectively ($p_i, k_i$ stand for the momentum)

for $f\overline{f} \to \phi h$ through $\phi, h$ propagator

$$\begin{aligned} M^2(f\overline{f} \ \to \ \phi h) = 4 \Big[ &\overline{g}^2_{ffh}\overline{g}^2_{\phi hh}\tfrac{1}{(q^2-m_h^2)^2} + \overline{g}_{ffh}\overline{g}_{ff\phi}\overline{g}_{\phi\phi h}\overline{g}_{\phi hh}\ \tfrac{1}{(q^2-m_h^2)(q^2-m_\phi^2)} + \\ &+ \overline{g}^2_{ff\phi}\overline{g}^2_{\phi\phi h}\tfrac{1}{(q^2-m_\phi^2)^2}\Big] \Big[(p_1p_2) - m_f^2\Big], \end{aligned} \tag{11}$$

for $f\overline{f} \to \phi\phi$ through $\phi, h$ propagator

$$\begin{aligned} M^2(f\overline{f} \ \to \ \phi\phi) = 4 \Big[ &\overline{g}^2_{ffh}\overline{g}^2_{\phi\phi h}\tfrac{1}{(q^2-m_h^2)^2} + \overline{g}_{ffh}\overline{g}_{ff\phi}\overline{g}_{\phi\phi h}\overline{g}_{\phi\phi\phi}\ \tfrac{1}{(q^2-m_h^2)(q^2-m_\phi^2)} + \\ &+ \overline{g}^2_{ff\phi}\overline{g}^2_{\phi\phi\phi}\tfrac{1}{(q^2-m_\phi^2)^2}\Big] \Big[(p_1p_2) - m_f^2\Big], \end{aligned} \tag{12}$$

for $f\overline{f} \to hh$ through $\phi, h$ propagator

$$\begin{aligned} M^2(f\overline{f} \ \to \ hh) = 4 \Big[ &\overline{g}^2_{ffh}\overline{g}^2_{hhh}\tfrac{1}{(q^2-m_h^2)^2} + \overline{g}_{ffh}\overline{g}_{ff\phi}\overline{g}_{\phi hh}\overline{g}_{hhh}\ \tfrac{1}{(q^2-m_h^2)(q^2-m_\phi^2)} + \\ &+ \overline{g}^2_{ff\phi}\overline{g}^2_{\phi hh}\tfrac{1}{(q^2-m_\phi^2)^2}\Big] \Big[(p_1p_2) - m_f^2\Big], \end{aligned} \tag{13}$$

where

$$\overline{g}_{\phi hh} =\frac{1}{\Lambda_\phi}[\{6b\xi(\gamma(ad+bc)+cd)+ad^2\}(k_1^2+k_2^2)$$
$$+d\{12ab\gamma\xi+2bc+ad(6\xi-1)\}k_3^2-4d(ad+2bc)m_{h_0}^2-3\gamma^{-1}cd^2m_{h_0}^2], \qquad (14)$$

$$\overline{g}_{hhh} = \frac{1}{\Lambda_\phi}\left[bd\left\{\left[12b\gamma\xi+d(6\xi+1)\right]\left(k_1^2+k_2^2+k_3^2\right)-12dm_{h_0}^2\right\}-3\gamma^{-1}d^3m_{h_0}^2\right], \qquad (15)$$

$$\overline{g}_{\phi\phi\phi} = \frac{1}{\Lambda_\phi}\left[ac\left\{\left[12a\gamma\xi+c(6\xi+1)\right]\left(k_1^2+k_2^2+k_3^2\right)-12cm_{h_0}^2\right\}-3\gamma^{-1}c^3m_{h_0}^2\right], \qquad (16)$$

$$\overline{g}_{\phi\phi h} =\frac{1}{\Lambda_\phi}[\{6a\xi\left(\gamma(ad+bc)+cd\right)+bc^2\}(k_1^2+k_2^2)$$
$$+c\{12ab\gamma\xi+2ad+bc(6\xi-1)\}k_3^2-4c(2ad+bc)m_{h_0}^2-3\gamma^{-1}c^2dm_{h_0}^2], \qquad (17)$$

$$\overline{g}_{f\overline{f}\phi} = -\frac{g}{2}\frac{m_f}{m_W}(c+\gamma a), \qquad (18)$$

$$\overline{g}_{f\overline{f}h} = -\frac{g}{2}\frac{m_f}{m_W}(d+\gamma b). \qquad (19)$$

Here, $q=p_1+p_2=k_1+k_2$ and $s=(p_1+p_2)^2$ is the square of the collision energy. We work in the center - of - mass frame: $p_1=(E_1,\overrightarrow{p}), p_2=(E_2,-\overrightarrow{p}), k_1=(E_3,\overrightarrow{k}), k_2=(E_4,-\overrightarrow{k})$ and denote the scattering angle by $\theta=(\overrightarrow{p},\overrightarrow{k})$.

From the expressions of the differential cross - section:

$$\frac{d\sigma}{d(cos\theta)} = \frac{1}{32\pi s}\frac{|\overrightarrow{k}|}{|\overrightarrow{p}|}|\overline{M}|^2, \qquad (20)$$

where M is the scattering amplitude, we will assess the number and make the identification, evaluation of the results obtained from the dependence of the differential cross - section (DCS) by $cos\theta$, the total cross - section fully follows $\sqrt{s}$ in Sec.IV.

*3.2. Decay of radion and Higgs*

We calculate the decay widths of the radion to the SM particles as follows:

$$;\ (\phi\to gg) = \frac{m_\phi^3\alpha_s^2}{8\pi(4\pi\Lambda_\phi)^2}(b_3a)^2, \qquad (21)$$

$$;\ (\phi\to\gamma\gamma) = \frac{m_\phi^3\alpha_s^2}{32\pi(2\pi\Lambda_\phi)^2}[(b_2+b_Y)a]^2, \qquad (22)$$

$$;\ (\phi\to \mathrm{f}\overline{\mathrm{f}}) = \frac{1}{32\pi \mathrm{m}_\phi}\mathrm{g}^2\frac{m_f^2}{m_{\mathrm{W}}^2}(c+\gamma a)^2(m_\phi^2-4m_f^2). \qquad (23)$$

Similarly, we calculate the decay widths of the Higgs to the SM particles as follows:

$$;\ (h\to gg) = \frac{1}{32\pi}\frac{(-\alpha_s)^2}{(4\pi v_0)^2}m_h^3\left(-2b_3\gamma b\right)^2, \qquad (24)$$

$$;\ (h\to\gamma\gamma) = \frac{1}{32\pi}\frac{\alpha^2}{(2\pi v_0)^2}m_h^3\left(b_2+b_Y\right)^2\left(\gamma b\right)^2, \qquad (25)$$

$$;\ (h\to f\overline{f}) = \frac{m_f^2}{32\pi m_{\mathrm{W}}^2}\frac{g^2}{m_h}\left(d+\gamma b\right)^2\left(m_h^2-4m_f^2\right). \qquad (26)$$

## 4. Numerical evaluation and discussion

We choose $\sqrt{s} = 3$TeV (CLIC), $m_e = 0.00051$ GeV, $m_\mu = 0.1057$ GeV, $m_\phi = 10$ GeV, $m_h = 125$ GeV (CMS), $\Lambda_\phi = 5$ TeV, $b_2 = 19/6$, $b_Y = -41/6$ , $b_3 = 7$, $g = \frac{2m_W}{v_0}$, $\xi = 1/6$, $v_0 = 246$ GeV [7, 10], to estimate for the cross - section as follows:

$$DCS(e^+e^- \xrightarrow{\phi, h} \phi\phi) = 8.63543 \times 10^{-18}(pbar), \quad (27)$$

$$DCS(e^+e^- \xrightarrow{\phi, h} \phi h) = 9.00065 \times 10^{-11}(pbar), \quad (28)$$

$$DCS(e^+e^- \xrightarrow{\phi, h} hh) = 2.16804 \times 10^{-13}(pbar), \quad (29)$$

$$DCS(\mu^+\mu^- \xrightarrow{\phi, h} \phi\phi) = 3.77275 \times 10^{-13}(pbar), \quad (30)$$

$$DCS(\mu^+\mu^- \xrightarrow{\phi, h} \phi h) = 3.93231 \times 10^{-6}(pbar), \quad (31)$$

$$DCS(\mu^+\mu^- \xrightarrow{\phi, h} hh) = 9.47199 \times 10^{-9}(pbar). \quad (32)$$

Therefore, the differential cross - section is independent on the $cos\theta$. DCS is minimum value in $e^+e^- \to \phi\phi$ collision and maximum value in $\mu^+\mu^- \to \phi, h$ collision. The total cross - section is shown as the function of the collision energy $\sqrt{s}$ in Fig.2 and Fig.3. We show that the total cross - section is approximately independent on $\sqrt{s}$ when $\sqrt{s} > 1$ TeV. Therefore, it is difficult to collect radion at very high energies.

The widths of the radion and Higgs decay channels dependence on the radion mass are shown in Fig.4 and Fig.5, respectively. The mass range is chosen as 10 GeV $\leq m_\phi \leq$ 100 GeV. The dominant decay modes are $\phi \to gg$ and $h \to gg$. The widths of the radion and Higgs decay channels increase or the liveable time of radion and Higgs decrease when the mass radion increases.

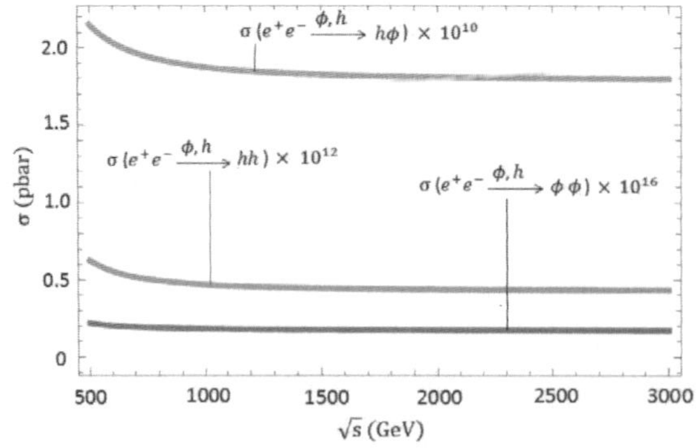


**Figure 2.** The total cross - section as a function of the collision energy $\sqrt{s}$ in $e^+e^-$ collision

## 5. Conclusion

In this work, the properties of radion and Higgs in their production and decay are evaluated in detail. At high energy values, the total scattering cross section does not depend on the collision energy $\sqrt{s}$. However, the total scattering cross section is very small. We also have studied the decay of radion and Higgs into the SM particles. The result shows that, the $h \to gg$ and $\phi \to gg$ modes dominate over the other decay channels.

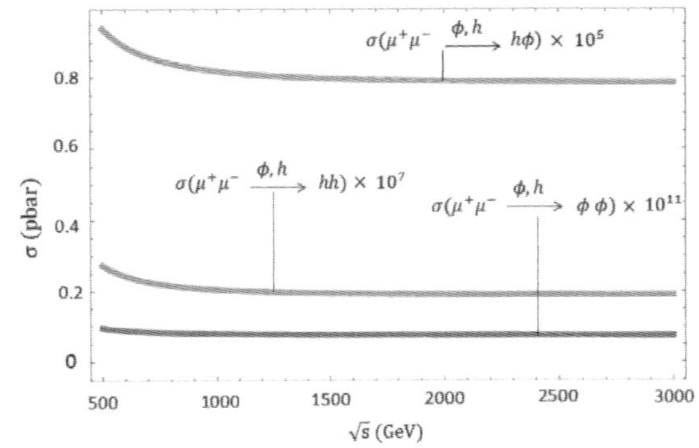


**Figure 3.** The total cross - section as a function of the collision energy $\sqrt{s}$ in $\mu^+\mu^-$ collision

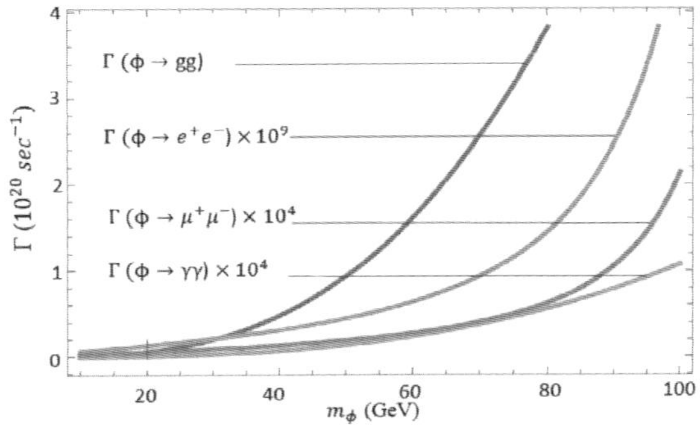


**Figure 4.** The widths of the radion decay channels as the function of the radion mass $m_\phi$

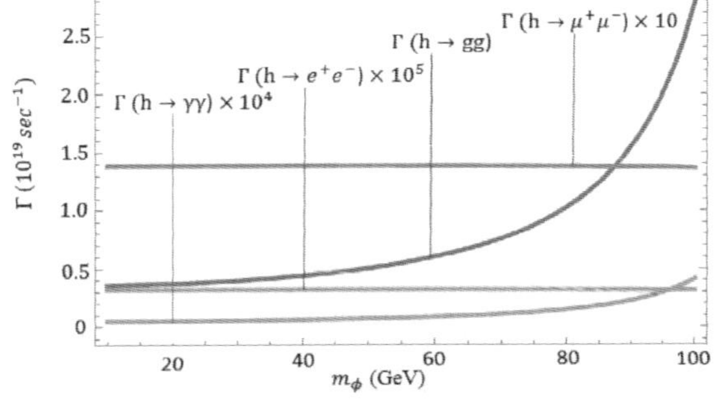


**Figure 5.** The widths of the higgs decay channels as the function of the radion mass $m_\phi$ with $m_h = 125GeV$

## References

[1] Randall L and Sundrum R 1999 *Phys. Rev. Lett.* **83** 3370
[2] Randall L and Sundrum R 1999 *Phys. Rev. Lett.* **83** 4690
[3] Soa D V, Thuy D T L, Thao N H, Tham T D 2012 *Mod.Phys.Lett.* A Vol.27 No.2 1250126
[4] Giang B T, Thuy D T L, Dai N V, Soa D V 2015 *Communications in Physics* Vol. 25 No. 1 45 - 50
[5] Csaki C, Hubisz J and Lee S J 2007 *Phys. Rev.* D **76** 125015
[6] Chaichain M, Datta A, Huitu K and Vu Z 2002 *Phys. Lett.* B **524** 161
[7] Bae S, Ko P, Lee H S and Lee J 2000 *Phys. Lett.* B **487** 299
[8] Csaki C, Graesser M L and Kribs G D 2001 *Phys. Lett.* D **63** 065002
[9] Giudice G F, Rattazzi R and Wells J D 2001 *Nucl. Phys.* B **595** 250
[10] Cheung K, Kim C S and Song J H 2003 *Phys. Rev.* D **67** 075017
[11] Kim C S, Kim J D ang Song J H 2003 *Phys. Rev.*D **67** 015001

# AUTHOR INDEX